Advances in Sol-Gel Derived Materials and Technologies

Series editors

Michel A. Aegerter
Michel Prassas

More information about this series at http://www.springer.com/series/8776

Suresh C. Pillai · Sarah Hehir
Editors

Sol-Gel Materials for Energy, Environment and Electronic Applications

Springer

Editors
Suresh C. Pillai
Nanotechnology and Bio-Engineering Research Group
Institute of Technology Sligo
Sligo
Ireland

Sarah Hehir
Nanotechnology and Bio-Engineering Research Group
Institute of Technology Sligo
Sligo
Ireland

ISSN 2364-0030 ISSN 2364-0049 (electronic)
Advances in Sol-Gel Derived Materials and Technologies
ISBN 978-3-319-50142-0 ISBN 978-3-319-50144-4 (eBook)
DOI 10.1007/978-3-319-50144-4

Library of Congress Control Number: 2016958479

Printed on acid-free paper

This Springer imprint is published by Springer Nature
The registered company is Springer International Publishing AG
The registered company address is: Gewerbestrasse 11, 6330 Cham, Switzerland

The International Sol-Gel Society (ISGS)

Dear Readers,

The International Sol-Gel Society (ISGS) was established in 2003 as an international, interdisciplinary, not-for-profit organization whose primary purpose and objective is to promote the advancement of sol-gel science and technology. ISGS's aims are both to represent the particular needs and aspirations of the international sol-gel community and to support this sol-gel community. The society's mission is threefold:

- to coordinate the promotion of sol-gel science and technology in the scientific and industrial community,
- to foster communication between researchers from different fields and geographical regions through the organization of conferences and the publication and circulation of technical papers,
- to encourage education, training, and research in the field of sol-gel science and technology.

To achieve these purposes, ISGS convenes the biannual International Sol-Gel Conference in many parts of the world. The XVIII edition of this International Conference was held in Kyoto, Japan in 2015. The next one will be held in Liège, Belgium in autumn 2017. These conferences play an important role to educate, federate, and disseminate scientific knowledge to people working in related fields. To initiate young researchers and engineers into the sol-gel field, a sol-gel summer school is also operated by ISGS every 2 years in addition to the International Sol-Gel Conferences.

In 2011, Springer and ISGS launched a new book series entitled "Advances in Sol-Gel Derived Materials and Technologies" to encourage education, training, and research in the field of sol-gel science and technology. Three volumes have already been published: Aerogels Handbook (2011), Sol-Gel Processing for Conventional and Alternative Energy (2012), and Sol-Gel Nanocomposites (2014). The present volume entitled "Sol-Gel Materials for Energy, Environment and Electronic Applications" is edited by Suresh C. Pillai and Sarah Hehir. This book focuses on a wide range of applications of sol-gel as well as covering the fundamentals of the relevant sol-gel processings. Therefore, this book is a very good guide to study the sol-gel processing for specific applications.

In Chap. 1, an introduction to sol-gel processings with a focus on aerogels is described in a systematic way so that it is easy to understand the applications described in the following chapters even for those who are beginners in the sol-gel approach. The realistic applications include varistor devices, functional coatings for optics, the role of diatoms in environmental applications, sensors, Li-ion batteries, corrosion protection, graphene incorporated in sol-gel materials for energy applications, titanium oxide and finally lanthanum phosphate ceramics for functional applications. Therefore, the readers obtain up-to-date knowledge on interesting applications of sol-gel materials.

Springer is also the publisher of the ISGS official journal the *Journal of Sol-Gel Science and Technology* (JSST), which is the best tool for readers of this book to learn about the latest research and developments on the science and technology of sol-gel related materials.

ISGS is very proud of having this productive relationship with Springer.

I wish you a very pleasant and educative reading!

Masahide Takahashi
President of the International Sol-Gel Society
http://www.isgs.org

Preface

In recent years sol-gel technology has become a hotbed for state-of-the-art developments in many diverse fields due to the accessibility of advanced materials with 'tailor-made' functionalities through inexpensive and environmentally viable processing routes. The focus of this book is on the use of such technologies and the resultant materials for applications in the energy, environmental and electronic sectors specifically providing a unique perspective. The current book comprises a broad scope encompassing basic knowledge, as well as high-level research advancements with the potential for commercialization and industrial use. Taking an integrated approach, a wide range of topics are covered, from electronic materials, photocatalysts, sensors and optics, to aerogels and materials for energy storage and conversion, consequently showcasing the combined use of chemistry, physics, materials science and engineering in the search for solutions to some of the most challenging issues of our time.

One of the reasons for the continued advancement of sol-gel technology is the ease of control of the nanoarchitecture of the resultant materials and the plethora of different material constructs which can be developed. The opening chapter focuses on introducing the core activities involved in sol-gel processing including the formation, ageing and drying of wet gels, with a particular focus on preparation of aerogels.

An important application of sol-gel processing in the area of electronics is outlined in Chap. 2, which presents an exploration of the full potential of this technology with regard to the preparation of high-performance varistors. Sol-gel methods can be easily integrated with any varistor processing method and examples of this and the advantages accrued are discussed. A detailed literature survey on the preparation, properties and advantages of sol-gel derived ZnO varistors is included along with recent work on the numerous sol-gel ceramic-polymer varistors.

Sol-gel technology has now reached a mature stage in the field of coatings for optics and photonics. Recent achievements in sol-gel derived functional optical coatings, specifically those dealing with reflection and anti-reflective phenomena such as photonic crystals and anti-reflective coatings are discussed in Chap. 3. Additionally, chemical strategies to modify the optical properties of sol-gel coatings are described together with a critical analysis of various liquid deposition techniques.

Chapter 4 evaluates the role of sol-gel modified unicellular microalgae, diatoms, in numerous environmental applications such as catalysis, separation science, filtration and emerging nanotechnologies. Much of the work presented in this chapter highlights the influence and importance of diatom species on the properties of sol-gel coated diatoms or diatom replicas.

The development of a diverse range of chemical sensors derived from sol-gel processing technology for a large range of applications, including determination of gases, forensic analysis and biosensing are reported in Chap. 5. The adaptability of the sol-gel process for the tailoring of material properties is key to the development of such a variety of sensors. The chapter also includes a discussion on the use of sol-gel films in sensor configuration to allow more flexibility.

In recent years, cathode materials prepared through sol-gel processing have exhibited improved electrochemical performance in rechargeable Li-ion batteries. Furthermore, some of the major drawbacks of current generation Li-ion battery cathodes have been efficiently alleviated by sol-gel technology processes. Chapter 6 outlines the synthesis, preparation and processing of these cathode materials in addition to some relevant modification procedures.

Sol-gel technology for the environmental-based application of developing an alternative to chromium-based corrosion control materials is the focus of Chap. 7. As well as the principles underlying the chemistry of the materials investigated, recent advances in the trends for preparing such corrosion protection systems are also presented. A number of strategies including sol-gel film design, introduction of nanoparticles, application of organic polymers and application of corrosion inhibitors are explored.

Due to its high specific surface area, excellent electrical and mechanical properties along with very good chemical stability, graphene is an ideal candidate for next generation energy devices. Chapter 8 gives an overview of the recent research on graphene incorporated sol-gel materials for energy conversion and storage applications including supercapacitors, solar cells, Li-ion batteries and fuel cells.

Chapter 9 details the preparation and applications of nanocrystalline titanium dioxide (TiO_2) via sol-gel processing. The various crystal forms of TiO_2 are presented and their application as photocatalysts is discussed.

In the final chapter, an overview of lanthanum phosphate ($LaPO_4$) derived from sol-gel processes is given. Properties such as hydrophobicity, metallophobicity, low thermal conductivity and machinability make $LaPO_4$ an ideal material for use in coating applications. A detailed case study of $LaPO_4$ nanocoating preparation and characterization is also included in this chapter.

The obvious strength of the sol-gel method lies in its inherent adaptability and flexibility while maintaining the integrity and consistency of the materials produced. The versatility of the method, as shown throughout this book, lies in the ease of integration of sol-gel technologies with other forms of processing, allowing multidisciplinary approaches to occur with very little effort. The application of sol-gel technologies to some of the work discussed is still at a relatively early stage of development and as such some authors highlight possible limitations to the commercialization of such methods in their respective fields. However, they also suggest strategies to overcome these perceived limitations in the future, thus allowing full commercialization and the benefits of such advancements to be felt by society at large.

Sligo, Ireland

Suresh C. Pillai
Sarah Hehir

Contents

Contributors

S. Ananthakumar Functional Materials Section, Materials Science and Technology Division, CSIR-National Institute for Interdisciplinary Science and Technology, Thiruvananthapuram, Kerala, India

S. Anas Functional Materials Section, Materials Science and Technology Division, CSIR-National Institute for Interdisciplinary Science and Technology, Thiruvananthapuram, Kerala, India; T. K. M. College of Arts and Science, Karicode, Kollam, Kerala, India

Killian Barton Dublin Institute of Technology, Focas Institute, Dublin 8, Ireland

Davide R. Ceratti Sorbonne Universités, UPMC Univ Paris 06, CNRS, Collège de France, Paris, France

Maikki Cullen Dublin Institute of Technology, Focas Institute, Dublin 8, Ireland

Francisco del Monte Instituto de Ciencia de Materiales de Madrid, Consejo Superior de Investigaciones Cientıficas, Madrid, Spain

Saoirse Dervin Nanotechnology and Bio-Engineering Research Group, Department of Environmental Sciences, Institute of Technology Sligo, Sligo, Ireland; Centre for Precision Engineering, Materials and Manufacturing Research (PEM), Institute of Technology Sligo, Sligo, Ireland

Peter Dockery School of Medicine, National University of Ireland, Galway, Ireland

Brendan Duffy Dublin Institute of Technology, Focas Institute, Dublin 8, Ireland

Vinodkumar Etacheri IMDEA Materials Institute, Getafe, Madrid, Spain

Marco Faustini Sorbonne Universités, UPMC Univ Paris 06, CNRS, Collège de France, Paris, France

U.S. Hareesh Materials Science and Technology Division, CSIR—National Institute for Interdisciplinary Science and Technology, Thiruvananthapuram, Kerala, India

M. Jeen Maria Functional Materials Section, Materials Science and Technology Division, CSIR-National Institute for Interdisciplinary Science and Technology, Thiruvananthapuram, Kerala, India

Honey John Department of Polymer Science and Rubber Technology, Cochin University of Science and Technology, Kochi, India

Maheswary Kavirajan Kavitha Department of Physics, Indian Institute of Technology, Chennai, India

Rajesh Komban Centrum for Angewandte Nanotechnologie (CAN) GmbH, Hamburg, Germany

Yvonne Lang School of Medicine, National University of Ireland, Galway, Ireland; School of Science, Institute of Technology, Sligo, Ireland

K.V. Mahesh Materials Science and Technology Division, Functional Materials Section, CSIR-National Institute for Interdisciplinary Science and Technology, Thiruvananthapuram, Kerala, India

Balagopal N. Nair R&D Centre, Noritake Co. Ltd, Miyoshi-Aichi, Japan

Shijina Nambiar Materials Science and Technology Division, CSIR—National Institute for Interdisciplinary Science and Technology, Thiruvananthapuram, Kerala, India

Guillaume Naudin Sorbonne Universités, UPMC Univ Paris 06, CNRS, Collège de France, Paris, France

M. Padmanabhan Department of Chemistry, Amrita Vishwa Vidyapeetham University, Kollam, India

Pradeepan Periyat Department of Chemistry, Central University of Kerala, Kasaragod, Kerala, India

Suresh C. Pillai Centre for Precision Engineering, Materials and Manufacturing Research (PEM), Institute of Technology Sligo, Sligo, Ireland; Nanotechnology and Bio-Engineering Research Group, Department of Environmental Sciences, Institute of Technology Sligo, Sligo, Ireland

Sankar Sasidharan Materials Science and Technology Division, CSIR—National Institute for Interdisciplinary Science and Technology, Thiruvananthapuram, Kerala, India

Sanjay Gopal Ullattil Department of Chemistry, University of Calicut, Thenhipalam, Kerala, India

Krishna G. Warrier Materials Science and Technology Division, CSIR—National Institute for Interdisciplinary Science and Technology, Thiruvananthapuram, Kerala, India

Aine M. Whelan School of Chemical and Pharmaceutical Sciences, Dublin Institute of Technology, Dublin 2, Ireland

Chapter 1
An Introduction to Sol-Gel Processing for Aerogels

Saoirse Dervin and Suresh C. Pillai

1.1 Introduction

The advancement of sol-gel processing is due to the ease of control of the nanoarchitecture throughout synthesis. The ability to tailor the structure of nanosystems from primitive processing phases advocates the preparation of pure materials with improved properties. The advanced properties of such compounds afford their use, in the form of fibres, films, fine powders and monoliths within a diverse range of applications. This chapter will introduce sol-gel processing with a particular focus on aerogels.

Sol-gel processing generally includes four key stages, additional procedures are often incorporated in an attempt to enhance mechanical properties and characteristics of the gel [1]. All steps involved in synthesis contribute to the framework of the gel and influence its properties and thus corresponding applications (Scheme 1.1) [2].

I. Sol preparation: A colloidal suspension is formed as a result of the dispersion of solid nanoscale particles, derived from a precursor material, within a solvent.
II. Sol to gel transition (gelation): Cross-linking and branching particles occurs due to the addition of an acid or a base catalyst, which initiates polymerisation forming an interconnected chain structure.

S. Dervin · S.C. Pillai (✉)
Nanotechnology and Bio-Engineering Research Group,
Department of Environmental Sciences, Institute of Technology Sligo, Sligo, Ireland
e-mail: Pillai.Suresh@itsligo.ie

S. Dervin · S.C. Pillai
Centre for Precision Engineering, Materials and Manufacturing Research (PEM),
Institute of Technology Sligo, Sligo, Ireland

S.C. Pillai and S. Hehir (eds.), *Sol-Gel Materials for Energy, Environment and Electronic Applications*, Advances in Sol-Gel Derived Materials and Technologies,
DOI 10.1007/978-3-319-50144-4_1

Scheme 1.1 Summary of sol-gel process

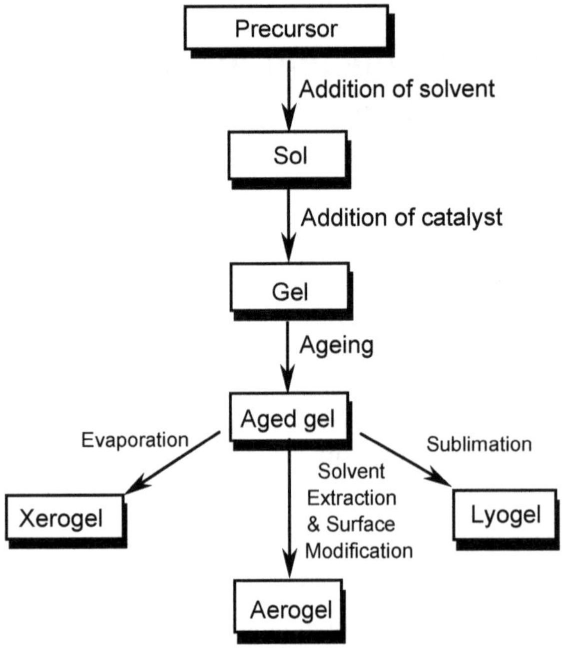

III. Ageing of the gel: The gel is aged in its mother solution in order to increase the backbone and mechanical strength of the gel.
IV. Drying of the gel: The solvent is removed from the pores of the gel in a manner deterrent of gel fracture.

Preparation begins with the formation of a colloidal suspension, a sol. Solid nanoparticles or precursor materials are dispersed within a solution of reactants and solvents. The addition of a catalyst promotes polymerisation, which includes hydrolysis and polycondensation reactions. Cross-linking and branching between polymeric species occurs resulting in the development of a 3D porous network within a wet, gel-like structure.

Sol-gel derived materials can be produced from various materials such as oxides, like silica, metal oxides, organic materials, such as polymers like cellulose and carbon materials, for example graphene and carbon nanotubes [2]. Composite materials have also become significant due to their ability to enhance gel strength and other material properties.

However, silica gels are the most widely used and greatly studied, also, possibly the simplest produced. As a result, the main focus of this chapter will be on the wet gel formation of silicates.

The final stage involved in the processing of the gels is drying (Fig. 1.1) and involves evaporation of solvent, extraction of solvent or sublimation during xerogel, aerogel and lyogel (cryogel) synthesis, respectively.

Aerogels are the lowest density solid materials produced with nanometer-scale pores [3]. Aerogels consist of microscopic beads which form a continuous network

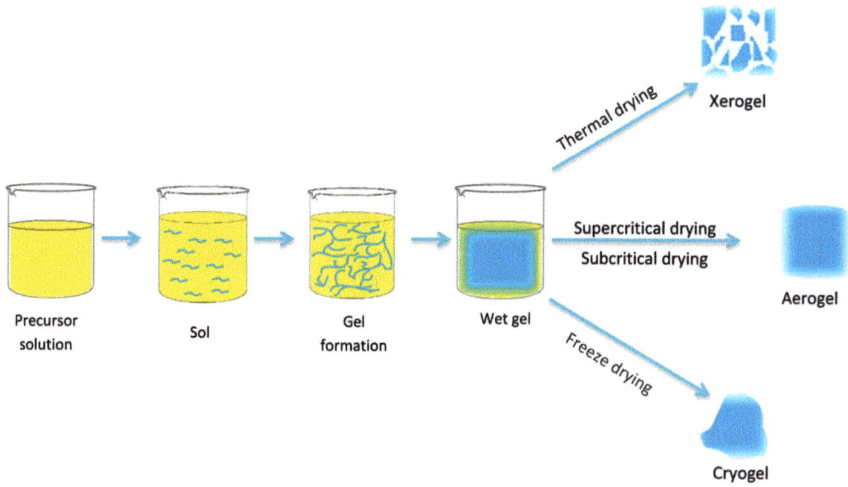

Fig. 1.1 Sol preparation and gel transformation [39]

possessing between 90 and 99% porosity, affording them very desirable qualities such as high specific surface areas, low densities, high thermal insulation values, ultra-low dielectric constant and low refractive indexes [4–7]. These unique properties paved the way for many future applications such as aerospace materials [8, 9], electronic devices [10, 11], ultrasonic and gas sensors [12, 13], materials for waste management [14, 15], nuclear particle detection (Cherenkov) [16, 17], optics and IR materials [18, 19], capacitors [20, 21], catalysts [22, 23] and thermal and acoustic insulation [3, 19, 24].

1.2 Silicon Alkoxide Sol-Gel Materials

A significant amount of research has been devoted to identifying novel precursors such as inorganic salts [25] and inexpensive organic materials [2]. However, the most common precursors used for the development of silica gels are silicon alkoxides (Table 1.1).

Dorcheh and Abbasi [1] provide an extensive review of the workings and findings of many authors who have compared and contrasted silicon alkoxide precursors. These findings demonstrate that the use of different precursors provides aerogels with different, desirable properties such as improved specific surface area, as well as pore size volume and distribution. For instance, Dorcheh [1] and Wagh et al. [26] investigated the influence of varying precursor material by preparing aerogels from TEOS, TMOS and PEDS. Their work concluded that silica aerogels of different physical properties were obtained by using various precursors, declaring that TMOS generated a gel of higher surface area, narrow, more uniform pores and

Table 1.1 Common Silica Precursors and their chemical structures

Precursor	Chemical structure
Tetramethylorthosilicate (TMOS)	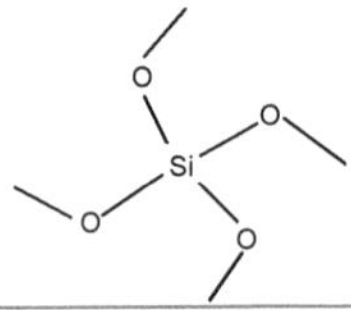
Tetraethylorthosilicate (TEOS)	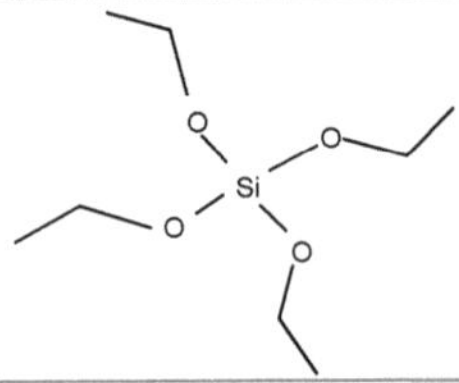
Polyethoxydisiloxane (PEDS)	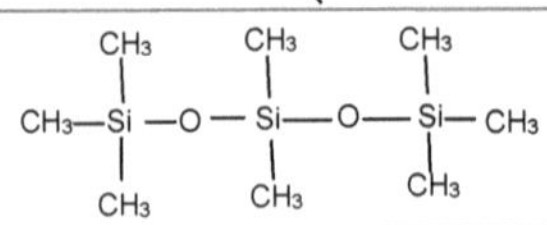
Methyltriethoxysilane (MTES)	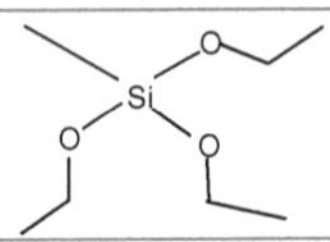

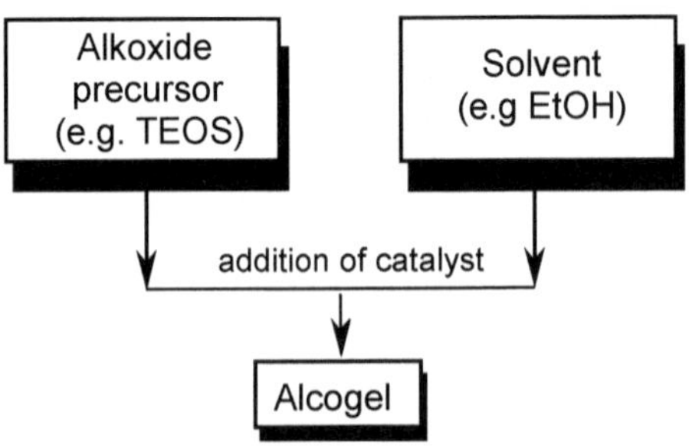

Scheme 1.2 Summary of silicon alkoxide gel formation

lower thermal conductivity than PEDS and TEOS [26]. Therefore, the choice of starting material depends predominantly on the required properties and applications of the aerogel. Although TEOS may yield less favourable material characteristics, this precursor is much more cost effective than other starting materials, as well as being a less harmful reagent, and so, has been widely used throughout aerogel research.

Structures are formed from silicon alkoxides as a result of the hydrolysis and polycondensation of alkoxides in solution; water and solvent, in the presence of a catalyst (Scheme 1.2) [27]. The hydrolysis and condensation reactions instigate polymerisation forming polymeric species composed of Si-O-Si cross-links and branches and are therefore the basis of the sol-gel transition [2, 28].

1.3 Sodium Silicate Sol-Gel Materials

The process of preparing sodium silicate gels differs from gels derived from alkoxide precursors. In order to prepare a sol from sodium silicate, sodium ions must first be replaced with hydrogen ions which produces silicic acid. If not removed, sodium ions, which do not dissolve in alcohol, are distributed throughout the gel network, forming opaque, weak, gels when dried (Scheme 1.3) [29].

Sodium ions can be removed from sodium silicate in one of the two ways:

I. Ion exchange [30].
II. Vapour passing method [31].

Ion exchange occurs before the sol is transferred to a gel state. Sodium silicate can be passed through strong acidic cation resins, commonly amberlite, within an ion exchange column [1]. This practice, however, is expensive and time consuming due to lengthy multiple wash steps.

In contrast, the vapour passing method occurs after gel formation. A vapour oven is employed, allowing water vapour to pass through the gel which is aged in an oven. The sodium ions are removed from the gel's pores as a result of temperature, and resulting pressure. This process however, can also be slow and monotonous.

Alternatively, in recent years, ion exchange and vapour passing methods have been replaced by washing and/or shaking the gels in the presence of water [32, 33].

Once sodium ions have been removed and silicic acid is formed, a catalyst, usually a base (NH_4OH), is added to the solution to modify the sol for gelation. Gelation of silicic acid occurs as a result of silica polymerisation also. Moreover, the rate of gelation of silicic acid can be controlled by acid, base or acid–base accelerated catalysis. The nature of the chosen catalyst affects the structure of sodium silicate gels in the same way as it affects the structure of gels prepared from alkoxide precursors.

Early stage sol-gel processing avoided the use of sodium silicate, due to the time consuming washing process which surrounded the use of the material. However,

Scheme 1.3 Summary of sodium silicate gel formation

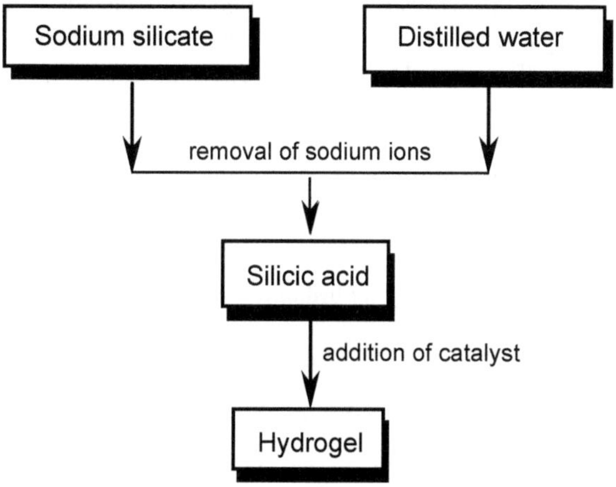

sodium silicate has once again become a desired precursor for industrial applications [32–35], due to the material's inexpensive nature, which affords its suitability for commercialisation [1, 36].

1.4 Wet Gel Formation

The transition from a sol to a wet gel commences with the initiation of a hydrolysis reaction. An oxygen group, found within water of a silicon atom is attacked by an electrophilic molecule, facilitated by the presence of homogenous mixture such as alcohol [28]. Through hydrolysis, hydroxyl groups or water molecules included on the precursor structure are replaced with alkoxy groups. Condensation then lengthens the chain of the molecule, releasing water or alcohol derived from such hydroxyl groups [37]. Polymerisation continues and a solid silicate network separates out at gel point [38]. Pierre [39] explains, in detail, the mechanisms of silicon precursor hydrolysis and condensation reactions (Schemes 1.4 and 1.5).

The rate, at which these reactions occur, has a significant impact on the structure of the final material. Consequently, controlling the conditions at which hydrolysis and condensation occur, such as pH, can influence the characteristics of the final material. Hydrolysis and condensation can take place simultaneously. However the rate and extent at which each reaction occurs can be controlled by the addition of a catalyst. Controlling these factors manipulates the structure and characteristics of the produced gel [40].

Common alkoxide precursors such as TEOS react slowly in alcoholic and aqueous solutions, resulting in prolonged gelation times. A reaction accelerator is therefore necessary, particularly where potential commercial products are concerned [37]. The rate at which hydrolysis and condensation occur can thus be accelerated with the aid of a catalyst; acid and/or base, in order to shorten the reaction process. There are three different courses (Scheme 1.6) which can be taken to control the reaction rates: Acid catalysis, base catalysis and acid–base catalysis.

It has previously been reported in much published work that the nature of the catalyst employed greatly influences the mechanical structure of aerogels due to their effects on pH of the gels environment [37].

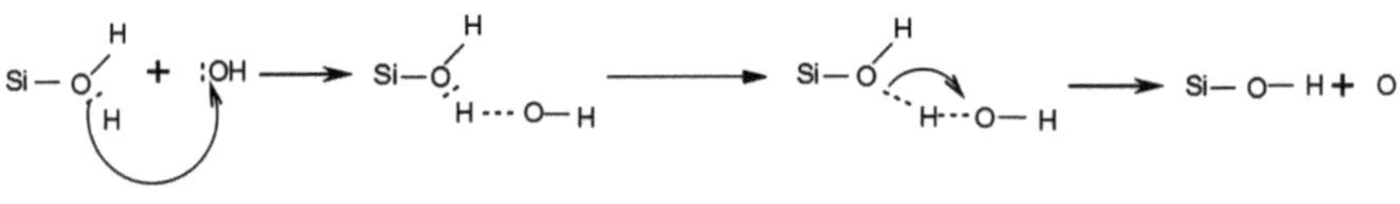

Scheme 1.4 Exchange of hydroxyl groups with alkoxy groups via hydrolysis reaction [39]

Scheme 1.5 Formation of siloxane bonds from condensation reactions [39]

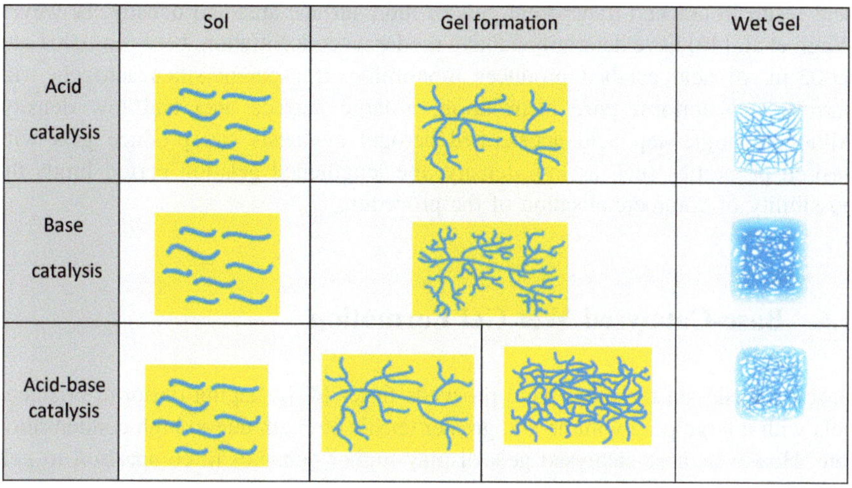

Scheme 1.6 The influence of the nature of catalyst on gel structure

1.5 Acid-Catalysed Wet Gel Formation

Acid catalysis can be performed with the use of various acids, such as hydrochloric, sulfuric, nitric, oxalic, formic and acetic acids, also, hydrogen fluoride. The class of acid itself does not notably influence structure or properties of gels [41, 42].

Under acidic conditions, hydrolysis is favoured and at first proceeds rapidly, meaning the rate of the condensation reaction is limited. The initial phase of hydrolysis, in which precursor materials are transformed to trialkoxy silanol groups, proceeds at a greater rate than the secondary phase of hydrolysis, due to unfavourable protonation of silanol species [43]. Consequently, the latter proceed at greater rates than the following condensation reaction which significantly impacts the resulting network.

Acid catalysis forms weakly branched, linear silica chains. Therefore highly porous, but weak gels are produced (Scheme 1.6), resulting in high levels of volume shrinkage during drying. Gels prepared with increased concentrations of acid (>0.001 m [44]), or using strong acids, also experience high volumes of shrinkage. However, gels synthesised in this manner tend to exhibit low optical transmission, higher densities and high porosity. Cross-linking may occur due to the accidental interlinking of chains after the hydrolysis of Si-OR side groups, resulting in weakly branched, dense silica networks with relatively small pores [43].

Moreover, acid-catalysed hydrolysis generally results in a lengthy gelation time due to reduced condensation process [26, 43, 45]. Wagh et al. [46] investigated the effects of acid catalysis on the physical properties of silica gels using citric acid as a new catalyst. The work demonstrated that low concentrations of acid catalyst result in low density opaque gels with a low surface area whereas high concentrations of acid result in cracked transparent gels of high surface area and density. However, Wagh et al. [46] have determined that a moderate concentration, between 0.001 and 0.005 m, of acid catalyst produces monolithic, transparent silica aerogels with narrow and uniform pore distribution, a large surface area and low density. Although single-step acid accelerated aerogel synthesis can produce gels with desired properties such as low density, the lengthened gelation period limits the possibility of commercialisation of the procedure.

1.6 Base-Catalysed Wet Gel Formation

Base catalysis, on the other hand proceeds faster [47], shaping uniform, branched gels with a large pore volume and greater tensile strength due to high condensation rate. However, base-catalysed gels display higher densities in comparison to gels prepared with acidic solutions.

Nadargi et al. [47] have reported solidification of wet gels, under basic conditions, in a matter of minutes. Under basic conditions, a quick hydrolysis reaction results in almost completely hydrolysed monomers. Rapid hydrolysis leads to cross-linking at an early stage in the process when unhydrolysed monomers are still present, which results in highly branched polymer chains [43]. The gels obtained by Nadargi et al. [47] were relatively low in density, displaying hydrophobic properties yet were completely opaque.

Brinker et al. [40] have thoroughly investigated the effect of pH and catalysis on the sol-gel transition in silicates showing that the nature of the polymer species is considerably different in acid- and base-catalysed gels. Under acidic conditions, weak, low density, porous, coarse textured gels were acquired. Slowing the rate of hydrolysis with the use of a base catalyst produces high density fine textured gels.

1.7 Acid–Base Catalysed Wet Gel Formation

Both an acid and a base catalyst can also be applied simultaneously in a two-step catalysis process. This two-step synthesis method allows for greater control of silica polymerisation as both the rate and extent of the hydrolysis and polycondensation reactions involved can be regulated by acid and base catalysis, respectively. Rao and Bhagat [44] compared; (i) acid, (ii) base and (iii) acid–base catalysed processes. The study reported that monolithic, low density, TEOS-based aerogels with low volume shrinkage and high optical transmission were obtained by a two-step (acid–base) sol-gel process [44]. Kim et al. [48] also detailed an advantageous two-step catalysis process. From their work, silica aerogels with an extremely high specific surface area of 1108 $m^2\ g^{-1}$ and a large pore volume of 4.7 $cm^3\ g^{-1}$ were produced from ambient pressure dried TEOS wet gels.

Both Rao and Bhagat [44] and Kim et al. [48] began their studies with the addition of an acid catalyst, oxalic acid, to the precursor solution. A base catalyst, ammonium hydroxide, was subsequently added to the sol at various time intervals, in order to observe the effects. The findings of both investigations illustrated that an increased time period, between the addition of acid and addition of base, resulted in increased optical transmission. However, Rao and Bhagat [44, 48] stated that incorporating the base catalyst, 24 h after the acid catalyst had been employed produced transparent aerogels with low density 0.1540 units, little volume shrinkage 17.48% and a porosity of 91.89%. However, delaying the addition of base further resulted in increased shrinkage and correspondingly, increased density.

Moreover, each investigation noted that the two-step process also resulted in a reduced gelation period. Rao and Bhagat [44] reported that acid accelerated gelation required a minimum of three days in order for the wet gel to stiffen. However, implementation of both an acid and base catalyst decreased the length of this process to as little as 2 min. Kim et al. [48] also described an impressively short gelation time, as low as 15 min.

1.8 Ageing of Gels

Once a wet gel has been formed, unreacted alkoxide groups may still remain within the gel structure. Consequently hydrolysis and condensation reactions continue once adequate time is provided. This prolonged period of polymerisation strengthens the silica network. The strength of a gel can also be improved by monitoring the pH, precursor concentration and water content of the solution which accommodates the ageing process [49, 50].

Low density silica aerogels have low mechanical strength and stiffness [51]. These gels comprise highly open structures in which secondary particles of silica are connected to each other with weak siloxane bonds. As a result, the fragile structure of these gels often fractures during drying [51–53]. In spite of this, bonding between secondary particles of silica can be reinforced with the appendage of additional siloxane bonds via an "ageing process" [54].

Ageing of a gel in its mother solution allows for incorporation of surplus silica monomers which strengthens the gel and prevents rupture during drying [54].

Prior to ageing wet gels are often washed in a solvent/water solution. Washing can aid the intensification of the gels strength and stiffness. Silica particles from the established silica network are dissolved and reprecipitated between the necks of the pre-existent silica network [51, 55]. However, this practice alone does not adequately enhance the robustness of the gel and cannot prevent damage to the gel structure. Thus the exercise of further reinforcing procedures is advantageous [56].

The ageing process strengthens the silica network of the wet gel due to; the dissolution and reprecipitation of silica particles to the gel network, addition of new silica monomers from the ageing solution, cross-linking and branching of additional silica monomers and enhanced polycondensation activity [51, 57].

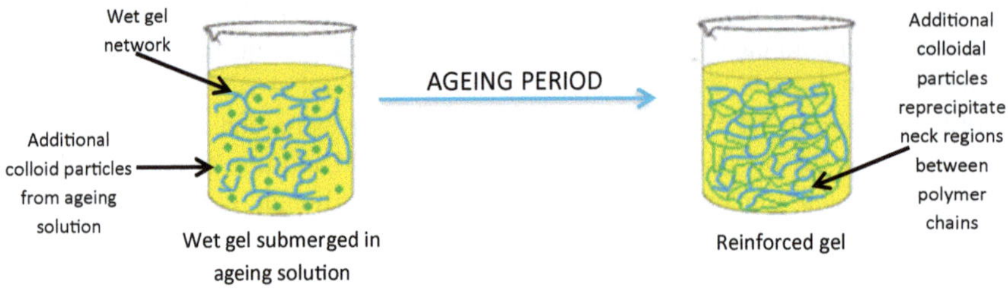

Fig. 1.2 Reinforcement of gel network by the addition of new Silica particles through an ageing process

Two particular mechanisms which take place during the ageing process have a strong impact on the structure of the gel; (1) Dissolution of small particles into larger ones, (2) Transport of material to the neck region. Small additional silica particles, resulting from the ageing solution, which are merged within silica network dissolve and reprecipitate between larger particles within the chain, reinforcing the gel complex (Fig. 1.2). Also, smaller pores within the gel structure are filled and pore volume is increased, permitting the preservation of porosity during drying and thus limiting the collapse of the gel structure. The solute may also precipitate the neck region between siloxane groups, as will supplementary silica particles, added to the established network from the ageing solution. Material which accumulates along the necks between silica polymers coarsens the silica network, increasing the gels durability [28, 50, 58].

Extensive studies have focused on the process of ageing aerogels, highlighting the effects the practice has on the solid framework of the dried gels. Hæreid et al. [51, 59] explored the effects of ageing time, temperature and pH on the stiffness and strength of wet gels, aged in TEOS/Ethanol. Their examination observed that TEOS proved to be more favourable ageing solution than ethanol. Also, an increase in pH of pore liquid, prior to ageing, results in a shorter gelation period. Finally, the nature and surface tension of solvent employed during the washing of the gel influences the linear shrinkage of the aerogel. The study found that shrinkage during drying in presence of various solvents were in the following order from n-heptane < ethanol < acetone.

The effects of ageing conditions on water glass (sodium silicate) based gels have also been studied [60]. An atmospheric drying method, using heptane as the solvent and hexamethyldisilazane (HMDZ) as the silylating agent, was employed in order to produce gels from Na_2SiO_3. Molar ratios of Na_2SiO_3/H_2O and gel-ageing solution were varied in order to observe the effects of such parameters. Monolithic, transparent silica aerogels of 3 cm diameter and 0.8 cm thickness were attained. These findings suggested that altering the molar ratios Na_2SiO_3/H_2O greatly influenced the optical transmittance in comparison to the adjustment of gel-ageing periods. An increase in molar ratio of Na_2SiO_3/H_2O from 2×10^{-3} to 17×10^{-3} resulted in a lower percentage of volume shrinkage with values decreasing from 35 to 0%, and also lowered gel densities which decreased from 0.370 to 0.140 g cm^{-3}. However, in order to increase the optical transmission of the gel from 20 to 70%, the molar ratio of Na_2SiO_3/H_2O was decreased from 17×10^{-3} to 2×10^{-3}. The research, also illustrated that increasing the gel-ageing period from 6 h to 48 h resulted in an increase in volume shrinkage, and an increase in density from 0.140 to 0.205 g cm^{-3}, as well as an increase in optical transmittance from 20 to $\sim$50%, for a 0.8 cm thick sample [60].

In another investigation, it was found that both sodium silicate and colloidal sol derived gels could be aged in a simple manner, in solutions of TEOS monomers, enhancing the strength and stiffness of each of the gels studied [56]. This research found that water glass based gels experienced an increase in shear (G) modulus of more than 10-fold within 27 h, due to high reactivity of the system. On the other hand, colloidal sol synthesised gels were exposed to a smaller increase in G modulus for the same increase in initial density. The larger pore size of these gels however, supports reduced capillary pressure during drying, resulting in xerogels, with densities as low as 0.22 g/cm^3 [56].

Anderson et al. [51] also analysed the effects of experimental parameters on strength and stiffness of gels. Their study included the ageing of aerogels, derived from TMOS precursors, in water, finding that the strength and stiffness of the gel was independent of temperature. The study determined that the shear modulus of the gel increased as a factor of 2 proportional to ageing time. However, unexpectedly, the research discovered a possible correlation between the variation in shear modulus and surface area, rather than ageing temperature. This discovery therefore established the possibility of similar influential ageing mechanisms between 40 and 100 °C. The findings concluded that optimum strength and stiffness of the gel was attained irrespective of temperature. In spite of this, an increase in temperature accelerated the ageing process, reducing the ageing time required to reach optimum strength from 132 h to 3 h [51].

A significant amount of research on the concept of ageing gels in organic materials has determined that ageing conditions such as nature and concentration of the ageing solution, pH of the pore liquid and ageing period strongly impact the structure and properties of silica aerogels [61].

Ageing a gel in water promotes the dissolution and reprecipitation of small silica particles onto the point of contact between siloxane groups, increasing the neck area thus enhancing the silica spine of the gel network. Ageing in mother solution also enhances the strength intensity of the gel by further constituting the silica network by extended hydrolysis and condensation reactions [61]. Kirkbir et al. [62] established that ageing a gel in a siloxane solution strengthens the structure of the gel by adding new monomers to the silica network, enriching the network with cross-linked silica polymers. Moreover, crack-free monolith silica aerogels have been prepared by ageing sodium silicate based gels in a TEOS/EtOH solution [33, 63, 64].

According to published results, increasing the concentration of the precursor solution present in the ageing solution enhances the bulk density of aerogels [37]. Sinkó [37] reported that gels prepared with dilute solvent volumes tend to have higher density. Saturated precursor solutions encourage condensation reactions, due to large amounts of solvent separating the reacting species. Consequently, lower precursor concentrations favour hydrolysis reactions over the condensation process.

Published works also indicate that silicon alkoxide gels require a lengthier ageing time compared to sodium silicate aerogels. For example, Bhangi et al. produced high quality silica aerogels from a sodium silicate precursor inclusive of a short ageing time of 3 h. The hydrophobic (146), and transparent (50%) gels presented favourable properties; a low density of 84 kg m^{-3}, a high porosity of 95%, and a low thermal conductivity of 0.090 W m K^{-1}. Smitha et al. [50] however, achieved high quality aerogels, prepared from TEOS and aged in 80% TEOS which required an ageing period of 48 h in order to exhibit preferred properties such as a surface area of 1098 m^2/g, pore volume of 1.3 cm^3 g^{-1} and an average pore size of 47.7 Å. Extending the ageing period improves the strength of the gel yet raising the ageing temperature shortens the process [51]. Elevated ageing temperatures also encourage the dissolution and reprecipitation process of silica, enhancing the strength of the silica framework, producing low density, monolithic, hydrophobic aerogels [61].

1.9 Drying of Gels

Drying of the gel is arguably the most significant and dominant process involved in preparing aerogels. Since the first upsurge of aerogel research, various methods of drying gels have been implemented (Fig. 1.1),

1. Supercritical drying
2. Ambient pressure drying (APD)—(Subcritical drying)
3. Evaporation
4. Lyophilisation or freeze-drying

Drying frees the pores of the solid structure from liquid which may otherwise cause the silica framework to cave in and succumb, and results in the complete collapse of the gel structure [1]. The drying process is controlled by capillary pressure (P_c) and thus its corresponding critical parameters such as surface tension ($-\gamma lv$), pore volume (V_p) surface area (S_p) and thickness of a surface adsorbed layer (θ) [28, 65, 66].

$$Pc = \frac{-\gamma lv}{(2(Vp/Sp) - \theta}$$

Capillary pressure and surface tension are proportional to one another. If a solvent with a relatively high surface tension is incorporated in the drying process the capillary pressure within the gel pores is increased. Upon removing the liquid, the pores will collapse causing the gel to shrink, resulting in a low surface area material.

Supercritical drying (SCD) and ambient pressure drying are common practices of aerogel drying which replace pore liquid with air. Supercritical drying is the most widely used route of transforming solvent filled gels into aerogels as it allows for the removal of liquid from the gel in a controlled manner, reducing the risk of gel fracture. This practice involves heating wet gels in closed containers causing the

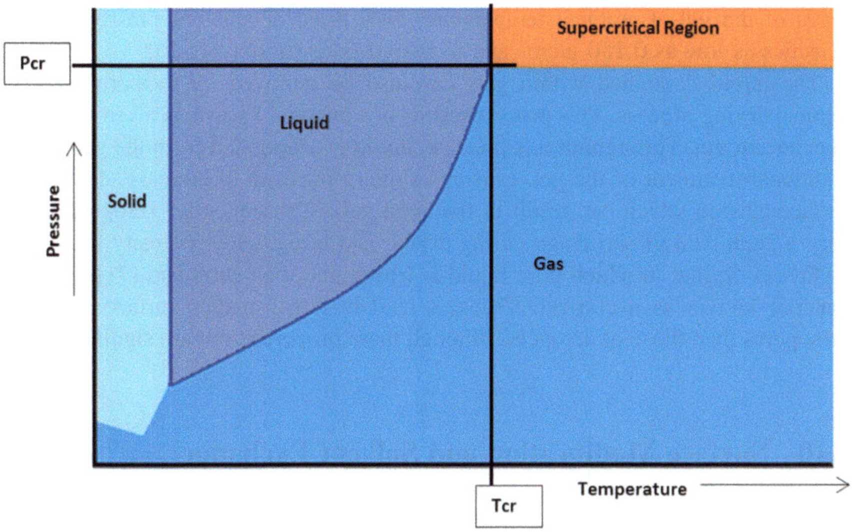

Scheme 1.7 Schematic representation of Supercritical drying process

pore liquid of the gel to reach its point of critical pressure Pcr and temperature Tcr (Scheme 1.7).

As a result the pore liquid converts to a gaseous phase and is evaporated [1, 39]. Supercritical drying can be completed at both ambient temperature (low temperature supercritical drying) and also at elevated temperatures (high temperature supercritical drying). Low temperature supercritical drying (LTSCD), often referred to as CO_2 exchange, involves replacing the precursor solvent, enclosed within the pores of a wet gel, by a liquid which displays a critical temperature and pressure similar to ambient temperature, such as CO_2 [67]. Gels dried with CO_2 however, tend to possess hydrophilic surfaces [1]. Drying gels at low temperature, with CO_2, results in the formation of hydroxyl groups on the surface of the aerogel [1]. On the other hand when SCD is completed at higher temperatures, the solvent which the gel is initially prepared in reaches its critical temperature and pressure before evaporating. The solvent reacts with the hydroxyl groups on the surface of the gel creating methoxyl groups, which results in a hydrophobic aerogel [1].

Ambient pressure, or subcritical, drying has proved resourceful and inexpensive in recent studies [50, 68, 69]. This form of drying is becoming increasingly popular as it is significantly less expensive than supercritical drying. Rather than evaporating the pore liquid using temperature and pressure ambient pressure drying relies on the processes of surface modification, hereafter drying the gel at ambient pressure. The liquid trapped within the pores of the gel is exchanged with a low surface tension solvent which is suitable for evaporative drying. Diffusion of a low surface tension solvent also chemically modifies the surface of the gel, replacing the polar surface groups with non-polar groups. The solvent is then evaporated from the pores of the gel at ambient temperature. Gels dried at ambient pressure present densities as low as 0.150 g/cm^3 and increased pore volume [1, 70].

The liquid contained within gels can also be removed by evaporation via a thermal drying process. This process results in a material known as a xerogel, rather than an aerogel. Throughout this practice, liquid is evaporated from the gel without additional treatment of the gel, causing capillary pressure to increase, due to high surface tension which can result in fractured gels. Consequently, xerogels experience a great deal of shrinkage during drying and possess low porosity [5].

Freeze drying, in which pore liquid is frozen and then sublimed [71], creates a material known as a cryogel. Cryogels tend to have smaller surface areas and mesopores than those of aerogels, although these properties remain significant [39].

1.10 Surface Modification and Solvent Exchange

The ageing process of aerogels allows extended hydrolysis and polycondensation reactions to be completed. Lingering particles dispersed throughout the silica chain are dissolved and incorporated into the gel network. However, excess water, alcohol, catalyst and even sodium ions may still remain within the pores of the newly formed gel. For instance, after gelation, sodium silicate gels contain a vast amount of water within their pores. The residue can be difficult to extract through mere employment of a drying processes, often leading to complications such as cracking of the gel [28]. Thus, additional measures must be implemented in order to remove the residue, such as surface modification and solvent exchange. Failure to remove deposits, such as water, from the gel pores, results in a decline of gel structure and will affect the surface nature of the gel, leading to hydrophilic aerogels.

Hydrogen bonding, due to polar alkoxy groups located on the surface of aerogels, and the presence of H_2O in the aerogels surrounding environment cause deterioration of aerogels structure overtime. Chemical surface modification or silylation is exercised in order to alter the silica surface network from hydrophilic to hydrophobic and also to remove organic compounds such as alcohol (Fig. 1.3). This transition therefore improves the surface composition of silica gels and eases the surface tension of the solvent contained within the pores of the gel, promoting ambient pressure drying [72].

Silylation replaces hydrogen of the protic material located on the gels surface with a substituted silicon atom (SiR_3), rendering a non-polar gel surface (Scheme 1.8).

Many silylating reagents are available for purpose of aerogel surface modification (Table 1.2). The most commonly used reagents are chemicals which contain a trimethlysilyl group (Me_3Si). However, many other silylating reagents are also available and have been recurrently employed in research, most of which are noted in Table 1.2.

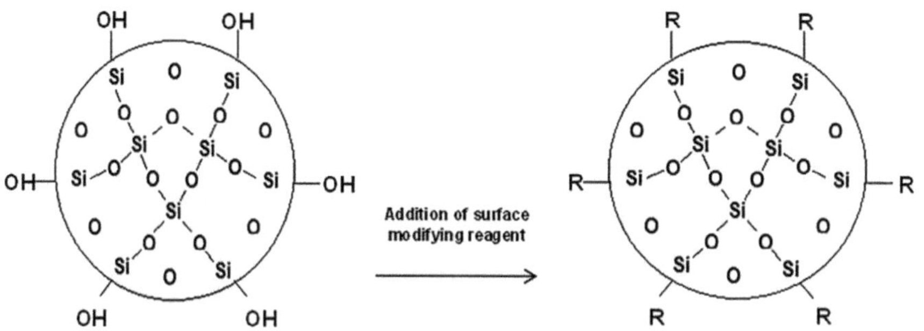

Fig. 1.3 Surface modification of hydrophilic silica gels

Scheme 1.8 Replacement of hydrogen with Silylating agent

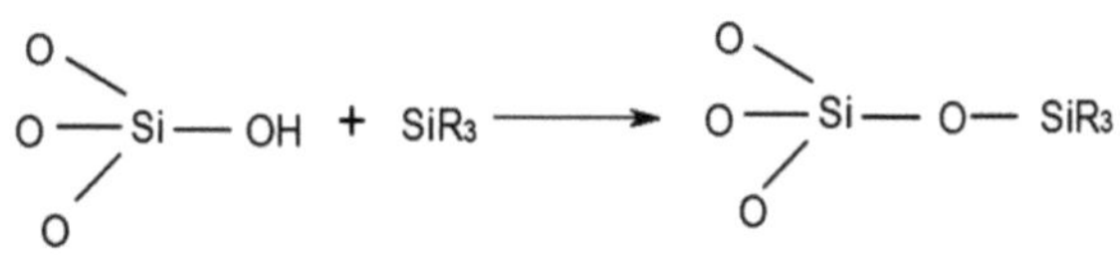

Table 1.2 Common silylating agents [73, 74]

Silyating agent	Abbreviation	Chemical structure
Methyltrimethoxysilane	MTMS	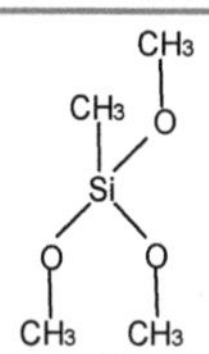
Methyltriethoxysilane	MTES	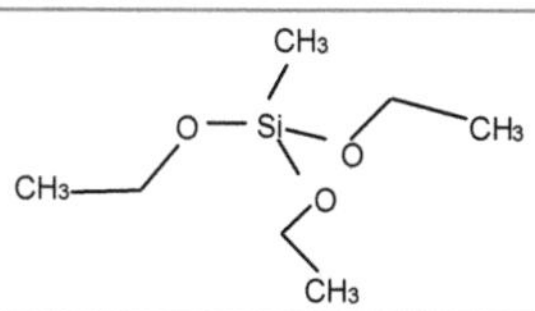
Dimethyldimethoxysilane	DMDMS	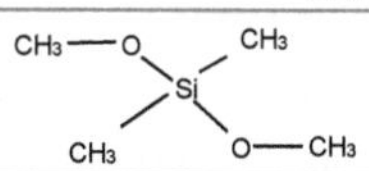
Dimethylchlorosilane	DMCS	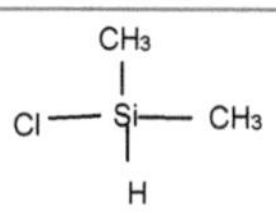
Trimethylethoxysilane	TMES	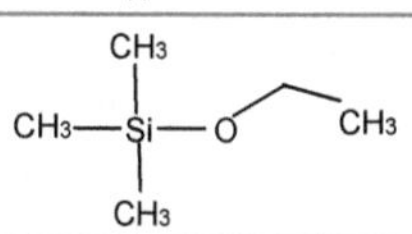
Trimethylchlorosilane	TMCS	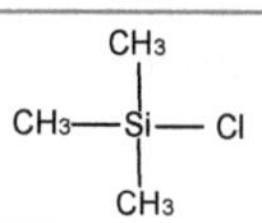

(continued)

Table 1.2 (continued)

Silyating agent	Abbreviation	Chemical structure
Vinyltrimethoxysilane	VTMS	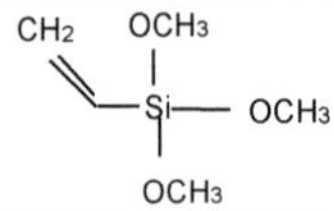
Phenyltrimethoxysilane	PTMS	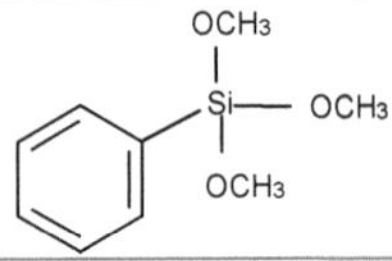
Phenyltriethoxysilane	PTES	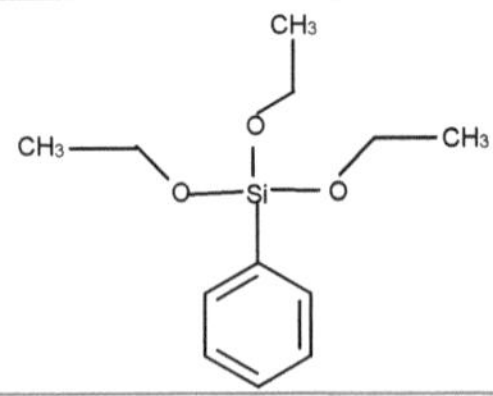
Ethyltriethoxysilane	ETES	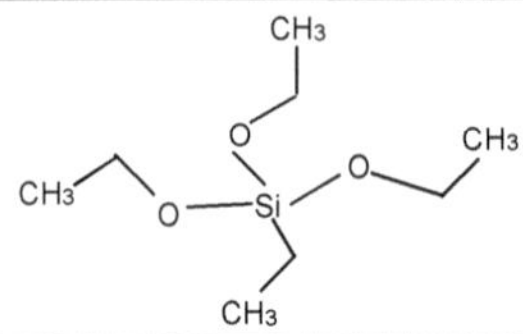
Bis(trimethylsilyl) acetamide	BTSA	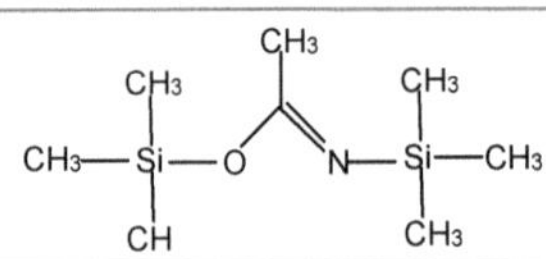
Hexamethyldisilazane	HMDZ	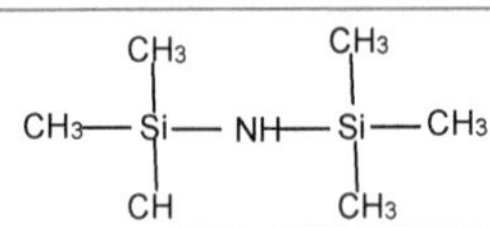

Wu et al. [72] accomplished a two-step surface modification employing methyltrimethoxysilane (MTMS) and trimethylchlorosilane (TMCS)/ethanol (EtOH)/n-Hexane of various molar ratios as silylating reagents, producing hydrophobic gels with properties comparable to those of subcritically dried gels. This route was applied in order to investigate a simple and effective means of synthesis exercising ambient pressure drying. The two-step modification process allowed modification of both the surface and properties of the aerogel. The study determined that increasing the molar ratio of MTMS/TEOS from 0.1 to 0.3 m increased the water contact angle of the produced aerogels from 137 to 156°. Furthermore the silica gels exhibited porosity of 92.27%, bulk density of 0.17 g cm^{-3}, and pore size ranging from 6.93 to 10.78 nm, depending on the molar ratio of reactants.

The effects of various modifying agents on the physical properties of aerogels were investigated by Rao et al. [74]. The results of this investigation are presented in Table 1.3.

Table 1.3 Physical and hydrophobic properties of the silica gels with various silylating agents [73, 74]

Silylating agent	Density	Porosity (%)	Pore volume (cm^3/g)	Transmission (%)	Thermal conductivity (W/mK)	Heat capacity per unit volume (kj/m^3K)	Refractive index (n)	Contact angle (n)
Methlytrimethoxysilane	0.152	92.05	6.05	85	0.164	375	1.029	60
Methyltriethoxysilane	0.164	91.45	5.51	88	0.170	380	1.031	65
Vinyltrimethoxysilane	0.157	91.94	5.84	86	0.169	382	1.030	70
Phenyltrimethoxysilane	0.145	92.56	6.37	75	0.154	340	1.027	80
Phenyltriethoxysilane	0.149	92.15	6.18	75	0.152	345	1.028	85
Dimethlychlorosilane	0.183	91.27	–	10	–	–	–	–
Dimethyldichlorosilane	0.195	90.69	–	8	–	–	–	–
Dimethyldimethoxysilane	0.131	93.32	7.10	70	0.134	330	1.025	88
Trimethylmethoxysilane	0.110	94.35	8.58	70	0.121	288	1.021	130
Trimethylethoxysilane	0.097	95.38		90	–	–	–	–
Trimethylchlorosilane	0.105	94.61	9.01	30	0.118	270	1.020	155
TrimethlySilyl-chloroacetate	–	–	–	5	–	–	–	–
Trimethylbromosilane	0.235	88.82	–	7	–	–	–	–
Tetraethylethoxysilane	0.097	95.39	–	91	–	–	–	–
Bis(trimethylsilyl) acetamide	0.090	95.38	10.60	60	0.102	260	1.017	150
Hexamethyldi silazane	0.060	96.90	16.15	65	0.091	204	1.011	165

Their examination concluded that tri alkyl silylating agents produced hydrophobic gels with lowest % of volume shrinkage, lowest density, lowest refractive index, large pore volume and highest percentage of porosity. Particularly HMDZ which produced an aerogel with a low density of 0.060 g/cm^3, low refractive index of 1.011, a high porosity of 96.9%, low thermal conductivity of 0.09 w/m K, good hydrophobicity (165°) and thermally stability up to 325 °C.

Another study was also performed by Rao et al. [73] in which different alkyl-alkoxy/chloro silanes were compared, observing their effects on aerogel properties. The group's findings reported that the use of chlorosilanes resulted in highly fractured acidic aerogels. However, alkyl-alkoxy silane modification produced hydrophobic, monolithic, transparent gels of low density as can be seen in Table 1.3.

As surface modification is generally completed with the use of organic solvents, solvent exchange is necessary. This process also purifies the remaining liquid within the pores of the gel by removing excess components, other than solvent, which may remain. For instance, Kistler employed solvent exchange using ethanol in order to remove sodium chloride from the pores of a waterglass based wet gel [75]. The gel is soaked in solvent causing impurities to diffuse out from the pores of the gel. The departed residue is then replaced with pure solvent.

The removal of water and other remaining components by solvent exchange and surface modification is therefore imperative [28, 76, 77]. Both solvent exchange and surface modification reduce capillary stresses [78]. Additionally liquids of low surface tension and high molecular volume, employed as surface modifiers produce high surface area gels [79].

These vital procedures can be completed separately, one after the other. However, each process is time consuming and would result in tedious processing steps. Recent research [68, 80, 81] has witnessed the implementation of a one-step solvent exchange, surface modification technique to remove excess water and other components at a reasonable cost and timeframe. Shi et al. [68] employed EtOH/TMCS/Heptane with EtOH/TMCS in a molar ratio of 1:1, to facilitate solvent exchange and act as a surface modifying reagent simultaneously. This synthesis technique resulted in crack-free silica aerogels during the reaction of EtOH/TMCS/Heptane solution with alcogels, by first adding ethanol and heptane. The procedure produced monolithic, hydrophobic microstructured gels with more uniform particle and pore size distribution, illustrating that one-step solvent exchange and surface modification of wet silica gels can be successfully, simultaneously accomplished.

1.11 Conclusions

The advancement of sol-gel processing is due to the ease of control of the nanoarchitecture throughout synthesis. The ability to tailor the structure of nanosystems from primitive processing phases advocates the preparation of pure materials with improved properties. The advanced properties of such compounds afford their use, in the form of fibres, films, fine powders and monoliths within a diverse range of applications.

References

1. Dorcheh, A.S., Abbasi, M.H.: Silica aerogel; synthesis, properties and characterization. J. Mater. Process. Technol. **199**(1–3), 10–26 (2008)
2. Du, A., Zhou, B., Zhang, Z., Shen, J.: A special material or a new state of matter: a review and reconsideration of the aerogel. Mater. **6**(3), 941–968 (2013)
3. Riffat, S.B., Qiu, G.: A review of state-of-the-art aerogel applications in buildings. Int. J. Low-Carbon Technol. **8**(1), 1–6 (2013)
4. Fricke, J.: Aerogels—highly tenuous solids with fascinating properties. J. Non-Cryst. Solids **100**(1–3), 169–173 (1988)
5. Fricke, J., Emmerling, A.: Aerogels. J. Am. Ceram. Soc. **75**(8), 2027–2035 (1992)
6. Carraher Jr, C.E.: Carraher's polymer chemistry, CRC Press (2013)
7. Schultz, J.M., Jensen, K.I., Kristiansen, F.H.: Super insulating aerogel glazing. Sol. Energy Mater. Sol. Cells **89**(2), 275–285 (2005)
8. Eisen, H.J., Wen, L., Hickey G., Braun, D.: Sojourner mars rover thermal performance, SAE Technical Paper (1998)
9. Fesmire, J.E.: Aerogel insulation systems for space launch applications. Cryogenics **46**(2), 111–117 (2006)
10. Ackerman, W.C., Changming, J., Cho, C.-C., Gnade, B.E., Johnston, G.C., Smith, D.M.: Porous dielectric material with improved pore surface properties for electronics applications. Google Patents (2000)
11. Cho, C.-C., Gnade, B., Levine, J.D.: Low density, high porosity material as gate dielectric for field emission device. Google Patents (1996)
12. Wang, C.-T., Wu, C.-L., Chen, I., Huang, Y.-H.: Humidity sensors based on silica nanoparticle aerogel thin films. Sens. Actuators B: Chem. **107**(1), 402–410 (2005)
13. Khuri-Yakub, B., Kim, J., Chou, C.-H., Parent, P., Kino, G. (eds.): A new design for air transducers. Ultrasonics Symposium, 1988 Proceedings, IEEE (1988)
14. Ahmed, M.S., Attia, Y.A.: Aerogel materials for photocatalytic detoxification of cyanide wastes in water. J. Non-Cryst. Solids **186**, 402–407 (1995)
15. Reynolds, J.G., Coronado, P.R., Hrubesh, L.W.: Hydrophobic aerogels for oil-spill cleanup? intrinsic absorbing properties. Energ Source **23**(9), 831–843 (2001)
16. Danilyuk, A., Kirillov, V., Savelieva, M., Bobrovnikov, V., Buzykaev, A., Kravchenko, E., et al.: Recent results on aerogel development for use in Cherenkov counters. Nuclear instruments and methods in physics research section A: accelerators. Spectrometers, Detectors and Associated Equipment **494**(1), 491–494 (2002)
17. Buzykaev, A., Danilyuk, A., Ganzhur, S., Gorodetskaya, T., Kravchenko, E., Onuchin, A., et al.: Aerogels with high optical parameters for Cherenkov counters. Nuclear instruments and methods in physics research section A: accelerators. Spectrometers, Detectors and Associated Equipment **379**(3), 465–467 (1996)
18. Plata, D.L., Briones, Y.J., Wolfe, R.L., Carroll, M.K., Bakrania, S.D., Mandel, S.G., et al.: Aerogel-platform optical sensors for oxygen gas. J. Non-Cryst. Solids **350**, 326–335 (2004)
19. Baetens, R., Jelle, B.P., Gustavsen, A.: Aerogel insulation for building applications: A state-of-the-art review. Energy Build. **43**(4), 761–769 (2011)
20. Zhang, Y., Feng, H., Wu, X., Wang, L., Zhang, A., Xia, T., et al.: Progress of electrochemical capacitor electrode materials: a review. Int. J. Hydrogen Energy **34**(11), 4889–4899 (2009)

21. Miller, J., Dunn, B., Tran, T., Pekala, R.: Deposition of ruthenium nanoparticles on carbon aerogels for high energy density supercapacitor electrodes. J. Electrochem. Soc. **144**(12), L309–L311 (1997)
22. Pajonk, G.: Aerogel catalysts. Applied Catalysis **72**(2), 217–266 (1991)
23. Reynolds, J.G., Hair, L.M., Coronado, P.R., Droege M.W., Wong, J. (eds.): Aerogel derived catalysts. MRS Proceedings, Cambridge Univ Press (1996)
24. Shukla, N., Kosny, J.: Aerogel thermal insulation—technology review and cost study for building enclosure applications **120**, 294–307 (2014)
25. Gash, A.E., Tillotson, T.M., Satcher Jr., J.H., Hrubesh, L.W., Simpson, R.L.: New sol–gel synthetic route to transition and main-group metal oxide aerogels using inorganic salt precursors. J. Non-Cryst. Solids **285**(1–3), 22–28 (2001)
26. Wagh, P.B., Begag, R., Pajonk, G.M., Rao, A.V., Haranath, D.: Comparison of some physical properties of silica aerogel monoliths synthesized by different precursors. Mater. Chem. Phys. **57**(3), 214–218 (1999)
27. Zhang, H., Hong C., Qiao, Y.: Synthesis, structural and thermal properties of nano-porous SiO2-based aerogels. INTECH Open Access Publisher (2011)
28. Brinker C.J., Scherer, G.W.: Sol-gel science: the physics and chemistry of sol-gel processing. Gulf Professional Publishing (1990)
29. Kim, P.B.S.J.-K., Park J.-K., Kim, H.-K.: Influence of solvent exchange on the physical properties of sodium silicate based aerogel prepared at ambient pressure (2006)
30. Schwertfeger, F., Frank, D., Schmidt, M.: Hydrophobic waterglass based aerogels without solvent exchange or supercritical drying. J. Non-Cryst. Solids **225**(0), 24–29 (1998)
31. Gurav, J.L., Rao, A.V., Rao, A.P., Nadargi, D.Y., Bhagat, S.D.: Physical properties of sodium silicate based silica aerogels prepared by single step sol–gel process dried at ambient pressure. J. Alloy. Compd. **476**(1–2), 397–402 (2009)
32. Rao, A.V., Bangi, U.K.H., Kavale, M.S., Imai, H., Hirashima, H.: Reduction in the processing time of doped sodium silicate based ambient pressure dried aerogels using shaker. Microporous Mesoporous Mater. **134**(1–3), 93–99 (2010)
33. Bangi, U.K., Rao, A.V., Rao, A.P.: A new route for preparation of sodium-silicate-based hydrophobic silica aerogels via ambient-pressure drying. Sci. Technol. Adv. Mater. **9**(3), 035006 (2008)
34. Sarawade, P.B., Kim, J.-K., Hilonga, A., Kim, H.T.: Production of low-density sodium silicate-based hydrophobic silica aerogel beads by a novel fast gelation process and ambient pressure drying process. Solid State Sci. **12**(5), 911–918 (2010)
35. Bangi, U.K.H., Jung, I.-K., Park, C.-S., Baek, S., Park, H.-H.: Optically transparent silica aerogels based on sodium silicate by a two step sol–gel process and ambient pressure drying. Solid State Sci. **18**(0), 50–57 (2013)
36. Shao, Z., Luo, F., Cheng, X., Zhang, Y.: Superhydrophobic sodium silicate based silica aerogel prepared by ambient pressure drying. Mater. Chem. Phys. **141**(1), 570–575 (2013)
37. Sinkó, K.: Influence of Chemical Conditions on the Nanoporous Structure of Silicate Aerogels. Mater. **3**(1), 704–740 (2010)
38. Norris P.M., Shrinivasan, S.: Aerogels: unique material, fascinating properties and unlimited applications. Annu. Rev. Heat Transf. **14**(14), (2005)
39. Pierre, A.C., Pajonk, G.M.: Chemistry of Aerogels and Their Applications. Chem. Rev. **102**, 4243–4265 (2002)
40. Brinker, C., Keefer, K., Schaefer, D., Ashley, C.: Sol-gel transition in simple silicates. J. Non-Cryst. Solids **48**(1), 47–64 (1982)
41. Rao, A.V., Pajonk, G., Parvathy, N.: Effect of solvents and catalysts on monolithicity and physical properties of silica aerogels. J. Mater. Sci. **29**(7), 1807–1817 (1994)
42. Karmakar, B., De, G., Ganguli, D.: Dense silica microspheres from organic and inorganic acid hydrolysis of TEOS. J. Non-Cryst. Solids **272**(2–3), 119–126 (2000)
43. Wright J.D., Sommerdijk, N.A.: Sol-gel materials: chemistry and applications. CRC press (2000)

44. Venkateswara Rao, A., Bhagat, S.D.: Synthesis and physical properties of TEOS-based silica aerogels prepared by two step (acid–base) sol–gel process. Solid State Sci. **6**(9), 945–952 (2004)
45. Kirkbir, F., Murata, H., Meyers, D., Chaudhuri, S.R., Sarkar, A.: Drying and sintering of sol-gel derived large SiO2 monoliths. J. Sol-Gel. Sci. Technol. **6**(3), 203–217 (1996)
46. Wagh, P., Rao, A.V., Haranath, D.: Influence of molar ratios of precursor, solvent and water on physical properties of citric acid catalyzed TEOS silica aerogels. Mater. Chem. Phys. **53**(1), 41–47 (1998)
47. Nadargi, D.Y., Kalesh, R.R., Rao, A.V.: Rapid reduction in gelation time and impregnation of hydrophobic property in the tetraethoxysilane (TEOS) based silica aerogels using NH4F catalyzed single step sol–gel process. J. Alloy. Compd. **480**(2), 689–695 (2009)
48. Sarawade, P.B., Kim, J.-K., Kim, H.-K., Kim, H.-T.: High specific surface area TEOS-based aerogels with large pore volume prepared at an ambient pressure. Appl. Surf. Sci. **254**(2), 574–579 (2007)
49. Hæreid, S., Dahle, M., Lima, S., Einarsrud, M.A.: Preparation and properties of monolithic silica xerogels from TEOS-based alcogels aged in silane solutions. J. Non-Cryst. Solids **186**(0), 96–103 (1995)
50. Smitha, S., Shajesh, P., Aravind, P., Kumar, S.R., Pillai, P.K., Warrier, K.: Effect of aging time and concentration of aging solution on the porosity characteristics of subcritically dried silica aerogels. Microporous Mesoporous Mater. **91**(1), 286–292 (2006)
51. Hæreid, S., Anderson, J., Einarsrud, M.A., Hua, D.W., Smith, D.M.: Thermal and temporal aging of TMOS-based aerogel precursors in water. J. Non-Cryst. Solids **185**(3), 221–226 (1995)
52. Estella, J., Echeverría, J.C., Laguna, M., Garrido, J.J.: Effects of aging and drying conditions on the structural and textural properties of silica gels. Microporous Mesoporous Mater. **102**(1–3), 274–282 (2007)
53. Strøm, R., Masmoudi, Y., Rigacci, A., Petermann, G., Gullberg, L., Chevalier, B., Einarsrud, M.-A.: Strengthening and aging of wet silica gels for up-scaling of aerogel preparation. J. Sol-Gel. Sci. Technol. **41**(3), 291–298 (2007)
54. Zarzycki, J., Prassas, M., Phalippou, J.: Synthesis of glasses from gels: the problem of monolithic gels. J. Mater. Sci. **17**(11), 3371–3379 (1982)
55. Haereid, S., Nilsen, E., Ranum, V., Einarsrud, M.-A.: Thermal and temporal aging of two step acid-base catalyzed silica gels in water/ethanol solutions. J. Sol-Gel. Sci. Technol. **8**(1–3), 153–157 (1997)
56. Einarsrud, M.-A., Nilsen, E.: Strengthening of water glass and colloidal sol based silica gels by aging in TEOS. J. Non-Cryst. Solids **226**(1–2), 122–128 (1998)
57. Einarsrud, M.-A., Haereid, S.: Preparation of transparent, monolithic silica xerogels with low density. J. Sol-Gel. Sci. Technol. **2**(1–3), 903–906 (1994)
58. Cuce, E., Cuce, P.M., Wood, C.J., Riffat, S.B.: Toward aerogel based thermal superinsulation in buildings: a comprehensive review. Renew. Sustain. Energy Rev. **34**(0), 273–299 (2014)
59. Hæreid, S., Nilsen, E., Einarsrud, M.-A.: Properties of silica gels aged in TEOS. J. Non-Cryst. Solids **204**(3), 228–234 (1996)
60. Rao, A.V., Rao, A.P., Kulkarni, M.: Influence of gel aging and Na2SiO3/H2O molar ratio on monolithicity and physical properties of water-glass-based aero-gel dried at atmospheric pressure. J. Non-Cryst. Solids **350**, 224–229 (2004)
61. He, F., Zhao, H., Qu, X., Zhang, C., Qiu, W.: Modified aging process for silica aerogel. J. Mater. Process. Technol. **209**(3), 1621–1626 (2009)
62. Kirkbir, F., Murata, H., Meyers, D., Chaudhuri, S.: Drying of aerogels in different solvents between atmospheric and supercritical pressures. J. Non-Cryst. Solids **225**, 14–18 (1998)
63. Cha, Y.C., Kim, C.E., Lee, S.H., Hwang, H.J., Moon, J.W., Han, I.S., et al.: Synthesis of silica aerogel thin film from waterglass. Solid State Phenom. **124**, 671–674 (2007)
64. Kim, C.E., Yoon, J.S., Hwang, H.J.: Synthesis of nanoporous silica aerogel by ambient pressure drying. J. Sol-Gel. Sci. Technol. **49**(1), 47–52 (2009)
65. Patel, P.R., Purohit, N.S., Suthar, M.A.: An Overview of Silica Aerogels. Int. J. ChemTech Res. **1**(4), 1052–1057 (2009)

66. Scherer, G.W., Hæreid, S., Nilsen, E., Einarsrud, M.-A.: Shrinkage of silica gels aged in TEOS. J. Non-Cryst. Solids **202**(1), 42–52 (1996)
67. Tewari, P.H., Hunt, A.J., Lofftus, K.D.: Ambient-temperature supercritical drying of transparent silica aerogels. Mater. Lett. **3**(9), 363–367 (1985)
68. Shi, F., Wang, L., Liu, J.: Synthesis and characterization of silica aerogels by a novel fast ambient pressure drying process. Mater. Lett. **60**(29–30), 3718–3722 (2006)
69. Smitha, S., Shajesh, P., Warrier, K.G.K.: Investigations on the effect of experimental parameters on the porosity features of silica aerogels synthesized at ambient drying conditions. Mater. Chem. Phys. **131**(1–2), 507–511 (2011)
70. Land, V.D., Harris, T.M., Teeters, D.C.: Processing of low-density silica gel by critical point drying or ambient pressure drying. J. Non-Cryst. Solids **283**(1), 11–17 (2001)
71. Shlyakhtin, O., Tretyakov, Y.: Recent progress in cryochemical synthesis of oxide materials. J. Mater. Chem. **9**(1), 19–24 (1999)
72. Wu, G., Yu, Y., Cheng, X., Zhang, Y.: Preparation and surface modification mechanism of silica aerogels via ambient pressure drying. Mater. Chem. Phys. **129**(1–2), 308–314 (2011)
73. Rao, A.V., Kulkarni, M.M., Amalnerkar, D.P., Seth, T.: Surface chemical modification of silica aerogels using various alkyl-alkoxy/chloro silanes. Appl. Surf. Sci. **206**(1–4), 262–270 (2003)
74. Rao, A.P., Rao, A.V., Pajonk, G.M.: Hydrophobic and physical properties of the ambient pressure dried silica aerogels with sodium silicate precursor using various surface modification agents. Appl. Surf. Sci. **253**(14), 6032–6040 (2007)
75. Kistler, S.S.: Method of producing aerogels. US Patent 2,093,454 (1937)
76. Ackerman, W.C., Vlachos, M., Rouanet, S., Fruendt, J.: Use of surface treated aerogels derived from various silica precursors in translucent insulation panels. J. Non-Cryst. Solids **285**(1–3), 264–271 (2001)
77. Rao, A.P., Rao, A.V.: Microstructural and physical properties of the ambient pressure dried hydrophobic silica aerogels with various solvent mixtures. J. Non-Cryst. Solids **354**(1), 10–18 (2008)
78. Deshpande, R., Hua, D.-W., Smith, D.M., Brinker, C.J.: Pore structure evolution in silica gel during aging/drying. III. Effects of surface tension. J. Non-Cryst. Solids **144**, 32–44 (1992)
79. Kumar, S.R., Pillai, P.K., Warrier, K.: Synthesis of high surface area silica by solvent exchange in alkoxy derived silica gels. Polyhedron. **17**(10), 1699–1703 (1998)
80. Wang, L.-J., Zhao, S.-Y., Yang, M.: Structural characteristics and thermal conductivity of ambient pressure dried silica aerogels with one-step solvent exchange/surface modification. Mater. Chem. Phys. **113**(1), 485–490 (2009)
81. Bhagat, S.D., Kim, Y.-H., Suh, K.-H., Ahn, Y.-S., Yeo, J.-G., Han, J.-H.: Superhydrophobic silica aerogel powders with simultaneous surface modification, solvent exchange and sodium ion removal from hydrogels. Microporous Mesoporous Mater. **112**(1), 504–509 (2008)

Chapter 2
Sol-Gel Materials for Varistor Devices

S. Anas, K.V. Mahesh, M. Jeen Maria and S. Ananthakumar

"Struck by lightning! Struck by lightning!"
— J.R.R. Tolkien
"Somewhere, something incredible is waiting to be known"
— Carl Sagan

2.1 Varistors—An Introduction

Varistors are known in academic, research and industrial sectors as, 'lightning arresters/surge suppressers/nonlinear resistors/voltage sensors/voltage arrestors/voltage regulators'. A varistor is a device which functions mainly by sensing and limiting the transient voltage during online operation [1]. Varistors protect circuits over a very wide range of voltages, from a few volts in semiconductor circuits to tens of kilovolts in electrical power distribution networks [2, 3]. They are also defined as nonlinear resistors due to their non-ohmic behaviour [4, 5]. The images of real varistors are provided in Fig. 2.1

Varistors are used from domestic appliances to industrial devices in both AC or DC power lines. They need only a very less response time of the order of less than 20 ns for clamping the transient surges. So, they are widely used to control the variable voltages in all modern electric or electronic devices like cell phones, laptops, tablets, notebooks, computers, cameras, watches, television, LCD and LED devices, multimedia devices, etc. [6–9]. Varistors can be connected directly across main supplies and across semiconductor switches for protecting transistors, MOSFET's and thyristor bridges. Unlike transient suppressor diodes, they absorb much higher transient energies and suppress both positive and negative transients.

S. Anas (✉) · K.V. Mahesh · M. Jeen Maria · S. Ananthakumar (✉)
Functional Materials Section, Materials Science and Technology Division,
CSIR-National Institute for Interdisciplinary Science and Technology,
Thiruvananthapuram, Kerala 695019, India
e-mail: anas.tkmchem@gmail.com

S. Anas
T. K. M. College of Arts and Science, Karicode, Kollam, Kerala 691005, India

S.C. Pillai and S. Hehir (eds.), *Sol-Gel Materials for Energy, Environment and Electronic Applications*, Advances in Sol-Gel Derived Materials and Technologies,
DOI 10.1007/978-3-319-50144-4_2

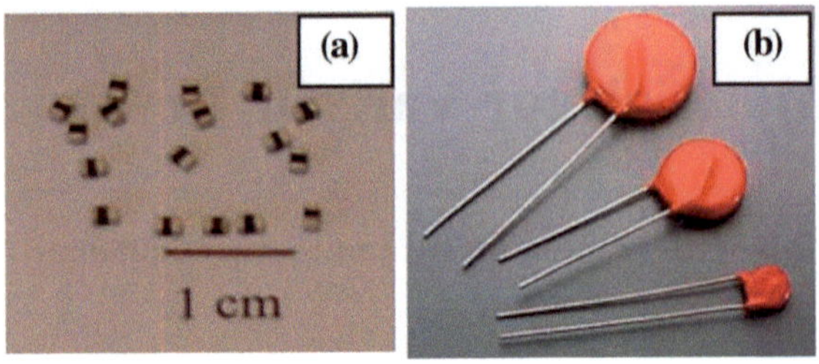

Fig. 2.1 Various types of ZnO varistors fabricated for different uses **a** varistors used in electronic circuits for ESD (electro static discharge) low-voltage suppression and protection **b** varistors used in industrial/AC line protection (reproduced with permission from [5])

Their energy absorption capability can also be tuned with respect to size of the component [10, 11]. They have wide range of voltage selection which allows for the easy selection of the correct component for any specific application. They exhibit low capacitance values, which makes them suitable even for the protection of the digital switching circuitry.

Varistors normally possess excellent surge withstanding capability because, unlike Zener diode which operates through a single junction, they generally functions as multijunction devices [12]. Furthermore, their special electrical properties are dominated by grain–boundary interface state [13]. A varistor consists of conductive grains and intergranular non-conductive layers. Existence of a potential barrier at grain boundaries results in nonlinear response of the varistor [14, 15]. This potential barrier, called Schottky barrier, blocks charge carriers by reducing the mobility of the carriers and increasing the effective resistivity of the grains. In fact, the Schottky barrier is formed due to trapping of electrons at grain boundaries. In other words, thin insulating intergranular layer between two ZnO grains creates back-to-back Zener diodes at grain boundaries. The typical property of a varistor can be better understood from the characteristic current–voltage (I/V) plot. The curve representing the current carrying capacity of a varistor as a function of the applied voltage is given as Fig. 2.2 [9].

Fig. 2.2 Schematic I–V characteristic *curve* of an idealized varistor **a** leakage or pre-switch region, I_L leakage current; V_c, breakdown voltage **b** non-ohmic or varistor operation region **c** high current or upturn region (reproduced with permission from [9])

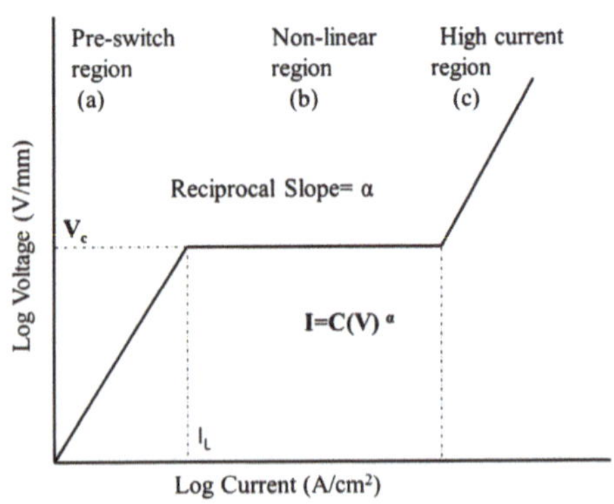

As illustrated in the figure, three major regions were identified for the varistors [4, 7]. The region 1 is the low-current region (<10^{-4} A/cm^{-2}). It is termed as 'pre-switch' or 'ohmic region'. Here, the applied current is a linear function of the voltage. The resistivity of the varistor is very high here (greater than 10^{10} Ω cm for a normal ZnO varistor). For a given varistor device, capacitance remains constant over a wide range of voltage and frequency in the pre-switch region and its dielectric characteristics are governed by the impedance of the ZnO grain boundaries. For AC applications, the total leakage current (IL) in the pre-breakdown region is composed of resistive (IR) and capacitive (IC) currents. In the design of a surge suppresser from a ZnO varistor, it is the resistive current (IR), which is of importance since this is responsible for Joule heating within the elements. The second region is associated with a nonlinear behaviour. This particular 'non-ohmic region', termed as the heart of the varistor device, decides the performance of a varistor. The I–V curve is nonlinear in this region wherein the varistor voltage remains approximately constant for a large change in current. This specific nonlinear response is due to its polycrystalline microstructure, which is primarily governed by the grain–grain interface architecture. The current (I)–voltage (V) relation of a varistor in this region can be expressed in power law as

$$I = C(V)^{\alpha} \tag{2.1}$$

or

$$J = (E/C)^{\alpha}, \tag{2.2}$$

where 'C' the proportionality constant corresponding to the resistance of an ohmic resistor (a material constant), 'α' is the nonlinearity coefficient, 'J' is the current density, 'E' the applied electric field.

The third region is the high-current region. Here, the varistor becomes highly conducting and draws the current through it. In this region, the current changes by many orders of magnitude for a small change in voltage. In the high-current region (>100 A cm^{-2}), the varistor approximates to a short circuit and the I–V curve exhibits linear characteristics similar to those in the low-current region. The dielectric characteristics of this upturn region are governed by the impedance of the grain in the ZnO microstructure.

There are three significant parameters that determine the varistor properties; the breakdown voltage 'V_b', the non- linear coefficient 'α', and the leakage current 'I_f' [2, 4, 16]. In detail, the breakdown voltage 'V_b' is defined as the voltage at which the device switches from resistive behaviour to conductive behaviour. The breakdown field is usually taken as the field applied when the current flowing through the varistor is 1 mA cm^{-2}. The breakdown field 'V_b' and the sample grain size 'D' are proportionally connected by the following relation:

$$V_b = V_{gb}/D, \quad (2.3)$$

where 'V_{gb}' is the barrier voltage. From the measured values of 'V_b' and 'D', 'V_{gb}' could be obtained.

The nonlinearity coefficient, α, is defined as

$$\alpha = \log[I_2/I_1]/\log[V_2/V_1], \quad (2.4)$$

where I_1 and I_2 are the currents corresponding to the voltages V_1 and V_2, respectively. It is practically measured from the reciprocal of the slope in the breakdown region. The nonlinearity coefficient represents protective power of a varistor device. The higher the value of nonlinearity or 'α', the better will be the protection power of the device. Because of its variation with current, 'α' values were taken in the range from 0.1 to 1 mA cm^{-2}, 1 to 10 mA cm^{-2} or in specific cases up to 100 mA cm^{-2}. Typical reported 'α' values for ZnO varistors varied from 20 to 100 over the current ranges mentioned above. The leakage current (I_L) is defined as the current flow at the steady operating voltage. The parameters, which are known to influence the leakage current, are (a) formulation of the materials (b) voltage applied and (c) time interval of the voltage applied. The most desirable characteristics of a varistor thus identified are, a high value of nonlinear coefficient (α), an acceptable rating of nonlinear voltage (V_b), a low value of leakage current (I_L), a long varistor life and a high-energy absorption capability.

2.2 Conventional Preparation of Varistors

The first work on varistor ceramics was reported on silicon carbide (SiC) and germanium (Ge) semiconductors [17, 18]. The strontium- and calcium-based titanates ($CaTiO_3$, $SrTiO_3$) are evolved later [19, 20]. However, these materials are outdated for the reason of poor nonlinear coefficients. For example, the nonlinear coefficient of only 10 was achieved in SiC varistor. In the 70s, Japanese physicist Matsuoka developed polycrystalline ZnO-based material as varistors [1]. Till today, his work is acknowledged as a great success in varistor ceramics. Based on the process identified by Matsuoka, the company Meidensha Corporation, a pioneer, developed ZnO surge arrestors commercially for the first time termed 'metal oxide varistors'. Now, ZnO varistors have become a commodity item and mass produced in several countries for applications ranging from power switching in electrical transmission systems to surge protection in automobile and semiconductor electronics.

Industrially, the ZnO varistors are produced in large quantities through a solid-state preparation method. The widely adopted solid-state preparation method for the varistors is provided as Fig. 2.3. Here, ZnO with the major additives and

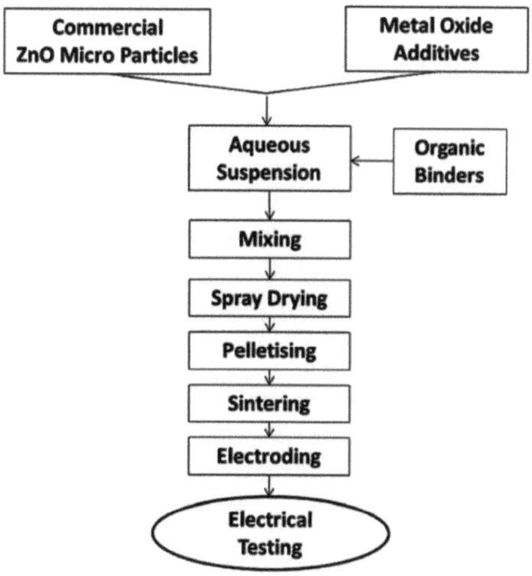

Fig. 2.3 Stages in conventional industrial processing of varistor ceramics (reproduced with permission from [9])

minor dopants were wet mixed and the resultant powder is mixed with an organic binder (e.g. polyvinyl alcohol) and plasticiser (e.g. polyethylene glycol) [9].

The selection of the dopants and additives in varistor is extremely important [21, 22]. In addition to Bi_2O_3, the various major additives and minor dopants used for preparing varistor compositions and their role, as seen from the literature, were summarized as Table 2.1. Typical varistor formulations consist of varistor former (oxides of elements of large ionic size Bi, Pr, La or Ba) [23, 24], performance enhancer (MnO_2, CoO) [25, 26] and performance highlighter (Na_2O, K_2O, CuO, etc.) [27].

The varistor powder containing more than 90% of high-purity ZnO powder, with the major additives and minor dopants, were initially spray dried to obtain a granulated powder. The resultant powder in the form of microspheres was then pressed uniaxially to form cylindrical blocks. Industrially, ZnO varistors were manufactured through a high-temperature reaction called sintering. Sintering involves a heat treatment of the ceramic powder compact at a temperature below the melting point of the main ceramic constituent. By means of sintering, a dense varistor product was normally obtained. Since the varistor performance depends on the final sintered microstructures, the sintering process must be carefully carried out. Liquid-phase sintering, the technique commonly employed for varistor processing was achieved by adding Bi_2O_3 [28, 29]. At ambient temperatures, Bi_2O_3 forms insoluble layers in ZnO and segregates at the grain boundaries. They behave as electrostatic barriers and increase the low-voltage resistivity.

For sintering, the varistor powder needs to be hard-pressed to ceramic discs/pellets and should be heated at a temperature in the range of 1100–1250 °C [8, 30]. Varistor sintering was normally achieved in three stages. The low-melting eutectic reaction between ZnO and Bi_2O_3 occurring at temperature 735 °C leading to the formation of a liquid phase was the first stage. The dissolution of the dopants occurs at this stage. The liquid-phase sintering associated with the diffusion and the

Table 2.1 Role of minor additives and dopants on the properties of varistors

Additives	Role	Additives	Role
Sb_2O_3	Inhibits ZnO grain growth, serves as pinning centres and retard ZnO grain boundary mobility [48, 100]	σ]λCoO, Cr_2O_3	Enhance powder sinterability, Improve electrical properties, prevent Bi_2O_3 evaporation [103, 114]
MnO_2, NiO	Decreases the average grain size, prevent the Bi_2O_3 evaporation and increase the non-ohmic properties [101, 102]	Al^{3+}, Ga^{3+}, In^{3+}, etc. WO_3, Nb_2O_5	Behave as donors and their addition (<0.1%) causes increase in the electrical and electronic conductivity of ZnO [35, 40, 104]
Y_2O_3, SiO_2	Grain growth inhibitors, increases the nonlinearity [34, 35, 105]	Li^+, Cu^{2+}	Doping of acceptors decrease the conductivity [102, 106]
Ag^+	Decrease the nonlinearity but increase the resistance to degradation [107]	Na^+, K^+	Improves the degradation resistance and controls the grain size of varistors [108]
σ]λTiO_2, SnO_2	Forms spinel phase and controls the grain size by triggering the formation of inversion boundaries (IBs) in ZnO grains [35, 109, 110, 113, 115]	Ta_2O_5	Small concentration results high grain conductivity. Excess Ta_2O_5 decreases both bulk conductivity and grain size [111]
La_2O_3, Er_2O_3, Dy_2O_3	Improves the nonlinearity and electrical stability [60, 112]	Pr_6O_{11}, V_2O_5, BaO, CeO_2	Varistor-forming oxides, mainly for low-voltage varistors [23, 103]

homogeneous distribution of the dopants and initiation of the grain growth are followed in the second stage. At the end of the second stage and also in the beginning of the third stage, the grain growth, crystallization of the secondary phases such as pyrochlore ($Zn_2Bi_3Sb_3O_{14}$) and spinel ($Zn_7Sb_2O_{12}$) phases, formation of the potential barriers, and retraction of the liquid phase from the two grain boundaries to the triple junctions are taking place [5, 16]. The new phases which are formed at elevated temperatures can be readily observed through the SEM (Fig. 2.4), EDAX (Fig. 2.5) and XRD (Fig. 2.6) analyses.

Finally, at the end of the sintering process, the varistor ceramics attain a microstructure-containing electrically conducting grains and insulating grain boundaries. The resultant sintered varistor discs were then silver electroded and leads were attached prior to encapsulating the disc in an epoxy-based polymer. This type of processing route is still the preferred method in varistor industries. Cost effectiveness, processing viability, needs of only basic instrumentation tools for powder preparation and varistor disc production, less risks and hazards associated are the main commercial attraction of this route.

Fig. 2.4 FESEM image of **a** ZnO grain, spinel ($Zn_7Sb_2O_{12}$) phase and bismuth-rich phases of a commercial varistor sample (reproduced with permission from [5])

2.3 Chemical Synthesis Strategies for Varistor Particles

Even though the solid-state preparative methods are commercially attractive, a major disadvantage of the solid-state route is the difficulty in obtaining a compositionally homogeneous microstructure, which is particularly important for the fabrication of the miniaturized devices required for modern electronic and communication equipments. Varistors prepared through the solid-state method requires high sintering temperature and extended sintering schedules. They also have the limitations like exaggerated grain growth, volatility of Bi_2O_3 liquid phase, formation of undesired reactive phases and poor stability [31–35]. Moreover, a typical solid-state-derived varistor component has the dimensions of 40 mm diameter; 40 mm height and ~450 gm weight (medium voltage varistors). As shown in Fig. 2.7, since ZnO undergoes excessive grain growth, the varistors will normally have an average sintered grain size of 10–20 μm [16]. For a miniaturized varistor, this has to be controlled well below 5 μm for achieving many-fold increase in energy handling capability.

In order to overcome the problems associated with solid-state varistor processing, researchers mainly suggest the development of various solution-based approaches for high-energy field varistors. Chemical techniques such as amine processing [36, 37], chemical combustion [38], citrate gel [31, 39–41], microemulsion [42], precipitation [43–46], microwave [47, 48]; precipitation and reflux [16, 49, 50], organometallic [51], polymeric [10, 36, 52–56], tape casting [57], thermal oxidation [58], templated grain growth [59], urea decomposition [33, 40, 60], and sol-gel [4, 26, 16, 49, 50, 61–78], methods are used for the synthesis of doped varistor grade multicomponent powders. Most of these precursor

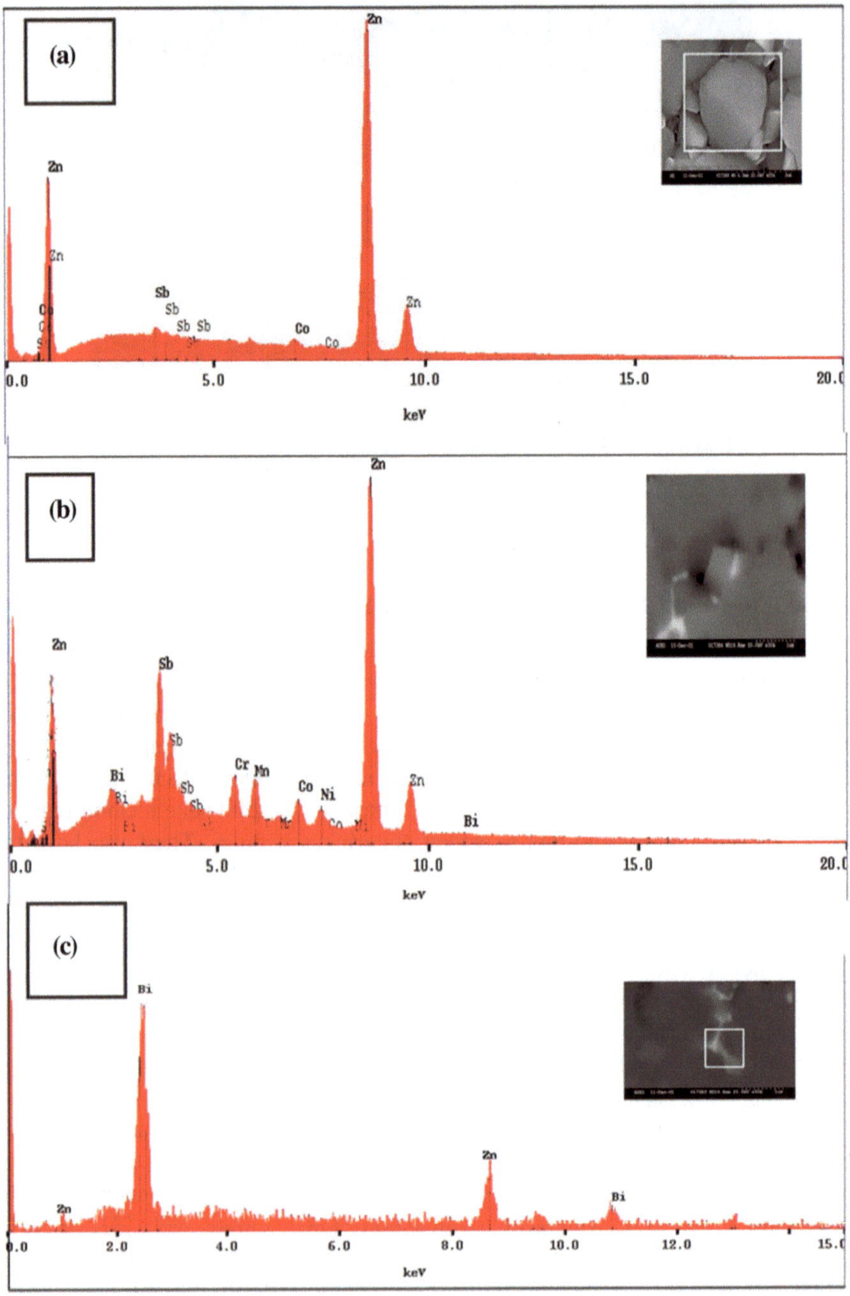

Fig. 2.5 EDAX images of **a** ZnO grains **b** spinel ($Zn_7Sb_2O_{12}$) phase **c** bismuth-rich phase (white areas in the microstructure) (reproduced with permission from [5])

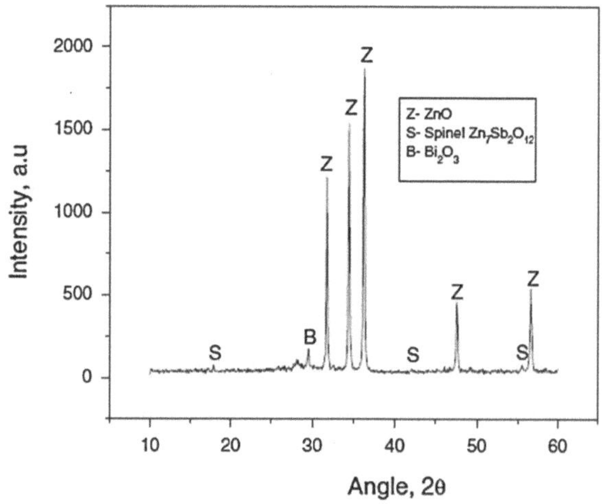

Fig. 2.6 X-ray pattern of a sintered ZnO varistor which contain various phases such as Z = ZnO; S = $Zn_7Sb_2O_{12}$; B = Bi_2O_3 (reproduced with permission from [16])

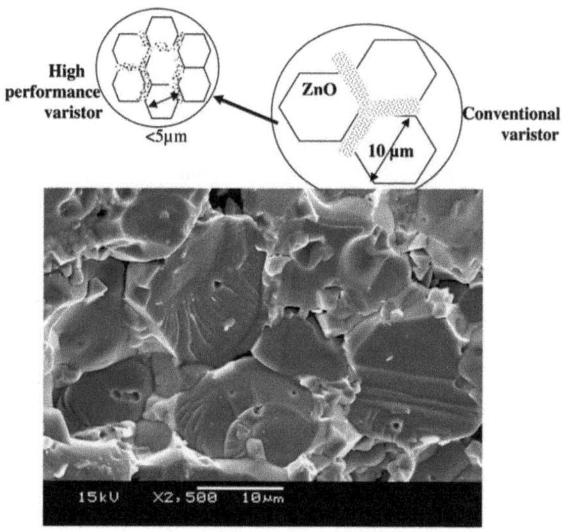

Fig. 2.7 Typical microstructure of a commercially available ZnO varistor (reproduced with permission from [16])

development methods are accompanied by a secondary sample heat treatment process like conventional, microwave, reflux and plasma methods for the conversion of sample hydroxides to corresponding oxides.

The advantages of wet chemical processing are primarily the higher purity, homogeneity and the lower processing temperature. Multicomponent dopant oxides can also be readily prepared by mixing relevant salt solutions of metal salts. By wet chemical methods homogeneity of the dopant ions among the ZnO grains can also be achieved at the atomic level. A number of liquid-phase approaches have been followed, are outlined in Table 2.2.

Table 2.2 Various wet chemical approaches for the syntheses of ZnO varistor powders and comparison of their electrical properties

Chemical process	Reported by	Sintering (°C)	V_b (V/mm)	α
Amine process	Hishita	1050	711	50
		1200	339	52
		1300	256	36
Citrate gel	Sinha	1150	515	35
	Lorenz	1000	460	72
		1000	510	75
Colloidal	Viswanath	750	3000	50
Emulsion	Hingorani	1200	450	83
Organometallic	Macary	550	1900	—
		600	1066	5.8
Pechini (polymeric complexation)	Duran	850	1500	70
	Duran	825	2000	270
	Lin	950	924	32
	Lin	950	895	48
Precipitation	Haile	1200	180	44
	Viswanath	750	3000	50
	Yang	1200	492	56
	Li	1050	540	50
	Shi	1060	218	55
	Cheng	1060	1610	15
Precipitation and reflux	Anas	1050	277	48
	Anas	1100	532	7.7
	Anas	1100	875	31
Sol-gel	Pillai	1050	786	34
	Pillai	1050	859	33
	Pillai	900	1192	—
	Westin	1100	249	37
	Tomandl	1100	375	60
	Ya	950	1000	45
	Lauf	1000	980	30
	Nobrega	1100	800	26
	Wang	1100	438	16.5
	Anas	950	557	34
	Hohenberger	1100	375	55
	Liu	1150	310	27.2
	Sedghi	1200	475	43
Sol-gel mixed route	Pillai	1050	941	33
Spray pyrolysis	Subasri	1100	960	70

2.4 Advantages of Sol-Gel Technology for Varistors

Among the various wet chemical methods, the sol-gel methods have definite advantages over other chemical methods in the production of varistor powders [9, 63, 79, 71]. Atomic level mixing of reactants, homogeneous distribution of low-level dopants and additives, hydrolysis and polycondensation reactions at room/low temperatures, preparation of particles with interaction through weak forces such as van der Waals/hydrogen bonded forces, formation of stable oxide network, size controlled particle formation through pH and concentration controls were the benefits realized through the sol-gel methods. Studies confirm that the sol-gel method yields the controlled production of 'varistor nanoparticles'. Unlike other solution-based approaches, the sol-gel methods exhibit specific control in the production of nanosized varistor particles. The control in the nanoparticle synthesis is reported to be difficult with other techniques such as chemical decomposition [48], thermal evolution [16, 41, 49, 50], solution growth [59, 79] and precipitation methods [55]. Most of these methods needed secondary heating of the precursor materials at elevated temperatures. They resulted in particles from nano-micro-macro ranges. Many of the thermal decomposition routes release unwanted byproducts such as CO, CO_2, NO, NO_2 gases which are hazardous to the safe environment. But, on the other hand, the sol-gel methods are reported to be environmentally safe. Most of the sol-gel syntheses adopt green chemical pathway. Multicomponent compounds can be prepared through sol-gel methods with a strict control in the stoichiometry by mere mixing sols of different compounds. There are several reports available in the literature, where sol-gel techniques have been utilized for varistor applications [9, 66, 67, 75]. In general two different approaches have been taken, the first involving, the sol-gel preparation of ZnO and the subsequent addition of solid dopant metal oxides and second, the preparation of the gel from the precursor solutions of the zinc and metal additives. On summarizing the processes involved, the currently adopted sol-gel methods for the production of varistors can be briefly represented as shown in Fig. 2.8.

It has been confirmed that the sol-gel processes result in the generation of mainly 'nanoparticles' [9, 68, 75]. Recent literature strongly supports the use of nanopowder in varistor industries to solve many limitations associated with this field [80, 51]. Due to the high surface area, nanopowders show benefits such as low-temperature sintering. They results in increased numbers of grains and grain boundaries per unit varistor volume. It has been confirmed that nanocrystalline varistor grain has large grain boundary volume that can generate more active grain boundaries per unit varistor volume [26, 36]. This is an important understanding about varistors which can be effectively utilized for controlling the varistor dimensions. Low-temperature densification, reduction in ZnO grain size and three to four times increase in the varistors performance were the main advantages realized when nanosized powders were used [26, 81–83].

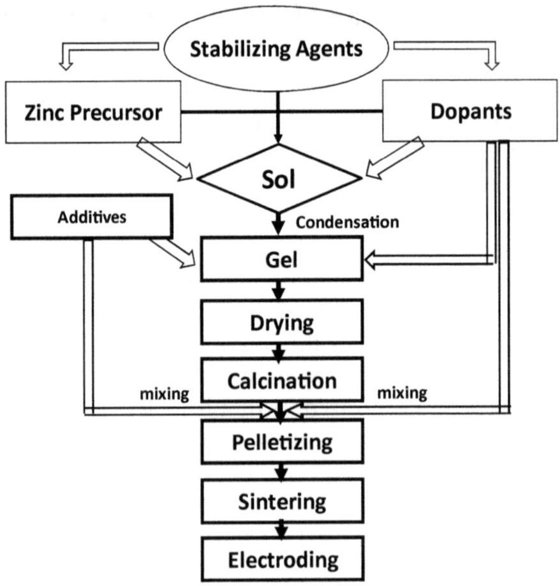

Fig. 2.8 Sol-gel synthesis of varistors with zinc oxide as the matrix

2.5 Sol-Gel Varistors

Many successful reports on the wet chemical sol-gel synthesis of varistor powders were addressed from early 1980s. The base work on the sol-gel varistors were acknowledged as the contribution from Lauf and Bond [67]. By following their processing methodology, Gert Hohenberger and Gerhard Tomandl prepared ZnO varistor powders using inexpensive source materials such as acetates and nitrates [64]. They prepared two sets of precursor solutions of ratio <3:1. The first one involved hot (90 °C) water solution containing the acetates of Zn, Co, and Mn, and H_3BO_3 and the second one contained Sb-acetate, Cr-nitrate, Bi-nitrate, Al-nitrate in ethylene glycol. By mixing and cooling these two solutions, a microcrystalline gel of the precursor material was resulted. As a result of the more homogeneous distribution of dopants, the sol-gel samples exhibited a higher breakdown field (375 V mm^{-1}) and nonlinear properties ($\alpha = 55$). By adopting similar processing strategies, improvement in the varistor properties were also achieved by Nobrega and his team [45].

Westin and co-workers [73, 84] have tried different sol-gel methods to obtain the varistor precursors. They used typical sol-gel precursors such as $Ni(OAc)_2.4H_2O$, $Co(OAc)_2.4H_2O$, $Mn(OAc)_2.4H_2O$, $Cr_3(OAc)_7(OH)_2$, $Al(NO_3)_3.9H_2O$, $Sb(OBun)_3$ and bi(2-ethylhexanoate)$_3$. A breakdown voltage of 249 V mm^{-1} and $\alpha = 37$ were successfully achieved in these varistor compositions. Puyane and his team [71, 72] reported the production of ZnO powders by a sol-gel route and showed that these powders were sintered at lower temperatures compared to the conventional varistor materials. In other studies, Ya and co-workers [75, 85] reported the preparation of varistors using sol-gel prepared ZnO nanopowders (20 nm). These materials displayed superior electrical properties with a high value of the nonlinear coefficient ($\alpha = 45$) and lower leakage current ($I_L < 1$ mA).

Sinha and Sharma [41] explained a most viable method for the preparation of varistor precursor gel by adding citric acid to nitrate solutions (pechini process) of varistor constituents. The gel on calcination resulted submicron-sized reactive varistor powder that exhibited homogeneity and stochiometric control. The citrate gel derived varistor exhibited better varistor properties ($V_b = 515$ V/mm and $\alpha = 35$) than the varistor derived from a urea route ($V_b = 280$ V/mm and $\alpha = 21$). Duran and his team [10, 36] introduced a metallorganic polymeric method for preparing homogeneous and nanosized (28 nm crystallite size) doped ZnO ceramic powders. Dense ceramic bodies (>99% of theoretical) were fabricated by normal liquid-phase sintering at 850 and 940 °C for 1–5 h. Ceramics so-fabricated showed a nonlinear coefficient, α, of >70, and a breakdown voltage, V_b (1 mA/cm^2), of >1500 V/mm.

Lorenz and co-authors [40] reported a modified citrate gel synthesis for the production of ZnO-based varistor powder. Doped ZnO powder was initially prepared by coating ZnO powder with a layer of citrates of the dopants using the citrate gel technique. Then, the doped ZnO powder was coated with a layer of the additives. ZnO-based varistors prepared by this modified citrate gel route showed higher values of characteristic field strengths [$V_b < 450$ V/mm and first nonlinear coefficient ($\alpha = 70$)]. Chu and his colleagues [86] proposed a new sol-gel processing method to prepare ZnO-based varistor powders, using inexpensive zinc acetate dihydrate, ethylene glycol, n- propyl alcohol and glycerol as starting materials. The procedure of the powder preparation is shown in Fig. 2.9. Compared to conventional oxide-mixing techniques, by adopting this route, the sintering temperature could be lowered by about 200 K to 1000 °C and the grain size as smaller as <2 mm. This proves very helpful for achieving high breakdown strength.

Huang and his team [65] illustrated the preparation of ZnO-based ceramic films, with thickness close to 2–4 μm. These films were deposited on Au/Si substrates by novel sol-gel process. The precursors were developed by dispersing doped ZnO ceramic nanopowders into the sols, which were prepared by dissolving zinc acetate dihydrate into 2-methoxyethanol and stabilized by diethanolamine and glacial acetic acid and doped with a concentrated solution of bismuth nitrate, phenylstibonic acid, cobalt nitrate, manganese acetate and chromium nitrate. The doped ZnO ceramic nanopowders were prepared by sol-gel processing. The samples were dried and calcined in air at 150 and 800 °C, respectively. The nonlinear V–I characteristics of the films for use as varistors were studied by transistor characteristics testing instrument. The studies confirm the production of varistor with nonlinear voltage as 2–10 V and the nonlinear coefficient (α) as 3–22. The highest nonlinearity coefficient with nonlinear voltage of 4 V was achieved at 750 °C by the films with a thickness of 2 μm.

Pillai and co-workers, the pioneers in the sol-gel high-energy field varistors [9], reported the production of zinc oxide precursor gels by the reaction of an ethanolic solution of zinc acetate and oxalic acid [69]. The gels were initially dried at 80 °C

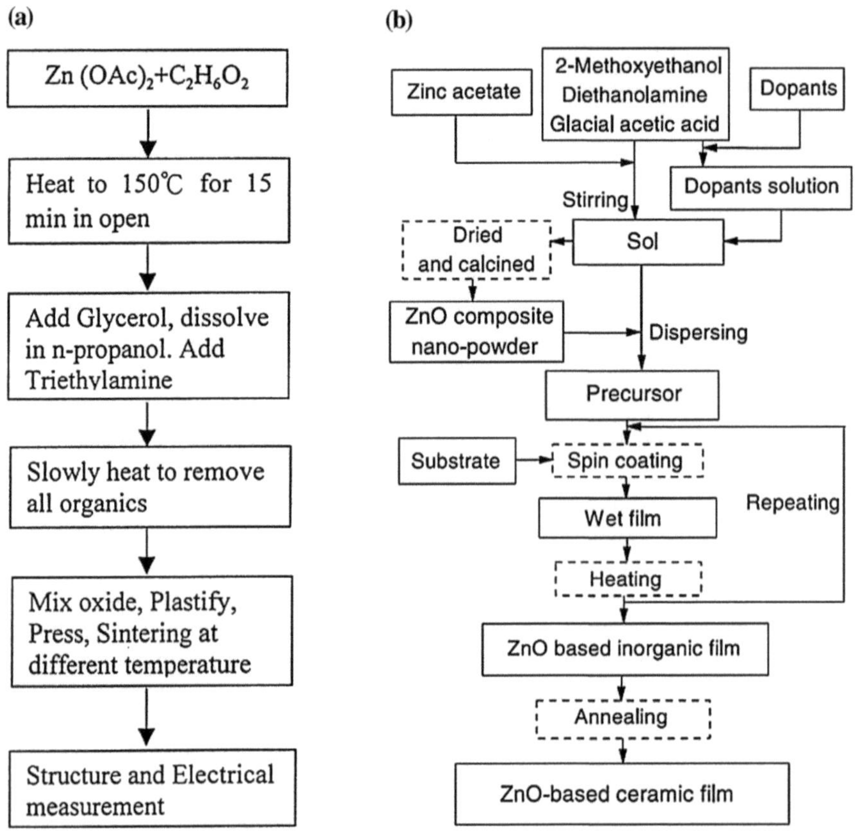

Fig. 2.9 **a** Flow diagram for the sol-gel preparation method of ZnO-based varistor powder (reproduced with permission from [86]). **b** Flowchart showing the procedure for preparing ZnO-based ceramic films (reproduced with permission from [65])

and calcined at 500 °C for 2 hours. This resulted ZnO nanoparticles with size of 30 nm. Subsequently, varistor materials were fabricated from this nanoparticular ZnO via two separate routes: (a) from a "core shell" material using metal salts as additives; (b) using a conventional solid-state mixing of metal oxides. As shown in Fig. 2.10, "Core shell" type varistor powders were prepared by coating the nano-ZnO with additive metal salts. To achieve a homogeneous coating of oxides (Sb, Bi and Co) around the ZnO particle, solutions of chloride salts of these elements were mixed with a dispersion of nano-ZnO (as prepared above) and reacted with ammonium carbonate. Sintering (1050 °C) and subsequent electrical studies were carried out for each of these samples and they were compared with commercial varistor samples prepared under similar conditions. "Core shell" type varistor material showed considerably higher breakdown voltage (V_c − 850 ± 30 V mm^{-1}) as compared to a sample prepared by mixing with metal oxides (V_c − 683 ± 30 V mm^{-1}) and commercial varistor discs (V_c − 507 ± 30 V mm^{-1}). The high breakdown voltage obtained is attributed to the formation of more varistor-active grain boundaries per unit area.

In a further experiment [26], linear arrays of ZnO nanoparticles have been successfully prepared by a simple sol-gel condensation reaction involving chemical modifiers [1,2-ethanediol (EG) and diethanolamine (DEA)]. The modifiers were added to the ethanolic solution of zinc acetate before the addition of the ethanolic solution of oxalic acid. The gel formed was subsequently treated and the powders were characterized. As shown in Fig. 2.11, the FESEM and TEM analyses revealed that the calcined material were composed of approximately spherical nanoparticles having average diameter 21 ± 3 nm, self-assembled to form arrays extending in length up to 2–4 mm. The morphology of the ZnO is found to depend sensitively on the amounts of chemical modifiers present. In their absence, the ZnO produced (nano-ZnO) is an unstructured agglomerate of nanoparticles. Varistors prepared from the nanoarray ZnO by sintering (1050 °C) with appropriate mixtures of metal oxides showed a breakdown voltage of 786 ± 30 V mm^{-1}, which is substantially higher than that of samples prepared under similar conditions from either micron-sized commercial ZnO (507 ± 30 V mm^{-1}) and from nano-ZnO (683 ± 30 V mm^{-1}).

Through a more sophisticated preparative method, high-performance varistor precursor gels were prepared from the precursor solution containing ethanol and diethanolamine solutions of zinc, cobalt, manganese, nickel, aluminium and ethylene glycol solution of bismuth, antimony and chromium salts, by adding with oxalic acid [5]. The obtained gels were dried at 80 °C and calcined at 500 °C. Ultrafine (~15 nm) doped ZnO nanoparticles were resulted. The obtained varistor powders were sintered at 1050 °C for 2 h. The xerogels, calcined powders and sintered materials were characterized. Full densification, high breakdown voltage (941 ± 30 V mm^{-1}; which is 85% higher than the commercial varistor at 1050 °C) and a good nonlinear coefficient (α = 33 ± 3) were observed for samples sintered

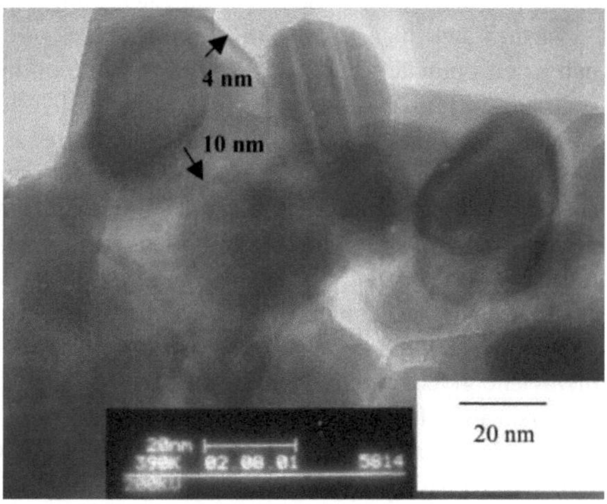

Fig. 2.10 HRTEM of ZnO nanoparticles coated with oxides of Sb, Bi and Co precursors (reproduced with permission from [69])

Fig. 2.11 Electron microscopy images of nanowire ZnO for varistor applications prepared by a sol-gel route. **a** and **b** FESEM images of samples calcined at 500 °C. **c** TEM image of a single array of ZnO nanoparticle separated by ultrasonication in ethanol. **d** FESEM image of ZnO nanoarray at 1000 °C (reproduced with permission from [26])

at 1050 °C for 2 h, Fig. 2.12. Highly dense varistor discs were prepared from the sintered material, with a breakdown voltage 85% higher than that of commercial varistors, making them suitable for use in miniaturized electronic circuitry. Improved performance of these materials has been attributed to the small grain size and better dispersion of additives on ZnO grains.

Zhang and his group [78] reported the preparation of a polycrystalline nanograin-boundary multi-doping ZnO-based nonlinear varistors with higher concentration additives, through a sol-gel method. The results showed the merit of sol-gel method in terms of high threshold voltage, at a lower sintering temperature, than the varistor prepared by a solid-state reaction method. The major conclusions are depicted in Fig. 2.13. The results showed that the grain size and the homogeneity of the material are the important factors which directly affect the characteristic of the varistor.

Anas and his team [50] reported the synthesis of varistor grade zinc oxide nanopowders through a non-hydrolytic sol-gel method. Thermo-gelation, using diethylene glycol and triethanolamine, was the method adopted for making varistor precursor gel. Hexagonal nanocrystalline varistor grade ZnO particles of size 50 nm were achieved by heating the sample precursors at 270 °C for 2 h and at 500 °C for 2 h. The varistor properties of the samples, from 850–1150 °C, revealed that the

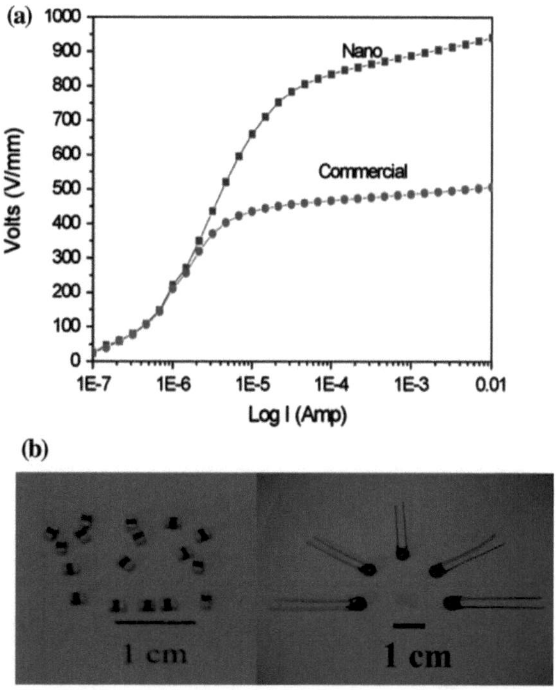

Fig. 2.12 Varistor samples prepared by sintering at 1050 °C: **a** I–V *curve*, **b** varistors fabricated using mixed precursor route (reproduced with permission from [5])

samples have narrow grain size distribution with higher breakdown voltage and nonlinearity, than any commercial varistor. As evident from the studies, Table 2.3, the breakdown voltage was found to be decreasing with an increase in the sintering temperature and grain size.

In another study, Cheng and co-workers [61] reported the preparation of Al_2O_3-doped ZnO varistors through sol-gel methods. Citric acid (CA) was added to ethylene glycol (EG) in a 3:2 CA/EG molar ratio, and this mixture was stirred at about 80 °C to form a transparent solution. This solution was added with the aqueous solution of the additive salt to form the sol. The resultant mixed oxide sols were subsequently evaporated. The obtained powder was calcined at 500 °C for 2 h and sintered at 900 °C for 2 h. Figure 2.14 shows the typical characteristics of the xerogels (a) and the powder (b) obtained by this method. The xerogels showed typical porous structure. The powder was loosely agglomerated with the particle size in the range of 30–40 nm. The studies confirms that the average grain size of the varistor decreased from 4.4 to 3.0 μm and the breakdown voltage increased from 720 to 1160 V/mm, as the amount of Al_2O_3 increased from 0.0 to 0.40 wt% which are incorporated in the varistor matrix through sol-gel pathway.

Very recently, Liu and his group [68] introduced xanthan gum, a polymerization agent, for preparing $ZnO–Co_2O_3–Bi_2O_3$ nanocomposites through sol-gel methods. As illustrated through Fig. 2.15, spherical- and hexagonal-doped nanocrystalline ZnO powders with particle size of about 20–45 nm were easily obtained in the

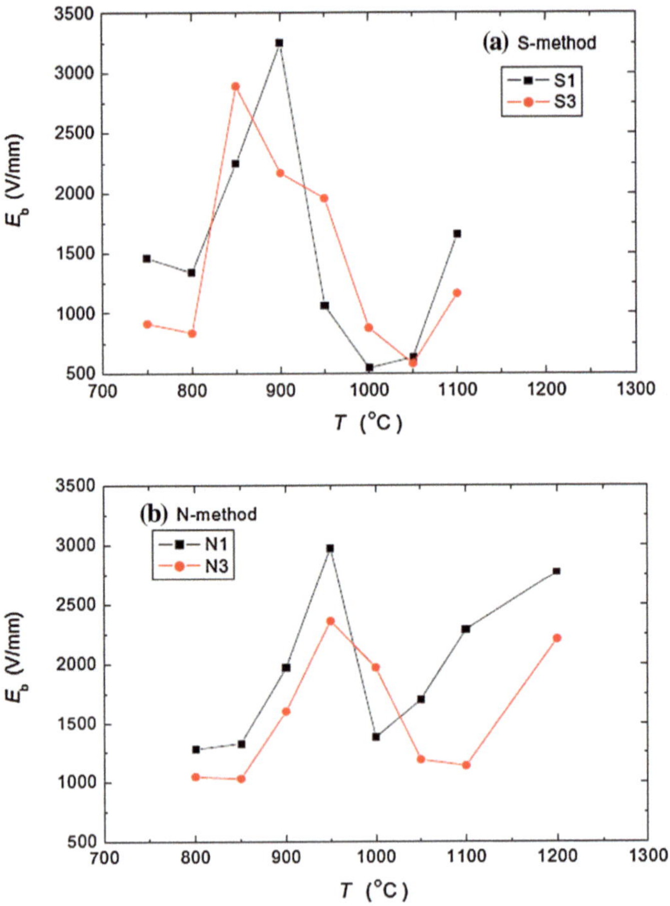

Fig. 2.13 Threshold voltage E_b versus sintering temperature: **a** sol-gel method; **b** solid-state reaction method (reproduced with permission from [78])

Table 2.3 Densification and electrical property studies on gel-derived varistors [50]

Sintering temperature	Grain size (μm)	Density (g/m^3)	Breakdown voltage (V_b)	Nonlinearity index, α
950 °C/2 h	<2	5.0	557	34
1050 °C/2 h	<4	5.1	493	36
1150 °C/2 h	<8	5.4	323	16

presence of xanthan gum. Xanthan gum was used as a polymerization agent to avoid hard agglomeration and then control the growth of the nanopowders. The metal elements (Co and Bi) were evenly distributed in the ZnO grain after calcination at 1150 °C for 2 h. The varistor ceramics sintered at 1150 °C for 2 h in air

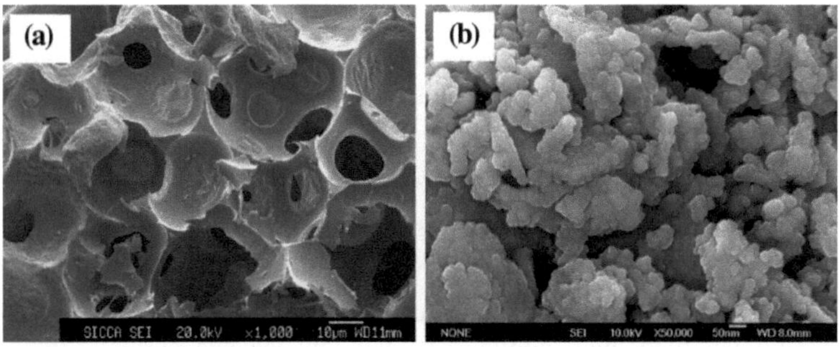

Fig. 2.14 SEM micrograph of the gel **a** and ZnO powder **b** prepared by sol-gel method (reproduced with permission from [61])

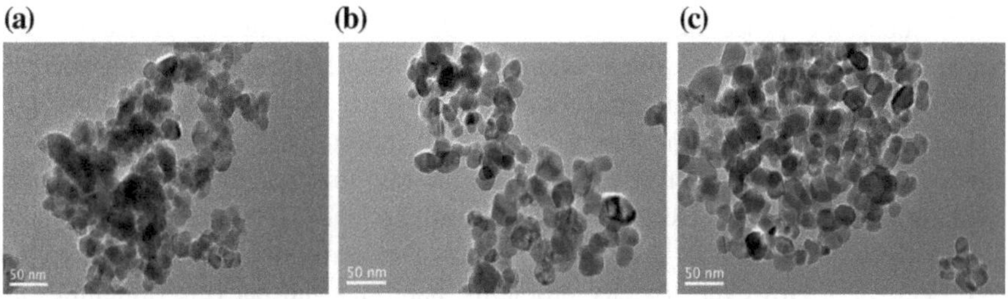

Fig. 2.15 TEM images for doped nanocrystalline ZnO powders with **a** 0.5% Xanthan gum, **b** 1% Xanthan gum, **c** 1.5% Xanthan gum (reproduced with permission from [68])

have a density of 5.52 g/cm^3 corresponding to 95.5% of the theoretical density with breakdown voltage of 310 V/mm and nonlinear coefficient of ~27.19. The experimental results showed the advantage of addition of the xanthan gum for avoiding hard agglomeration and improving electrical performance of the varistors.

Of late, the effect of transition metals and rare earth ions on sol-gel derived varistors has also been studied. Dong and his colleagues [62] published Y_2O_3-doped ZnO-Bi_2O_3 thin films on silicon substrates by sol-gel process and annealing in air at 750 °C for 1 h. The article confirms the development of thin-film varistors having nonlinear *V–I* characteristics with the leakage current of 0.46 mA, the threshold field of 110 V/mm and the nonlinear coefficient of 3.1, with the films contain 0.2% (mole fraction) yttrium ion. With increasing Y_2O_3 content, the grain size of ZnO was found to be decreased. In a similar way, Wang and co-workers [77] reported pure and Pr-doped ZnO (0.5 at.%, 1 at.%,1.5 at.% and 2 at.%) nanoparticles synthesized through a sol-gel method. The samples exhibited hexagonal ZnO lattice structure and another Pr_6O_{11} phase was detected in the 2 at.% Pr-doped ZnO nanoparticles. The calculated average crystalline size decreases from 29.73 to 25.5 nm when Pr content increased from 0 to 2 at.%. The energy dispersive X-ray spectroscopy (EDS) analysis for 2 at.% Pr-doped ZnO confirms the evidence of Pr in ZnO nanoparticles, and the varistor ceramics has a sintered density of 5.42 g/cm^3 corresponding to 96.8% of its theoretical density, with breakdown voltage of 438 V/mm and nonlinear coefficient of 16.5.

2.6 High-Performance Varistors: Through Nanoinclusions

High-performance varistors are miniaturized varistors, which generally possess high breakdown filed, high nonlinearity and low leakage currents, that can be commonly used for the purpose of arresting high voltages in high-energy fields and for smoothening variable voltages in modern electric and electronic devices such as laptops, mobile phones, etc. In the current scenario, the major thrust in the varistor field is the production of this miniaturized lightweight varistors via a safer and much economic processing route [5, 9, 39, 87]. This greatly demands novel approaches and strategies which requires large number of conducting grains and higher fractions of insulating grain boundaries within a small varistor volume [10, 51]. It is evident from the scientific journals/reports that the 'sol-gel derived nanopowders' can be effectively utilized as the raw material for the production of miniaturized varistors. Detailed information on this and the fabrication of high-performance varistors using nanopowders were provided in the earlier section (see Sect. 2.5). However, toxicity problems, handling and packing difficulties are the drawbacks involved in the use of only the nanopowders. At the same time, the industries also call up on new developments in the varistor field for attaining miniaturized devices at low production cost. These issues received most attention in the design and development of state-of-the-art processes or smart techniques for the development of miniaturized varistors.

2.6.1 Grain Size Reduction and Grain Boundary Area Improvement

2.6.1.1 Addition of Nanofillers

High-performance varistor blocks can be achieved by effective grain size reduction [9, 34, 51, 83, 88, 89]. This can be achieved by the inclusion of nanosized varistor powder to the commercial micron-sized varistor powder [16, 87]. Rather than using nanopowder as a single source for varistor manufacture, scientific reports authenticate the use of nanopowder as 'nanofiller' for improving the varistor properties. The microstructure modifications as expected in the nanofiller-based composite varistor approach can be depicted schematically as Fig. 2.16.

As shown in Fig. 2.16, before sintering, the average diameter of an industrially processed spherical varistor powder granule is in the range of <50 μm. This varistor

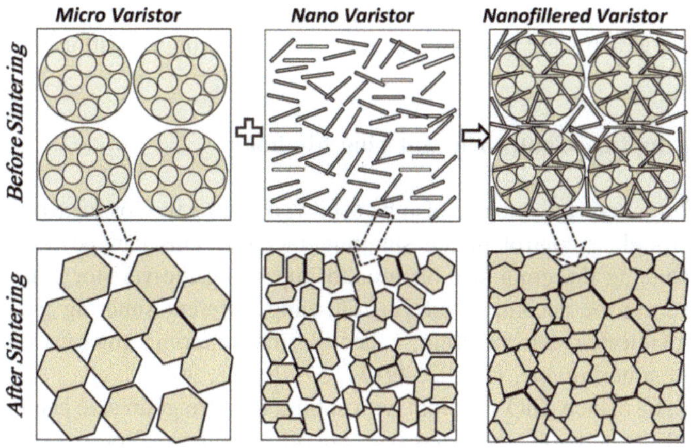

Fig. 2.16 Schematic representation of the 'nanofiller' approach for the high-energy field varistors (reproduced with permission from [87])

granule contains loosely adhered spherical particles having average size 3–5 μm, with distributions of nano- to micro-scale interstitial porosities. The large amounts of nano- and micropores on the powder granule and the interstitial macropores produced during compaction cause sintering problems in varistors and hence they need to be eliminated. Usually varistors require high sintering temperatures to remove these pores. But high sintering temperatures causes excessive grain growth in ZnO varistors. By the use of nanopowders as nanoadditives or nanofillers to the commercially available spray-dried micron-sized granules, the nanopowder enters into the pores of the micropowder and fills the varistor interstitial voids. This will not only increase the particle packing but also yields high particle contact area which favours accelerated densification at comparatively low temperatures. The resultant composite varistor powder produces sintered varistor ceramics with significant grain size reduction and possesses sintered microstructures with distributions of nano- and micrograins, say a bimodal grain size distribution. In either case, an effective increase in the number of ZnO grains within a specific varistor volume can be achieved.

In addition to the grain size reduction, the nanofiller addition also helps the varistor industries to have many practical advantages. The blending of nanofillers with a commercial micron-sized varistor powder simply by dry or wet milling can easily be introduced as an intermediate unit operation during varistor manufacturing. Varistors made out of such nanofillers not only promises high-energy field varistors but also provide a huge potential to miniaturize varistors at low production cost. Hence, this method can be taken as a 'smart technique' for the manufacturing of varistors. Reports on the nanofiller varistor studies shows that only a 5 wt% addition of the nanofiller resulted a varistor with theoretical density = 99%, average sintered grain size = 1 μm, $V_b = 875 \pm 30$ V mm^{-1} and $\alpha = 31 \pm 3$, when compared to the commercial sample with average grain size = 6 μm, V_b = 421 ± 30 V mm^{-1} and α = 17 ± 3, which were processed under similar sintering conditions at ≤1100 °C.

2.6.1.2 Adopting Controlled Sintering Methods

High-performance varistors can also be achieved from nano- or microsized varistor powder through a control in the sintering method. This can be possible either through reducing the grain size or through improving the varistor grain boundary area. These can be accomplished via following different sintering processes like multistage sintering/step sintering, spark plasma sintering, microwave sintering, constrained sintering, etc. [16, 47, 49–51, 57, 70, 82, 90].

High-performance ZnO varistors with submicrometre grain size can be achieved through a strategic two-stage, one-cycle, pressureless thermal processing method at temperatures as low as 825 °C in air [10]. Densification in a two-stage sintering is believed to be based on a rapid rearrangement of the solid ZnO nanoparticles in the presence of a liquid phase at the first stage up to 900 °C. In the second stage, on cooling to 825 °C, the liquid-phase recedes, thus, improving the ZnO–ZnO direct contacts and suppressing the grain boundary migration while maintaining densification by very active grain boundary diffusion. Full density with a close-packed arrangement of the grains was obtained at the second stage without additional grain growth. The nonlinear coefficients 'α' of such fully dense varistors were as high as 270, and the breakdown field 'V_b' of about 20 kV/cm at 1 mA/cm^2 can be achieved. Recently, Pillai and co-workers [70] reported the production of nano-zinc oxide varistor material by a sol-gel route. Varistor discs fabricated from these materials were subjected to a two-step sintering schedule. In a typical experiment, the samples were heated to 1000 °C, then allowed to cool for over 30 min to 900 °C and held there for 6 h. As evident from Table 2.4, the studies showed that the nanovaristor samples prepared through the stepwise sintering conditions would normally exhibit higher breakdown voltages (1192 ± 30 V mm^{-1}) than any commercially available varistor.

Spark plasma sintering (SPS) is another method preferred for the preparation of high-performance nanostructured varistors. Through SPS, grain size reduction can be achieved by sintering the sample at very low temperatures. An enhancement in the varistor property can thus be expected at low-sintering temperatures. For example, Macary and his team [51] used 8-nm-sized zinc oxide nanoparticles, which were synthesized through an organometallic approach, for SPS. As seen in Fig. 2.17, nanostructured varistors were obtained by SPS by sintering the sample at

Table 2.4 Breakdown voltages of varistor samples sintered at different temperatures [70]

Sample	V_b (V/mm)
Commercial varistor (1000 °C/0 h; 900 °C/6 h)	723
Nanovaristor (1000 °C/0 h; 900 °C/6 h)	1192
Commercial varistor (1000 °C/2 h)	584
Nanovaristor (1000 °C/2 h)	951

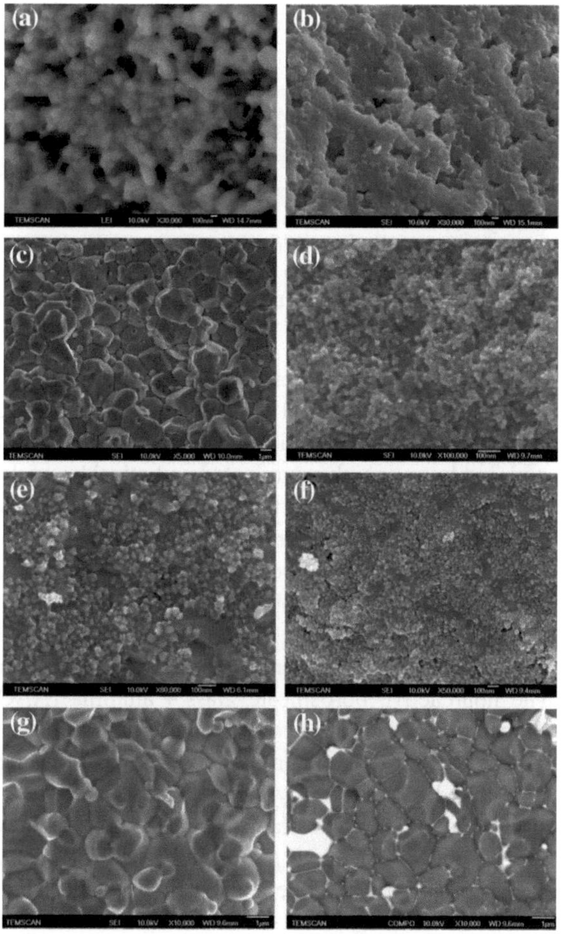

Fig. 2.17 SEM SEI micrographs of fracture surface of: **a** Sample 1 sintered at 1000 °C in a furnace; **b** Pellet 2, Sample 1 sintered by SPS at 800 °C; **c** Pellet 3, Sample 1 sintered by SPS at 900 °C; **d** Pellet 1, Sample 1 sintered by SPS at 600 °C; **e** Pellet 5, Sample 2 sintered by SPS at 550 °C for 15 min; **f** Pellet 6, Sample 2 sintered by SPS at 600 °C for 4 min; **g** Pellet 10, Sample 2 sintered by SPS at 800 °C; **h** SEM BEI micrograph of the same image as **g** (reproduced with permission from [51])

sintering temperatures as low as 550–650 °C. At 600 and 650 °C, the grain size was evaluated for these varistors and it was found to be equal to 300–500 nm and to 1 mm, respectively, with the breakdown field values above 1000 V/mm.

Microwave sintering is another promising method for obtaining varistors with smaller and uniform grain size distribution [16, 47–50, 83]. When compared to the conventionally sintered samples which exhibit non-homogeneous Bi_2O_3 vaporization, the microwave-sintered samples can be produced with homogenous Bi_2O_3 vaporization. Thus more uniform microstructures can be resulted with this method. Also, the time required for the microwave sintering was only less than half the time that required for conventional sintering. For example, Cong [91] and Subasri [83] prepared microwave-sintered samples and compared their properties with respect to the conventional ones. The study proves that microwave-sintered samples hold good varistor properties that can be even modulated for developing miniaturized varistors at a very low cost.

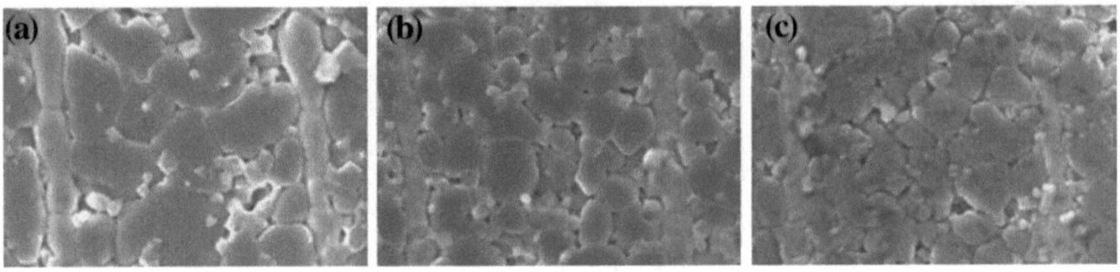

Fig. 2.18 Microstructure of ZnO-based MLV obtained by free sintering (**a**) at 925 °C for 2 h (**b**), and (**c**) microstructure of ZnO-based MLV obtained by constrained sintering with a constraining layer thickness of 24 and 44 mm at 925 °C for 2 h (reproduced with permission from [80])

Another promising approach for the fabrication of ZnO-based high-performance varistor is the preparation of multilayer varistor by constrained sintering [80, 90]. Here, nonreactive borosilicate glass and alumina (Al_2O_3) can be used as the constraining layer, which is laminated on both sides of the multilayer varistor (MLV). It is obvious from Fig. 2.18 that the mean grain size and the distribution of grain size of ZnO-based MLVs are greatly reduced because of constrained sintering. A sandwich structure of borosilicate glass + alumina/ZnO-based MLV/borosilicate glass + alumina can be fabricated. The in-plane shrinkage of the ZnO-based MLV tape can be reduced using a nonreactive borosilicate glass + alumina (Al_2O_3) tape, which densifies in a different temperature region than the ZnO-based MLV tape. The nonreactive borosilicate glass + alumina (Al_2O_3) tape will densifies at low temperature, where it is constrained by the non-shrinkage green ZnO-based MLV tape. At higher temperature, the ZnO-based MLV tape densifies and the shrinkage is inhibited by the already dense nonreactive borosilicate glass + alumina (Al_2O_3) tape. Hence, a self-constrained sintering can be accomplished, because each tape acts beside its functional performance as a constraining layer. It is clear from the studies performed by Lee and co-workers [80] that the mean grain size and the distribution of the grain size of ZnO-based MLVs can be remarkably reduced when constrained sintering is used instead of free sintering. The mean grain size of ZnO-based MLVs decreases with increase in the thickness of the constraining layer. Thus the property of a multilayer varistor can be effectively tuned with respect to the thickness of the constraining layer.

2.6.2 *Homogeneous Dopants Distribution at the Grain Boundaries*

2.6.2.1 Conducting Wet Chemical Syntheses

High-performance varistors can also be accomplished by performing wet chemical syntheses. Detailed discussion on the synthesis of nanoparticles through wet chemical methods and the resultant varistor with homogeneous distribution of

Fig. 2.19 SEM photographs of microstructure of ZnO ceramics for **a** 1 wt% Bi_2O_3, **b** 3 wt% Bi_2O_3 and **c** 5 wt% Bi_2O_3 sintered at 1150 °C (reproduced with permission from [92])

dopants are given in earlier sections (please refer Sects. 2.3, 2.4 and 2.5). Among the many methods employed, the solution coating methods stands out as the best one for the effective distribution of dopants at the varistor grain boundaries [34, 35, 44]. In this method, the ZnO particles were coated with dopants such as Bi_2O_3, Sb_2O_3, Co_2O_3, Cr_2O_3 and other minor additives via liquid nanocoating technique. For example, Li and his team [44] suggests the preparation of an additive solutions in alcohol media (ethylene glycol and ethanol in the volume ratio is 5:3) as the initial step for developing a nanolayer-coated ZnO grains. The method involves the ultrasonic treatment of the pure ZnO, which is dispersed in the additive solution, for about 1 h. The solution was then heated up to 90 °C and stirred until it changed into slurry. The slurry was dried in an oven at 150 °C for 8 h. It was then pulverized into powder and heated at 450 °C. The microscopic analyses of the resultant powder showed that the ZnO composite powder was homogeneously coated and ultrafine. The resultant varistors prepared through the nanocoating method showed higher threshold voltage (540 V/mm) and the nonlinear coefficient ($\alpha = 50$). Likewise, different solution coating methods were reported by many researchers to coat various minor additives and dopants over ZnO surfaces. For example, Yuan and his collaborator [92] tried to deposit a homogeneous layer of Bi_2O_3 over ZnO through nanocoating. As seen in Fig. 2.19, through solution coating methods, effective distribution of dopants at the grain boundaries were achieved.

2.6.2.2 Performing Temperature Controlled Step-Sintering Processes

Another method to control the distribution of dopants at the varistor grain boundaries is to sinter the material at temperatures near the liquid-phase formation or the varistor key-phase formation steps [5, 10, 28]. By performing the liquid-phase sintering, dopants and minor additives can be uniformly distributed along the varistor grain boundaries. In fact many researchers have reported that in $ZnO–Bi_2O_3$ varistor systems, the Bi_2O_3 additive forms the liquid phase at 850 °C, which is actually the driving force for grain boundary diffusion and rapid densification. For example, the effect of microstructure modification by step-sintering of varistor discs was studied by Anas and co-workers [16]. As illustrated through Fig. 2.20, by step-sintering (heating the sample to a higher temperature and cooling

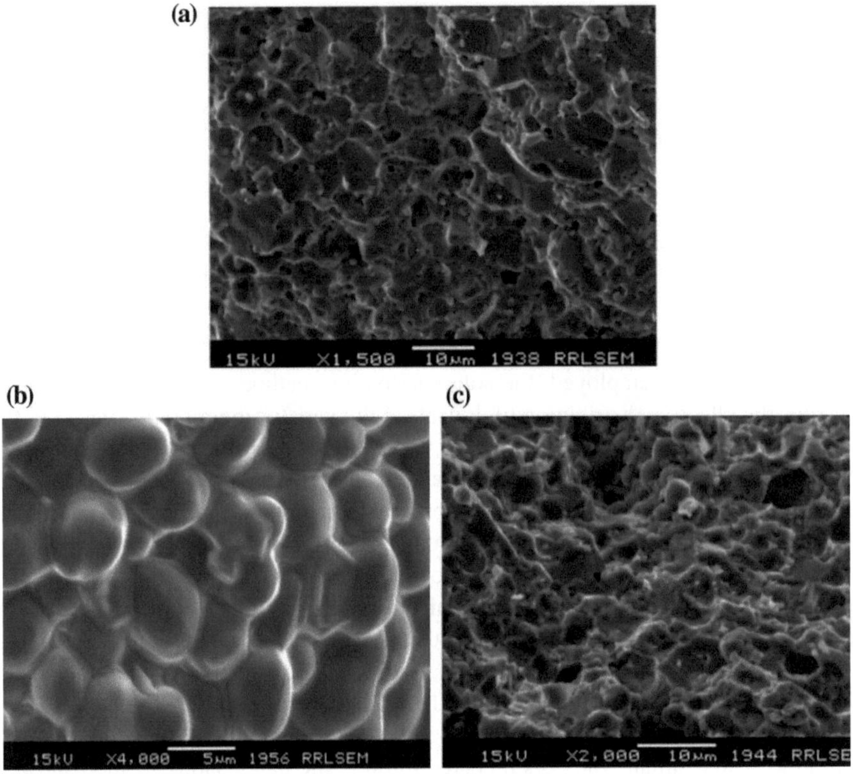

Fig. 2.20 Microstructures of step-sintered varistors at 1050–850 °C **a** commercial varistor disc **b** nanopowder varistor disc **c** composite varistor disc (powder **a** + 5 wt% powder **b**) (reproduced with permission from [16])

down to a low temperature drastically), homogeneous distribution of additives at the grain boundaries were achieved. Uniformly coated secondary additives on the varistor grain boundaries can be seen from these images. As a result of the liquid-phase sintering, the commercial micron-sized varistor discs were resulted with an average grain size of 8 µm, nanopowder varistor discs with 4 µm and nanofillered varistor powder sample with an average grain size of 2 µm.

2.6.3 Use of Economic and High-Quality Raw Materials

For any varistor device applications, one must not compromise with the quality of the components. It should be very high [4, 59, 72]. At the same time, the raw materials should be available abundantly, at low cost. Zn dust is a readily available and a low-cost source material which can be effectively utilized for developing high-performance varistors. For example, very recently, Maria et al.

(2016) introduced a new thermal oxidation method for the much economic production of high-performance varistors from Zn dust. The ceramic–metal composite (Cermet) varistor mixture was processed by mechanical blending of micron-sized metallic Zn dust with rare earth oxides (La_2O_3, CeO_2) and other varistor-forming minor additives (CoO, Cr_2O_3, Al_2O_3, MnO_2 and SiO_2). The cermet varistor mixture was uniaxially pressed into cylindrical pellets and sintered at various temperatures to obtain ZnO varistors. In this method, the zinc dust was converted to different zinc oxide nanostructures like nanorods, mutipods and nanowires at elevated temperatures through sequential thermal oxidation. The cermet-derived varistors yields dense, nano-micro-mixed, fine-grained varistor microstructures with good current–voltage properties. At 1250 °C, the Zn dust derived cermet composition showed 99% theoretical density, average sintered grain size <2 μm, breakdown voltage $V_b = 480$ V mm^{-1} and nonlinear coefficient $\alpha = 46$.

2.7 Sol-Gel Ceramic–Polymer Composite Varistors

2.7.1 Varistor–Epoxy Composites

Polymer composites containing varistor fillers exhibit nonlinear electrical–thermal switching properties [52–54, 56, 93]. Varistor–polymer composite can exhibit two nonlinearities in electrical conductivity. In the conducting state above the varistor filler breakdown, the polymer matrix reduces the conductivity, if a critical temperature is reached. Here, the overvoltage is limited by the fillers (i.e. varistor effect) and the overcurrent is limited by the polymer matrix. Above the breakdown field, the nonlinear effect dominates and then melting takes place. These materials act as positive thermal resistance coefficient (PTCR) due to microcontact separation in the insulating polymer matrix and they acts as varistors mainly due to the nonlinear properties of the grain boundaries.

Figure 2.21 shows the electrical–thermal dependence of a varistor–polymer composite [56, 79]. The high-current paths are visualized by the increase in temperature caused by Joule heating. In the nonlinear regime (B), the current increases exponentially and is then limited to a certain value in the upturn regime (C). The nonlinear phenomenon can be explained by the semiconducting behaviour of the varistor filler. Starting at around 100 °C, the epoxy matrix expands strongly due to the melting of crystalline areas. This leads to a separation of the varistor fillers and causes an increased resistivity. This is known as the electrical–thermal switching effect. Thus the nonlinear electricity determined by the quality of effective grain boundaries of the varistor fillers dominate the overall behaviour only as long as the microcontact resistances are smaller than the filler resistance. The resistance effects are always accompanied by losses in the form of heat. The PTCR effects can be used to limit the dissipated energy because the melting polymer does not allow a further increase in the electrical field. When the applied field is larger than the

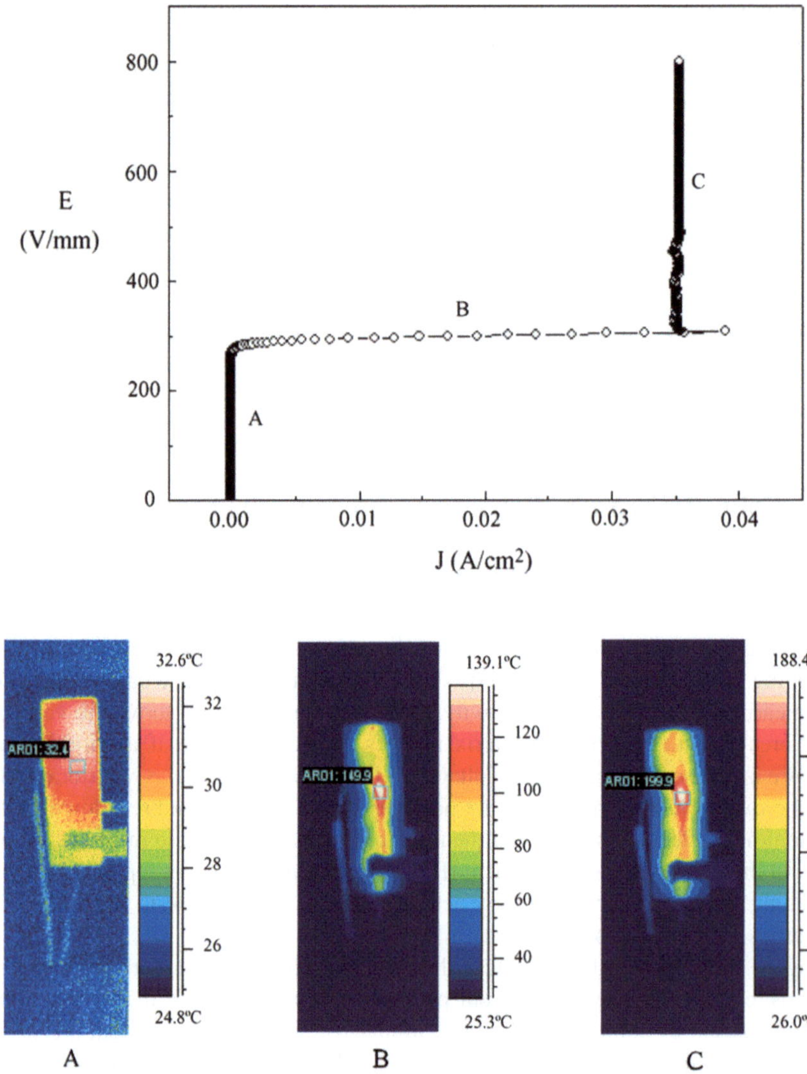

Fig. 2.21 Composite nonlinear electrical–thermal dependence of the epoxy polymer composite varistors. Here, the fillers (45 vol%) were sintered at 1100 °C for 5 h and the composite was cured at 120 °C for 10 min. **'a'** is pre-breakdown region with very low current; **'b'** is nonlinear region with high-current paths which can be visualized by the increase in temperature caused by Joule heating; **'c'** is upturn region with limited current which is controlled by the melting point of epoxy (reproduced with permission from [56])

breakdown field and melting does not occur, there is a relatively wide distribution of current paths, i.e. the nonlinear effect. On the other hand, when the applied field is further increased and local melting occurs, the melting eliminates a portion of the connections and creates new tunnelling paths at the same time.

2.7.2 ZnO–PANI Composites

ZnO/Polyaniline nanostructure heterojunction devices show characteristic room temperature I–V properties. The heterojunction devices can be developed using a PANI thin film (200 nm, on top of a conducting Au film) [94]. On top of the PANI film, a ZnO film should be deposited (high-purity ZnO powder, 70–400 nm) along with a polarizing Al contact (200 nm, deposited using thermal evaporation). At the microscopic level, the structure of PANI should consist of highly conducting islands embedded in a non-conductive sea of the polymer [95, 96]. The electrical properties of these devices can be varied from a standard diode (rectifying in forward polarization) to those of a varistor (rectifying in both forward and reverse polarizations) by tuning the thickness of the ZnO thin film or the polymer doping state. The nonlinear I–V characteristics for the ZnO/PANI heterostructures are likely due to intrinsic surface and interface effects, which are a common feature between grain boundaries of PANI and ZnO.

2.7.3 ZnO–PANI–PVA Composites

Low-voltage varistors with good nonlinearity can be achieved using ZnO–polyaniline–polyvinyl alcohol composites. The polymer composite thin films can be easily prepared by solution-casting technique. Mixing PANI with ZnO inorganic oxide and pressing the mixture resulted in brittle discs which were not suitable for use in electrical devices. So, to prepare uniform and mechanically stable films, solution-casting method can be used. To increase the elasticity of final films, and to prevent them from tearing, polyvinyl alcohol (PVA) can be effectively employed in the varistor matrix.

As shown in Table 2.5, an increase in the polymer content leads to a decrease in the breakdown voltage [52–54, 79]. On the other hand, by increasing the amount of ZnO powder or, alternatively, decreasing the amount of polymer matrix, the number of back-to-back Zener diodes in a unique thickness increases and this leads to a higher breakdown voltage. Since the devices with high nonlinearity coefficients (α) and low breakdown voltages are useful in protecting sensitive electrical circuits, samples with higher amount of polymer content are suggested for modern sophisticated devices.

2.7.4 GaAs–PANI–PE Composites

Low-voltage varistors can be made out of a low-band gap semiconductor, gallium arsenide—as filler and polymers—e.g. polyaniline and polyethylene—as matrix

Table 2.5 Parameters of composite varistors [52–54]

Film No.	ZnO content (%)	Polymer matrix (%)	Breakdown voltage (V)	α	Band gap (eV)
1	20	80	120	4.084	3.23
2	40	60	200	3.632	3.21
3	60	40	280	2.920	3.17
4	80	20	350	2.700	3.16

[52–54, 93, 97]. The efficacy of this composite as low-field varistor depends completely on the material properties of each of the individual constituents. Gallium arsenide, the key III–V semiconductor used for these varistor formulations, normally possess higher saturated electron velocity, direct band gap transition and higher electron mobility. Due to these benefits, they are commonly used in devices for high-frequency and low-power applications. PANI is a rigid polymer with weak mechanical stability. It functions as a semiconducting polymer, due to the probability of electron movement along the polymer chain [52–54, 94]. Conductivity of emeraldine salt form of PANI is about 1 S cm^{-1} which is comparable to the conductivity of any semiconductors. The composite samples without PANI do not show nonlinearity and its I–V characteristic is, unlike varistors, i.e. linear. Addition of PANI reduces resistance of the second phase and causes samples to exhibit nonlinearity [96]. In spite of its unique properties, low mechanical stability of PANI is a major disadvantage. To overcome its frangibility, one method followed is to combine it with a thermoplastic polymer, e.g. polyethylene, which consists of long hydrocarbon chains. Polyethylene normally exhibits excellent thermal and electrical insulating properties. Within the framework, they show high resistance to chemicals and possess good mechanical and optical properties.

The composite varistor with GaAs–polyaniline–polyethylene compositions can be easily prepared as thin discs using hot pressing method [52–54]. GaAs–PANI–PE composite varistors with higher content of PANI possesses lower breakdown voltages, nonlinearity coefficients and optical impurity band gaps. They protect circuits against over voltages ranging from $\sim$55 to 70 V. The composite varistor has hysteresis which increases through increase in its polyaniline content. However, samples with higher percentages of GaAs (>50%) display the varistor properties. They can be used to protect circuits from $\sim$60 V up to over 90 V voltages. The higher the content of GaAs in the varistor, the lower the breakdown voltage would be. Their nonlinearity coefficients, hysteresis, optical impurity bandgaps and electrical bandgaps would also be higher. The composite varistor sample exhibits hysteresis which decreases through the increase in sintering temperature. This causes the varistors to have longer lifetime due to their low degradation. Changing the sintering temperature affects properties of GaAs–PANI–PE composite varistors in such a way that its increment increases their breakdown voltages, nonlinearity coefficients and impurity bandgaps as well as their lifetime; whereas leakage current and barrier heights of these varistor normally decrease.

2.7.5 Si–Polymer Composite Varistors

Silicon has some unique properties that make it important in varistor industry. It is a hard, relatively inert metalloid which possesses a low-band gap value of 1.1 eV. It is a good thermal conductor, which enables very dense packing of transistors that need to get rid of their heat of operation. As a varistor, the breakdown voltages of Si–polymer composite varistors are low in comparison with similar ZnO–polymer composite varistors [98, 99]. By Si content exceeding 65% of the whole Si-polymer mixture, samples exhibit varistor behaviour with nonlinear coefficient of about 10. Studies on (I–V) characteristics of samples show that by increasing Si content in the Si-polymer mixture, nonlinear coefficient and hysteresis loop increase whereas breakdown voltage and potential barrier height decreases.

There are two reasons to observe hysteresis loop on I–V diagrams, which have effect on lifetime of a varistor: (a) Joule heating and (b) existence of dipole moments. The former is very important because of lower band gap of Si that causes higher leakage current compared with ZnO varistors. For these varistors, leakage current is in mA range, which increases operating temperature and results in lifetime reduction. One of the most important effects of Joule heating is that it can directly affect polymer structure. Heating could change electrical parameters such as intergranular resistivity. Polymer may be decomposed during long inactive operating time resulting in more degradation. The latter is less important with respect to the first one. On the other hand, in the absence of an external electric field, dipole moments of superficial conditions of two contacting surfaces are randomly oriented; they can even be antiparallel. By increasing the applied voltage, majority of dipole moments orient along the applied electric field. Apparently, by decreasing the applied field, a limited number of dipole moments remain along the field direction; however, most of them do not align with the field. Therefore, a hysteresis on (I–V) curves will be observed. As shown in Fig. 2.22 [98, 99], the hysteresis loops shrink by increasing temperature because of the mechanisms' equalization.

For Si-polymer varistors, the leakage current is in mA range; as a result, it is one of the most probable reasons for instability of grain boundaries, especially for samples with higher amount of Si content. The temperature dependence of current–voltage characteristic shows that the increase in ambient temperature lowers breakdown voltage as well as nonlinearity coefficient. It should be noted that Si-PANI–PE composite varistors cannot be used at higher temperatures because of the following two reasons:

1. PE starts to melt at 130 °C, so the discs would become spoiled.
2. Nonlinear behaviour of the sample is not good enough for varistor application even at 100 °C. At temperatures between 100 °C and 130 °C, the nonlinear conductivity of varistor gradually turns to Ohmic conductivity.

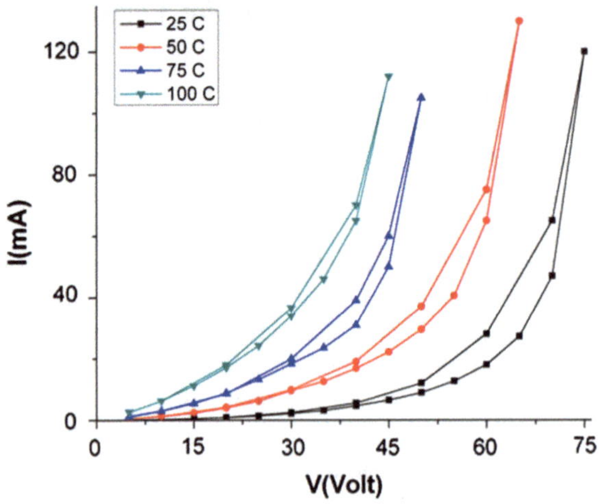

Fig. 2.22 Hysteresis loops for the Si-polymer sample at different temperatures (reproduced with permission from [98, 99])

Si-PANI–PE composite varistor, with lower breakdown voltage, can be a suitable substitute for ZnO-based varistors for the purpose of protection from low overvoltage.

2.8 Conclusions

The chapter presented an overview on ‘sol-gel derived varistors’. The full potential of sol-gel method is explored in the current chapter by not only presenting simple sol-gel processing routes for preparing multicomponent nanoparticles system, but also by analyzing their derived products for high-performance miniaturized varistor applications. The unique characteristics of sol-gel syntheses which make them useful for varistor applications are their capability to do atomic level mixing of reactants, homogeneous distribution of low-level dopants and additives, stable oxide network formation through controlled hydrolysis and polycondensation reactions at room/low temperatures, control in nanoparticles preparation and the capability to produce high surface area particles with greater homogeneity. The advantage of sol-gel method is that it can be easily integrated with any varistor processing methods. Scalable, multicomponent particles can be prepared through sol-gel methods with a strict control in the stoichiometry by mere mixing sols of different compounds. Miniaturized varistors can be processed through these promising methods which yields higher breakdown voltage, low leakage current and good nonlinear characteristics. By integrating these simple processing routes and sophisticated sintering methods, high-performance or next-generation varistor devices can be achieved.

References

1. Matsuoka, M.: Nonohmic properties of ZnO ceramics. Jpn. J. Appl. Phys. **10**, 736–746 (1971)
2. Levine, J.D.: Theory of varistor electronic properties. Crit. Rev. Solid State Mater. Sci. **5**, 597–608 (1975)
3. Mahan, G.D., Levinson, L.M., Philipp, H.R.: Theory of conduction in ZnO varistors. J. Appl. Phys. **50**, 2799–2812 (1979)
4. Levinson, L.M., Philipp, H.R.: Zinc Oxide varistors-a review. Am. Ceram. Soc. Bull. **65**, 639–646 (1986)
5. Pillai, S.C., Kelly, J.M., McCormack, D.E., et al.: High performance ZnO varistors prepared from nanocrystalline precursors for miniaturised electronic devices. J. Mater. Chem. **18**, 3926–3932 (2008)
6. Bok-Hee, L., Sung-Man, K.: Properties of ZnO varistor blocks under multiple lightning impulse voltages. Curr. Appl. Phys. **6**, 844–851 (2006)
7. Clarke, D.R.: Varistor Ceramics. J. Am. Ceram. Soc. **82**, 485–502 (1999)
8. Gupta, T.K.: Application of zinc oxide varistors. J. Am. Ceram. Soc. **73**, 1817–1840 (1990)
9. Pillai, S.C., Kelly, J.M., Ramesh, R., et al.: Advances in the synthesis of ZnO nanomaterials for varistor devices. J. Mater. Chem. C **1**, 3268–3281 (2013)
10. Duran, P., Capel, F., Tartaj, J., et al.: A Strategic two-stage low-temperature thermal processing leading to fully dense and fine-grained doped-ZnO varistors. Adv. Mater. **14**, 137–141 (2002)
11. Ezhilvalavan, S., Kutty, T.R.N.: Low-voltage varistors based on zinc antimony spinel Zn-Sb-O. Appl. Phys. Lett. **68**, 2693–2695 (1996)
12. Clarke, D.R.: The microstructural location of the intergranular metal oxide phase in a zinc oxide varistor. J. Appl. Phys. **49**(4), 2407–2411 (1978)
13. Kingery, W.D.: Distribution and influence of minor constituents on ceramic formulations. Mater. Sci. Res. **21**, 281–294 (1986)
14. Alim, M.A., Li, S., Liu, F., et al.: Electrical barriers in the ZnO varistor grain boundaries. Phys. Stat. Sol. **203**, 410–427 (2006)
15. Clarke, D.R.: Grain boundary segregation in a commercial ZnO-based varistor. J. Appl. Phys. **50**(11), 6829–6832 (1979)
16. Anas, S., Mangalaraja, R.V., Mukundan, P., et al.: Direct synthesis of varistor-grade doped nanocrystalline ZnO and its densification through a step-sintering technique. Acta Mater. **55**, 5792–5801 (2007)
17. Dienel, H.F.: Silicon carbide varistors: properties and construction. Bell Lab. Rec. **34**, 407 (1956)
18. Frosch, C.J.: Improved silicon carbide varistors. Bell Lab. Rec. **32**, 336–340 (1954)
19. Hsiang, H.-I., Wang, S.-S.: Cooling rate effects on the electrical properties of TiO_2-based varistor. Mater. Sci. Eng. B **128**, 25–29 (2006)
20. Nakano, Y., Ichinose, N.: Oxygen adsorption and VDR effect in (Sr, Ca) TiO-based ceramics. J. Mater. Res. **5**, 2910–2921 (1990)
21. Brankovic, Z., Brankovic, Bernik S., et al.: ZnO varistors with reduced amount of additives prepared by direct mixing of constituent phases. J. Eur. Ceram. Soc. **27**, 1101–1104 (2006)
22. Sedky, A., Abu-Abdeen, M., Almulhem, A.A.: Nonlinear I–V characteristics in doped ZnO based-ceramic varistor. Phys. B **388**, 266–273 (2006)
23. Mukae, K.: Zinc oxide varistors with praseodymium oxide. Am. Ceram. Soc. Bull. **66**, 1329–1331 (1987)
24. Takemura, T., Kobayashi, M., Takada, Y., et al.: Effects of bismuth sesquioxide on the characteristics of ZnO varistors. J. Am. Ceram. Soc. **69**(5), 430–436 (1986)
25. Onreabroy, W., Sirikulrat, N., Brown, A.P., et al.: Properties and intergranular phase analysis of a ZnO–CoO–Bi_2O_3 varistor. Solid State Ionics **177**, 411–420 (2006)

26. Pillai, S.C., Kelly, J.M., McCormack, D.E.: Self-assembled arrays of ZnO nanoparticles and their application as varistor materials. J. Mater. Chem. **14**, 1572–1578. Pillai, S.C., Kelly, J. M., McCormack, D.E., et al.: Microstructural analysis of varistors prepared from nanosize ZnO. Mater. Sci. Tech. Ser. **20**, 964–968 (2004)
27. Gardner, T.J., Doughty, D.H., Lockwood, S.J., et al.: The effect of low level dopants on chemically prepared varistor materials. Ceram. Trans. **3**, 84–92 (1989)
28. Kingery, W.D.: Densification during sintering in the presence of a liquid phase I. Theory and II. Experimental. J. Appl. Phys. **30**(3), 301–310 (1959)
29. Shaw, T.M.: Liquid redistribution during liquid-phase sintering. J. Am. Ceram. Soc. **69**(1), 27–34 (1986)
30. Senda, T., Bradt, R.C.: Grain growth in sintered ZnO and ZnO–Bi_2O_3 ceramics. J. Am. Ceram. Soc. **73**, 106–114 (1990)
31. Fan, J., Sale, F.R.: Citrate gel route processing of ZnO varistors. Br Ceram Proc **52**, 151–157 (1994)
32. Lihong, C., Guorong, L., Liaoying, Z., et al.: Analysis of high-voltage ZnO varistor prepared from a novel chemically aided method. J. Am. Ceram. Soc. **93**, 2522–2525 (2010)
33. Sonder, E., Quinby, T.C., Kinser, D.L.: ZnO varistor made from powders produced using a urea process. Am. Ceram. Soc. Bull. **65**(4), 665–668 (1986)
34. Yuke, L., Guorong, L., Qingrui, Y., et al.: Preparation of ZnO varistors by solution nano-coating technique. Mat. Sci. Eng. B **130**, 264–268 (2006)
35. Wang, Q., Qin, Y., Xu, G.J., et al.: Low-voltage ZnO varistor fabricated by the solution-coating method. Ceram. Int. **34**, 1697–1701 (2008)
36. Duran, P., Capel, F., Tartaj, J., et al.: Sintering behavior and electrical properties of nanosized doped-ZnO powders produced by Metallorganic Polymeric Processing. J. Am. Ceram. Soc. **84**, 1661–1668 (2001)
37. Hishita, S., Yao, Y., Shirasaki, S.-I.: Zinc oxide varistors made from powders prepared by amine processing. J. Am. Ceram. Soc. **77**(2), 338–340 (1989)
38. Riahi-Noori, N., Sarraf-Mamoory, R., Alizadeh, P.: Synthesis of ZnO nano powder by a gel combustion method. J. Ceram. Process. Res. **9**, 246–249 (2008)
39. Banerjee, A., Ramamohan, T.R., Patni, M.J.: Smart technique for fabrication of Zinc Oxide varistor. Mater. Res. Bull. **36**, 1259–1267 (2001)
40. Lorenz, A., Ott, J., Harrer, M., et al.: Modified citrate gel routes to ZnO-based varistors. J. Eur. Ceram. Soc. **21**, 1887–1891 (2001)
41. Sinha, A., Sharma, B.P.: Novel route for preparation of high voltage varistor powder. Mater. Res. Bull. **32**, 1571–1579 (1997)
42. Hingorani, S., Shah, D.O., Multani, M.S.: Effect of process variables on the grain growth and microstructure of ZnO–Bi_2O_3 varistors and their nanosize ZnO precursors. J. Mater. Res. **10**, 461–467 (1995)
43. Haile, S.M., Johnson Jr., D.W., Wiseman, G.H., et al.: Aqueous precipitation of spherical ZnO powders for varistor applications. J. Am. Ceram. Soc. **72**, 2004–2008 (1989)
44. Li, Y., Li, G., Yin, Q.: Preparation of ZnO varistors by solution nano-coating technique. Mater. Sci. Eng. B **130**, 264–268 (2006)
45. Nobrega, M.C.S., Zolotar, M.S., Mannheimer, W.A., et al.: ZnO varistors produced using colloidal-gel powders. J. Non-Cryst. Solids **147–148**, 803–807 (1992)
46. Viswanath, R.N., Ramasamy, S., Ramamoorthy, R., et al.: Preparation and characterization of nanocrystalline ZnO based materials for varistor applications. Nanostruct. Mater. **6**, 993–996 (1995)
47. Varma, H.K., Ananthakumar, S., Perumal, P., et al.: Microwave processing of ZnO varistor. Ceram. Trans. **36**, 115 (1993)
48. Varma, H.K., Ananthakumar, S., Warrier, K.G.K., et al.: Synthesis of zinc oxide varistors through microwave-derived precursor. Ceram. Int. **22**, 53–56 (1996)
49. Anas, S., Metz, R., Sanoj, M.A., et al.: Sintering of surfactant modified ZnO–Bi_2O_3 based varistor nanopowders. Ceram. Int. **36**, 2351–2358 (2010)

50. Anas, S., Mukundan, P., Sanoj, M.A., et al.: Synthesis of ZnO based nanopowders via a non-hydrolytic sol gel technique and their densification behaviour and varistor properties. Proc. Appl. Ceram. **4**, 7–14 (2010)
51. Macary, L.S., Kahn, M.L., Estournes, C.: Size effect on properties of varistors made from Zinc Oxide nanoparticles through low temperature spark plasma sintering. Adv. Funct. Mater. **19**, 1775–1783 (2009)
52. Bidadi, H., Aref, S.M., Ghafouri, M., et al.: Effect of changing Gallium arsenide content on Gallium arsenide–polymer composite varistors. J. Phys. Chem. Solids **74**, 1169–1173 (2013)
53. Bidadi, H., Aref, S.M., Ghafouri, M., et al.: The effect of sintering temperature on varistor characteristics of galliumarsenide–polyaniline–polyethylene composite varistors. Mat. Sci. Semicon. Proc. **16**, 752–758 (2013)
54. Bidadi, H., Olad, A., Parhizkar, M., et al.: Nonlinear properties of ZnO-polymer composites prepared by solution-casting method. Vacuum **87**, 50–54 (2013)
55. Hashimov, A.M., Hasanli, S.M., Mehtizadeh, R.N., et al.: Zinc oxide-and polymer-based composite varistors. Phys. Stat. Sol. C. **3**, 2871–2875 (2006)
56. Lin, C.-C., Lee, W.-S., Sun, C.-C., et al.: A varistor–polymer composite with nonlinear electrical-thermal switching properties. Ceram. Int. **34**, 131–136 (2008)
57. Lanyi, W., Guoyi, T., Zheng-Kui, X., et al.: Preparation and electrical properties of multilayer ZnO varistors with water-based tape casting. Ceram. Int. **35**, 487–492 (2009)
58. Maria, M.J., Balanand, S., Anas, S., et al.: Zn-dust derived ultrafine grained ZnO non-linear ceramic resistors via in-situ thermal oxidation of cermet reactant mixture. Mater. Des. **92**, 387–396 (2016)
59. Suvaci, E., Ozgur, I.O.: Processing of textured zinc oxide varistors via templated grain growth. J. Euro. Ceram. Soc. **25**, 1663–1673 (2005)
60. Rahul, S.P., Mahesh, K.V., Sujith, S.S., et al.: Processing of La_2O_3 based rare earth non-linear resistors via combustion synthesis. J. Electroceram. **32**, 292–300 (2014)
61. Cheng, L.H., Zheng, L.Y., Meng, L., et al.: Electrical properties of Al_2O_3-doped ZnO varistors prepared by sol–gel process for device miniaturization. Ceram. Int. **38S**, S457–S461 (2012)
62. Dong, X., Bin, J., Lei, J., et al.: Sol-gel synthesis of Y_2O_3-doped ZnO thin films varistors and their electrical properties. Trans. Nonferrous Met. Soc. China **22**, 110–114 (2012)
63. Gert, H., Gerhard, T.: Sol-gel processing of varistor powders. J. Mater. Res. **7**, 546–548 (1992)
64. Hohenberger, G., Tomandl, G.: Sol-gel processing of varistor powders. J. Mater. Res. **7**(3), 546–548 (1992)
65. Huang, Y.Q., Meidong, L., Yike, Z., et al.: Preparation and properties of ZnO-based ceramic films for low-voltage varistors by novel sol-gel process. Mater. Sci. Eng. B: Solid-State Mater. Adv. Technol. **86**(3), 232–236 (2001)
66. Jincang, Z., Shixun, C., Ruiying, Z., et al.: Effect of fabrication conditions on I–V properties for ZnO varistor with high concentration additives by sol–gel technique. Current Appl. Phys. **5**, 381–386 (2005)
67. Lauf, R.J., Bond, W.D.: Fabrication of high-field zinc oxide varistors by sol–gel processing. Am. Ceram. Soc. Bull. **63**(2), 278–281 (1984)
68. Liu, T.-T., Wang, M.-H., Zhang, H.-P., et al.: Sol–gel synthesis of doped nanocrystalline ZnO powders using xanthan gum and varistor properties study. J. Mater. Sci.: Mater. Electron. **26**, 9056–9062 (2015)
69. Pillai, S.C., Kelly, J.M., McCormack, D.E., et al.: The effect of processing conditions on varistors prepared from nanocrystalline ZnO. J. Mater. Chem. **13**, 2586–2590 (2003)
70. Pillai, S.C., Kelly, J.M., McCormack, D.E., et al.: Effect of step sintering on breakdown voltage of varistors prepared from nanomaterials by sol gel route. Adv. Appl. Ceram. **105**, 158–160 (2006)
71. Puyane, R., Toal, F., Hampshire, S.: Production of doped ZnO powders for varistor applications using sol-gel techniques. J. Sol-Gel. Sci. Technol. **6**, 219–225 (1996)

72. Puyane, R., Guy, I., Metz, R.: High performance varistor discs obtained from chemically synthesized doped Zinc Oxide powder. J. Sol-Gel. Sci. Technol. **13**, 575–578 (1998)
73. Westin, G., Ekstrand, A., Nygren, M., et al.: Preparation of ZnO-based varistors by the sol–gel technique. J. Mater. Chem. **4**, 615–621 (1994)
74. Xu, D., Jiang, B., Jiao, L., et al.: Sol-gel synthesis of Y_2O_3-doped ZnO thin films varistors and their electrical properties. Trans. Nonferrous. Mat. Soc. China **22**, 110–114 (2012)
75. Ya, K.X., Diao, W.T., Yin, H., et al.: Sol-gel process doped ZnO nanopowders and their grain growth. Mater. Res. Bull. **32**, 1165–1171 (1997)
76. Yanqiu, Q.H., Liu, M., Zeng, Y., et al.: Preparation and properties of ZnO-based ceramic films for low-voltage varistors by novel sol-gel process. Mater. Sci. Eng., B **86**, 232–236 (2001)
77. Wang, M.-H., Zhao, Z.-Y., Liu, T.-T.: Synthesis of Pr-doped ZnO nanoparticles by sol–gel method and varistor properties study. J. Alloys Comp. **621**, 220–224 (2015)
78. Zhang, J., Cao, S., Zhang, R., et al.: Effect of fabrication conditions on I–V properties for ZnO varistor with high concentration additives by sol–gel technique. Curr. Appl. Phys. **5**, 381–386 (2005)
79. Ozgur, U., Alivov, Y.I., Liu, C., et al.: A comprehensive review of ZnO materials and devices. J. Appl. Phys. **98**, 041301 (2015)
80. Lee, W.-H., Chen, W.-T., Lee, Y.-C.: Performance enhancement of ZnO-based multilayer varistor by constrained sintering. J. Am. Ceram. Soc. **91**, 2938–2942 (2008)
81. Jianzhang, S., Quanxi, C., Yunge, W., et al.: ZnO varistor manufactured by composite nano-additives. Mat. Sci. Eng. B **99**, 344–347 (2003)
82. Mehdi, M., Zahedi, A.M., Sadrnezhaad, S.K.: Two-step sintering of nanocrystalline ZnO compacts: effect of temperature on densification and grain growth. J. Am. Ceram. Soc. **91**, 56–63 (2008)
83. Subasri, R., Asha, M., Hembram, K., et al.: Microwave sintering of doped nanocrystalline ZnO and characterization for varistor applications. Mater. Chem. Phys. **115**, 677–684 (2009)
84. Westin, G., Nygren, M.: Sol-gel preparation of M-Sb oxides from Sb (OBun)3-M- Acetate precursors with M = Mn Co, Ni. J. Mater. Chem. **3**, 367–371 (1993)
85. Ya, K.X., Yin, H., De, T.M., et al.: Analysis of ZnO varistors prep ared from nanosize ZnO precursors. Mater. Res. Bull. **33**, 1703–1708 (1998)
86. Chu, S.-Y., Yan, T.-M., Chen, S.-L., et al.: Analysis of ZnO varistors prepared by the sol-gel method. Ceram. Int. **26**, 733–737 (2000)
87. Anas, S., Mahesh, K.V., Jobin, V., et al.: Nanofillers in ZnO based materials: a 'smart' technique for developing miniaturized high energy field varistors. J. Mater. Chem. C. **1**, 6455–6462 (2013)
88. Peiteado, M., Fernandez, J.F., Caballero, A.C.: Processing strategies to control grain growth in ZnO based varistors. J. Eur. Ceram. Soc. **25**, 2999–3003 (2005)
89. Peiteado, M., Fernandez, J.F., Caballero, A.C.: Varistors based in the ZnO–Bi_2O_3 system: microstructure control and properties. J. Eur. Ceram. Soc. **27**, 3867–3872 (2007)
90. Lee, W.-H., Chen, W.-T., Su, C.-Y.: Attaining a homogeneous microstructure of a ZnO-based multilayer varistor by constrained sintering. J. Am. Ceram. Soc. **90**, 3296–3298 (2007)
91. Cong, L., Zheng, X., Hu, P., et al.: Bi_2O_3 Vaporization in microwave-sintered ZnO varistors. J. Am. Ceram. Soc. **90**(9), 2791–2794 (2007)
92. Yuan, F., Ryu, H.: Microstructure of varistors prepared with zinc oxide nanoparticles coated with Bi_2O_3. J. Am. Ceram. Soc. **87**, 736–738 (2004)
93. Aref, S.M., Olad, A., Parhizkar, M., et al.: Effect of polyaniline content on electrophysical properties of gallium arsenide polymer composite varistors. Solid State Sci. **26**, 128–133 (2013)
94. Felix, J.F, da Silva Jr., E.F., de Vasconcelos, E.A., et al. Tailoring the electrical properties of ZnO/Polyaniline heterostructures for device applications. J Korean Phys. Soc. **58**(5), 1256–1260 (2011)

95. MacDiarmid, A.G., Epstein, A.J.: Poly anilines: a novel class of conducting polymers. Faraday Discuss Chem. Soc. **88**, 317–332 (1989)
96. MacDiarmid, A.G., Chiang, J.C., Richter, A.F.: Polyaniline: a new concept in conducting polymers. Synth. Met. **18**, 285–290 (1987)
97. Greuter, F., Siegrist, M., Kluge-Weiss, P., et al.: Microvaristors: functional fillers for novel electroceramic composites. J. Electroceram. **13**, 739–744 (2004)
98. Ghafouri, M., Parhizkar, M., Aref, S.M., et al.: Effect of temperature on the electrophysical properties of Si–polymer composite varistors. Microelectron. Reliab. **54**, 965–971 (2014)
99. Ghafouri, M., Parhizkar, M., Bidadi, H., et al.: Effect of Si content on electrophysical properties of Si-polymer composite varistors. Mater. Chem. Phys. **147**, 1117–1122 (2014)
100. Daneu, N., Recnik, A., Bernik, S.: Grain growth control in Sb_2O_3-doped zinc oxide. J. Am. Ceram. Soc. **86**, 1379–1384 (2003)
101. Shi, J., Cao, Q., Wei, Y., et al.: ZnO varistor manufactured by composite nano additives. Mater. Sci. Eng. B **99**, 344–347 (2003)
102. Yano, Y., Takai, Y., Morooka, H.: Interface states in ZnO varistors with Mn Co, and Cu impurities. J. Mater. Res. **9**, 112–118 (1994)
103. Nahm, C.W.: Effect of sintering temperature on nonlinear electrical properties and stability against DC accelerated aging stress of (CoO, Cr_2O_3, La_2O_3)-doped ZnO–Pr_6O_{11}-based varistors. Mater. Lett. **60**, 3311–3314 (2006)
104. Wang, M.H., Hu, K.A., Zhao, B.Y., et al.: Electrical characteristics and stability of low voltage ZnO varistors doped with Al. Mater. Chem. Phys. **100**, 142–146 (2006)
105. Nahm, C.W., Shin, B.C.: Highly stable nonlinear properties of ZnO–Pr_6O_{11}–CoO– Cr_2O_3–Y_2O_3 based varistor ceramics. Mater. Lett. **57**, 1322–1326 (2003)
106. Binks, D.J., Grimes, R.W.: Incorporation of monovalent ions in ZnO and their influence on varistor degradation. J. Am. Ceram. Soc. **76**, 2370–2372 (1993)
107. Fan, J., Freer, R.: The roles played by Ag and Al dopants in controlling the electrical properties of ZnO varistors. J. Appl. Phys. **77**, 4795–4800 (1995)
108. Gupta, T.K., Miller, A.C.: Improved stability of the ZnO varistor via donor and acceptor doping at the grain boundary. J. Mater. Res. **3**, 745–754 (1988)
109. Daneu, N., Recnik, A., Bernik, S., Kolar, D.: Microstructural development in SnO_2-Doped ZnO Bi_2O_3 ceramics. J. Am. Ceram. Soc. **83**, 3165–3171 (2000)
110. Santos, P.A., Maruchin, S., Menegoto, G.F., et al.: The sintering time influence on the electrical and microstructural characteristics of SnO_2 varistor. Mater. Lett. **60**, 1554–1557 (2006)
111. Silva, I.P., Simoes, A.Z., Filho, F.M., et al.: Dependence of La_2O_3 content on the nonlinear electrical behaviour of ZnO, CoO and Ta_2O_5 doped SnO_2 varistors. Mater. Lett. **61**, 2121–2125 (2007)
112. Houabes, M., Metz, R.: Rare earth oxides effects on both the threshold voltage and energy absorption capability of ZnO varistors. Ceram. Int. **33**, 1191–1197 (2006)
113. Wang, J.-F., Chen, H.-C., Su, W.-B., et al.: Effects of Sr on the microstructure and electrical properties of (Co, Ta)-doped SnO_2 varistors. J. Alloys Comp. **413**, 35–39 (2006)
114. Margionte, M.A.L., Simoes, A.Z., Riccardi, C.S., et al.: Nonlinear characteristics of Cr_2O_3, WO_3, ZnO and CoO doped SnO_2 varistors. Mater. Lett. **60**, 142–146 (2006)
115. Yang, H.G., Zeng, H.C.: Synthetic architectures of $TiO_2/H_2Ti_5O_{11}$ H_2O, $ZnO/H_2Ti_5O_{11}$ H_2O, $ZnO/TiO_2/H_2Ti_5O_{11}$ H_2O and ZnO/TiO_2 nanocomposites. J. Am. Chem. Soc. **127**, 270–278 (2005)

Chapter 3
Sol-Gel Derived Functional Coatings for Optics

Guillaume Naudin, Davide R. Ceratti and Marco Faustini

3.1 Introduction

3.1.1 Sol-Gel Chemistry and Optical Coatings: An Industry-Driven Development

Because of its exceptional versatility, sol-gel chemistry was identified as one of the key technologies for optical applications since the beginning of its long history [1]. The great potential of the sol-gel process was described more than two decades ago by Livage et al. as the ability to go all the way from the molecular precursor to the product, allowing a better control of the whole process and the synthesis of "tailor-made" materials [2]. However, differently from many other fields, the sol-gel "technology" (driven by industrial needs) had developed before the sol-gel "science".

In 1845, the French chemist Ebelman reported for the first time that transparent hydrated silica could be obtained from the slow hydrolysis of a silicic ester in presence ambient moisture [3]. The first communication on sol-gel optical coatings dates back to 1930; the patent deposited from Geffcken and Berger from Schott Glaswerke Company describes the fabrication of oxide layers from metal oxide precursors by dip-coating [4]. The first products appeared on the market in 1953 and large-scale production in 1959 with automotive rear-view mirrors, later antireflection coatings, and sun-shielding windows were introduced using TiO_2 and SiO_2 layers. In the 60s, titanium butoxide was used by Schoroeder in order to adjust the refractive index of optical layers [5]. At the end of this period, almost any mixture of metal oxide could be industrially deposited as thin films from the

G. Naudin · D.R. Ceratti · M. Faustini (✉)
Sorbonne Universités, UPMC Univ Paris 06, CNRS, Collège de France,
UMR 7574, Chimie de la Matière Condensée de Paris, 75005 Paris, France
e-mail: marco.faustini@upmc.fr

S.C. Pillai and S. Hehir (eds.), *Sol-Gel Materials for Energy, Environment and Electronic Applications*, Advances in Sol-Gel Derived Materials and Technologies,
DOI 10.1007/978-3-319-50144-4_3

solutions containing various metal alcoxides, dissolved metal oxides, or metal salts. The second wave of interest in fundamental sol-gel chemistry came with the first International Workshop on Glass Ceramics from Gels that took place in Padova (Italy) only in October 1981. With the aim of finding inexpensive alternatives to physical deposition of films, sol-gel applications was revitalized in the 90s and applied in numerous application fields such as ultrafiltration membranes [6]. Since then, sol-gel chemistry was developed as a soft-chemistry method and was coupled with other fields to gave birth to more general approaches of hybrid organic-inorganic materials aiming at finding the best possible synergies between the inorganic and organic chemistry worlds. In this regard, a milestone in the development of functional coatings was the fabrication of mesoporous films by evaporation-induced self-assembly [7, 8]. These new and multidisciplinary approaches to materials science developed extremely fast and are now widely applied for the preparation of functional films in many fields including micro-electronics, optics, surface protection, energy harvesting, isolation, conductive layers, smart windows, etc. [9]. The success of the soft-chemistry strategies for optical coatings is due to the unique characteristics of the sol-gel process:

(i) The sol-gel rheology is ideally suited for the shaping of materials as thin films.
(ii) The homogeneous mixing of several components (organic and inorganic) at the molecular level allows tailoring the sub-wavelength optical properties.
(iii) This inorganic polymerization can be naturally adapted to form macro and mesostructured materials by a wide range of templating strategies (soft templating, hard templating, emulsion templating, phase separation).

In addition to the modulation of the optical properties, the versatile sol-gel chemistry is also particularly suited to assembling/integrating other components in order to fabricate real multifunctional optical devices.

A significant number of excellent reviews and chapters have summarized the main activities on the field of sol-gel derived materials in optics [10–14]. This chapter will give (non-exhaustive) focus on sol-gel functional optical coatings and, in particular, those dealing with reflection and antireflection phenomena, such as photonic crystals and antireflective coatings. The following sections will describe some chemical strategies to modulate the optical properties of sol-gel coatings by adapted liquid deposition techniques. Moreover, some recent examples of functional (and multifunctional) antireflective coatings and photonic crystals will be highlighted.

3.1.2 Sol-Gel Strategies Toward Original Optical Properties

Light interactions with matter are the foundation for many useful applications in optics and photonics. Light can be absorbed, transmitted, reflected, scattered, or emitted depending on the intrinsic properties of the material or its organization in space. When shaped as a thin film, the optical performances result from a combination of the inherent properties of the material, the surface/interface characteristics, its thickness and its morphology. While sol-gel processing does not produce completely new optical phenomena, it does offer new ways to assemble materials. Some general strategies associated with the molecular design and construction of functional sol-gel derived coatings having tunable optical properties are illustrated in Fig. 3.1. The scheme gives a general conceptual overview about the strategies employed to tune the optical response of the materials at a certain wavelength such as in the visible range.

Initially the modulation of the refractive index of a "transparent" dielectric material,which is not absorbing or scattering/diffracting light. The simplest case is illustrated in Fig. 3.1i; it corresponds to a sol-gel derived dense film, made of silica for instance, transparent in the visible range. The refractive index of the material depends on its composition and can be easily tuned by modifying the initial synthetic conditions. Depending on the nature of the inorganic precursor, the composition of the final materials (and its refractive index) can be modulated, as illustrated in Fig. 3.1ii, without affecting the optical transparency in the visible range. Table 3.1 presents some of the most common materials used in sol-gel optical coatings and the corresponding refractive index at 550 nm [15, 16]. Among the sol-gel materials, SiO_2 is considered a low refractive index material (n = 1.45–1.47 at 550 nm) while TiO_2 is the most used as high refractive index component (n = 2.2–2.4 at 550 nm) in optical devices. A mixture of two (or more) sol-gel oxides such as SiO_2 and TiO_2 allows precise tuning of the refractive index of the coatings. As shown in Fig. 3.1iii, this can be achieved by the formation of mixed-oxides network or by the formation of nanocomposite materials in which,

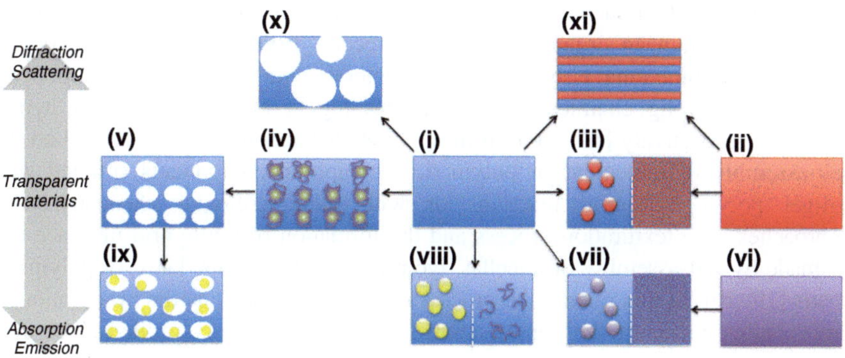

Fig. 3.1 Conceptual illustration of the some general strategies associated to the molecular design and construction of functional sol-gel derived coatings with tunable optical properties: **i** low refractive index material, **ii** high refractive index material, **iii** mixed-oxides material, **iv** hybrid organic/inorganic material, **v** porous material, **vi** absorbing material, **vii** absorbing mixed-oxides system, **viii** absorbing/emitting component-embed in dense material, **ix** absorbing/emitting component-embed in a porous material, **x** scattering material, **xi** diffracting/photonic system

Table 3.1 Examples of materials commonly used in sol-gel coatings and their corresponding refractive indices [15, 16]

Materials	Refractives index (550 nm)
SiO_2	1.45–1.47
Al_2O_3	1.60–1.65
MgO	1.72–1.74
MgF_2	1.37–1.38
ZrO_2	1.95–2.05
Ta_2O_5	2.00–2.14
TiO_2	2.20–2.40

the two components, phase separate at the nanoscale (sublight wavelength to avoid scattering). In a recent report, amorphous, and crystalline optical films made by binary SiO_2–TiO_2 network were obtained by co-condensation of Si and Ti oligomers [17]. By varying the Si/Ti ratio, the refractive index could be tuned from 1.45 to 2.20. Thermal treatment of the films at temperatures lower than the crystallization temperature, the binary network was amorphous and the Si and Ti atoms were homogenously dispersed at the molecular level. Above the crystallization temperature of TiO_2, the system evolved toward the formation of discrete anatase nanocrystals embedded into an amorphous silica network [17]. Alternatively, these nanocomposite networks with tunable optical density have been obtained by co-deposition of preformed TiO_2 nanocrystals and SiO_2 molecular precursors or particles [18].

Optical modulation of the refractive index can also be achieved by adding an organic component which results in hybridization of the inorganic matrix at the molecular level or by creating a hybrid mesophases as illustrated in Fig. 3.1iv. Many chemical approaches have been used to achieve those hybrid materials [19]. Via sol-gel pathways, amorphous hybrid networks can be obtained through hydrolysis–condensation of various metal species (alkoxides/salts) that could be organically modified such as in the case of bridged silsesquioxanes [19]. The sol could also contain an organic component (specific organic molecule, a biocomponent or polyfunctional polymers) that can interact with the inorganic components by weak or strong chemical interactions leading to class I or class II hybrid materials, respectively [19]. Hybrid mesophases formation process is based on the self-assembly of molecular or polymeric amphiphilic components coupled with sol-gel polycondensation reactions or with other colloidal chemistry-based approaches. The texturation process and the formation of the hybrid interface can be made by self-assembly of micelles, lyotropic liquid crystals, latex, or polymeric beads [19, 20]. In this regard, the processing conditions such as the evaporation of the volatile species or the condensation degree need to be controlled since they affect the kinetics of nano-segregation and formation of hybrid organic/inorganic interfaces, allowing the design of tailored structuration [21]. From the practical point of view, the formation of hybrid organic/inorganic films is often an intermediate step toward the fabrication of micro- and mesoporous inorganic materials as illustrated in Fig. 3.1v. In a conventional procedure, the porosity is obtained after the removal of the organic template by washing or by thermal treatment. Alternatively, nanoporous layers can be obtained by nanoparticles films, in which the void is localized in the interparticles space. A micro- or mesoporous layer can thus be seen as composite material composed by the inorganic phase and by air. Optically speaking, adding porosity into a sol-gel matrix leads to a decrease in the refractive index. Porosity is thus another powerful lever to tune (toward lower values) the refractive index of the media as in the case of antireflective layers that will be widely discussed later in this chapter.

A number of different strategies have been developed to fabricate colored, absorbing and/or emitting sol-gel-based coatings. As schematized in Fig. 3.1vi, the composition of the sol-gel matrix can be chosen to be optically absorbing itself such as in the case of hematite films [22]. Following a similar approach previously reported for SiO_2/TiO_2 films, mixed-oxides amorphous, or nanocomposite films can be synthetized to be absorbing. As shown in Fig. 3.1vii, the requirement is that at least one of the sol-gel derived component should be absorbing in the considered range of wavelengths. One recent remarkable example of sol-gel absorbing layer for solar energy conversion was composed of RuO_2 nanocrystals in situ generated into a SiO_2 matrix during thermal curing [23]. The layer showed outstanding absorbance properties and was capable of withstanding high operating temperatures for several thousands of hours.

The soft conditions of sol-gel processing have been identified as an important opportunity to introduce organic species within an inorganic glassy network. The pioneering work of David Avnir's group remains a milestone in this domain [24]. They showed for the first time that sol-gel chemistry allows preparation of inorganic glasses doped with organic dyes that preserved their optical activity after incorporation within the matrix. Following this strategy, as shown in Fig. 3.1viii, colored molecules, fluorescent dye, [25] quantum dots, [26] metallic particles [27] and so on can be embedded into a sol-gel dense films. Mesoporous films can also be used to host and shape absorbing components. (Fig. 3.1ix) In particular, the unique feature of the absorbing objects, (such as plasmonic gold particles) coupled with the high surface area and accessibility of the porous network has opened great opportunities in various fields, including biosensing and nonlinear optics. Recently, the film mesoporosity was exploited to "shape" gold nanoparticles (and thus their optical properties) by in situ seeded growth and branching mechanism [28]. This aspect concerning absorbing/emitting sol-gel systems will not be discussed further here.

Instead of transparency or absorbance, another target property of optical coating is opacity or translucency. These optical effect can be attributed to light scattering which is described of as the deflection of a ray of light from a straight path, for example by irregularities in the propagation medium, particles, macropores or in the interface between two media. Scattering coatings can be realized by the formation of heterogeneous structures such as macropores, having sizes in the hundreds of nanometer—micrometric scale (Fig. 3.1x). The micrometer range structuring can be easily reached by various templating strategies using, for example, controlled phase separation phenomena, emulsion droplets, bigger latex beads, bacterial threads, colloidal templates, or organogelators [19]. Macroporous gel systems have been infiltrated with liquid crystals to develop smart-windows switching reversibly from transparent to opaque state under electric field application [29].

A unique feature of sol-gel materials is the capability to be shaped/combined periodically in order to "mold the flow of light" [30] by inducing diffraction and photonic phenomena through the use of diffraction gratings, waveguides or photonic crystals. The most reported examples concerns the fabrication of sol-gel-based Bragg reflectors (Fig. 3.1xi), opal or inverse opals films [30]. The sol-gel photonic structures will be discussed in depth in Sect. 3.

3.1.3 Liquid Deposition Techniques

The first requirement for a coating to be applied in optics concerns its optical quality: the film needs to be optically homogeneous, of controlled thickness, and not scattering light (at least in the wavelength range of interest). A key of step for preparing optical coating is thus the deposition step. Liquid deposition processes are very versatile methods to produce homogeneous coatings, with thickness typically ranging from few nanometers to few micrometers, through spreading of a solution onto a substrate and evaporation of the volatile compounds. They are seen as serious alternatives to dry deposition techniques, such as PVD, CVD, or PECVD, because they combine excellent control of the inner layer structure and composition with rapidity and low cost [31]. Due to the peculiarities of the sol-gel chemistry, sol-gel derived optical coatings are usually applied onto the substrate by liquid deposition processes. Amongst the various techniques available, some of them, such as dip coating or spin-coating, are well suited for the preparation of optical quality coatings and are already commonly used in various R&D and production industries. Other liquid deposition techniques, such as spray, ink-jet, or roll to roll coatings, are still in their infancy and are in the process of been better understood and mastered for the fabrication of optical coating. Recently, both spin and dip coating have been combined with sol-gel processes in more than 3000 articles each, while the other techniques appear in less than 200 sol-gel-related articles. Despite the fact that dip-coating and spin-coating are well known and easy to apply, the deposition and, eventually the self-assembly, is a complex process that involves evaporation of the volatile species and dynamic evolution of the physicochemical parameters of the system including concentration, viscosity, etc. The accurate control of all the evaporation-related parameters is thus crucial for achieving good quality layers in term of thickness homogeneity and eventually self-assembly [32].

In the dip-coating method, the substrate is successively dipped into the solution and withdrawn at a constant speed [31]. Appropriate solvents are usually alcohols because they have low surface tension and are fairly volatile to promote relatively fast evaporation. After evaporation of the solvent, a layer of the nonvolatile species is obtained above the drying line. The withdrawal speed of the substrate and the evaporation rate are the two factors that allow preparation of a film with desired thickness and quality from a given solution. The effect of the dipping rate on thickness has been extensively studied [7, 33, 34]. Two different regimes were reported, namely the capillary regime and the draining regime [33]. The capillary regime was found at speed below 0.1 mm s^{-1} where the deposition is governed by the combination of capillary filling and evaporation of the solvent at the triple point. In this configuration the final thickness is proportional to the withdrawal speed at the power −1. The deposition speed and thus the final thickness in the capillary regime is also directly dependent on evaporation rate. Increasing the atmospheric temperature induces faster deposition in the capillary regime due to faster evaporation rate. The capillary deposition can be used to prepare thick sol-gel film (>1000 nm) from aqueous solutions at high temperature [33]. A major drawback exists when coating in the capillary regime for optical application. Because of the slow meniscus motion, a stick-and-slip effect may occur in certain conditions, which produces periodical fluctuation in thickness. While these fluctuations are typically of only a few tens of nanometer, they are not critical for the film cohesion but only affect the aesthetics of the layer [31]. Optical quality coatings are usually obtained at speeds above 1 mm s^{-1} in the draining regime. It was first described by the Landau–Levich model, which relies on the gravity induced viscous drag opposing the adhesion of the fluid on the substrate for Newtonian and non-evaporating fluid [34]. It predicts that the equilibrium thickness depends on the density, the surface tension, and the viscosity of the fluid and is proportional to the withdrawal speed at the power 2/3. This tendency is only verified in a narrow range of speed centered around a few mm s^{-1} because most of the dip-coated sol-gel solutions are non-Newtonian as a result of their viscosity and surface tension being continuously modified during solvent evaporation. Between the capillary and draining regime, an intermediate regime is located between 0.1 and 1 mm s^{-1} where the minimal thickness is obtained. Dip-coating conformation allows also to process films with one-directional thickness gradients by programming withdrawal speed along the dipping direction, which is very useful for sampling the influence of the thickness on a single substrate sample [35]. The main drawbacks of dip-coating lays in the fact that a great amount of chemical solution is needed to fill the reservoir and that all faces of the substrate will be coated, which is not often required when manufacturing optical devices. However, recently a new bi-phasic method was developed to overcome these issues [36]. The dip-coating method is a highly convenient and fast technique to apply a solution onto a substrate, with the highest degree of control, making it the process the most appropriated for R&D. It can also be used to deposit optical coatings on large surfaces as in industrial production or for high technology requirement [37].

Spin-coating is one of the most widespread liquid deposition methods [38]. While its development was closely associated to that of the electronic industries, it is widely used in R&D for the fabrication of thin films for optics. In addition, it is predominantly used to deposit a control thickness of the various types of lithographic resins on silicon wafers along the multistep top-down process leading to 2D photonic devices [38]. It is thus an adaptable method that is well appropriated for large-scale production as well as for R&D or also to prepare high added value coatings. A typical spin-coating protocol involves maintaining a flat substrate onto a horizontal rotating sample stage using vacuum [38]. The solution is then dropped on the center of the substrate and rotation is started. In some cases, the rotation is launched before solution dispensing. The rotation speeds are usually comprised between 1000 and 10,000 rev s^{-1} depending on the solution viscosity, dilution, and the final desired thickness. The rotation is ideally maintained up to the stabilization of a homogeneous thickness on top of the substrate. The typical time depends mainly on the evaporation of the solvent and on the condensation rate of the nonvolatile species, and is usually found between 20 s and a few minutes. The system is then removed from the stage to be processed further (imprint, lithography, curing, etc.). The first evident restriction is that the shape of the substrate must allow fast rotation, limiting the viability to substrates having a rotational symmetry axe normal to the surface to coat. In addition, very bulky and cumbersome substrates cannot be rotated. Second, to make sure that the whole surface is coated a great excess of initial solution has to be dispensed. In addition, while evacuating the solvent, turbulences are formed above the substrate, leading to thickness local fluctuation especially in the first stage of the film thinning with loss of the optical homogeneity.

Another alternative deposition method called "capillary coating" (or meniscus coating) was developed by Lawrence Livermore National Laboratory [39] and by CEA/Limeil-Valenton [40]. The technique is somewhat similar to dip coating in capillary regime but the substrates are coated in a horizontal upside-down position. The liquid coating medium is pumped into a slot applicator tube, flows out in a laminar motion to the surface through the slot, and forms a continuous and homogeneous liquid film on the outside [39]. When the substrate is placed in contact with the liquid film, a narrow meniscus is created between it and the applicator tube, and this latter is then smoothly moved horizontally relative to the substrate. A liquid film is transferred to the substrate as the applicator tube or the substrate translates and the coating is formed after solvent evaporation. The capillary coating technique was developed and exploited for the preparation of sol-gel and nanoparticles-based multidielectric mirrors and diffractive optics on very large scale [41].

As mentioned other liquid deposition techniques are still under development [31]. Conventional spray-coatings, where a sol-gel solution is atomised in relatively small droplets before been impacted on the substrate surface, is seen as the most appropriate process for industrial production. However, it is barely used due to the difficulties in controlling the deposition and the coalescence of the liquid droplets on the surface to yield homogenous coatings, a critical issue for optical applications. Alternatively, electro-assisted generation of sol-gel films was also proposed for the fabrication of optically homogeneous coatings such as mesoporous silica films [42]. It is base on an electro-induced modification of the chemical composition of a sol at the contact of an electrode. The aim is to locally induce a chemical modification of stable cations (or metal complex) for creating metal hydroxides suitable for sol-gel condensation. However, a conductive (optically absorbing) substrate is always needed to electro-generate the coatings.

3.2 Sol-Gel-Based Multifunctional Antireflective Coatings

3.2.1 Fundamentals of Antireflective Coatings

Reflection occurs when light reaches an interface between two media with different refractive indices. The main goal of an antireflective coating is to control this refractive index contrast and thus to avoid or limit reflection and maximize transmission. Depending on the optical properties of the media or the interface morphology, a fraction of the light intensity could also be scattered or absorbed. To achieve high transmittance these additional phenomena have to be reduced, using well-adapted materials and coating morphologies. In case of single layer thin films, light travels as depicted in Fig. 3.2a.

Following the configuration of Fig. 3.2, Fresnel Eq. (1) can be used to calculate the reflectance *R*:

$$R = \left[\frac{n_{\mathrm{air}} n_{\mathrm{s}} - n^2}{n_{\mathrm{air}} n_{\mathrm{s}} + n^2}\right]^2 \tag{1}$$

According to this equation, zero reflectance is reached if $n = \sqrt{n_{\mathrm{air}} n_{\mathrm{s}}}$. In case of a glass substrate ($n_s = 1.5$), this requirement is fulfilled when the refractive index of the thin film n is equal to 1.22. More generally, the reflected rays R_1 and R_2 (Fig. 3.2a) must be out of phase, thus the optical thickness ($d \times n_{\mathrm{film}}$) of the monolayer has to be equal to $\lambda/4$, λ being the desired wavelength with complete transmission. With an incident light at $\lambda = 550$ nm, an optimal antireflective film on

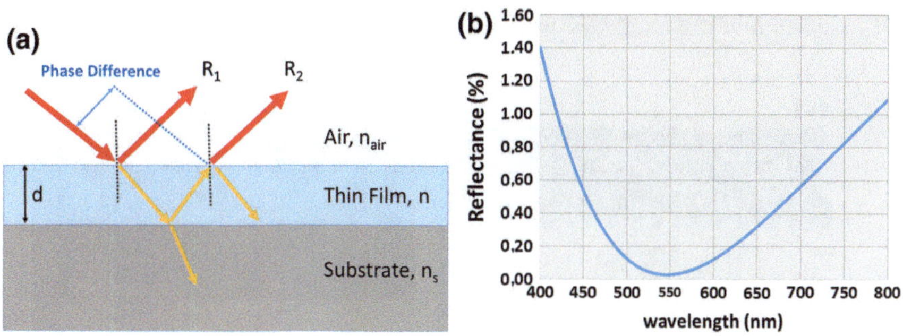

Fig. 3.2 **a** Propagation of light rays through a thin film (refractive index n and thickness d) on a substrate (with $n_s > n$). The different media are considered as isotropic, homogenous, and non-absorbing and **b** simulated reflectivity spectrum of a single layer coating on a glass substrate ($d = 112.7$ nm and refractive index $n = 1.22$) at normal incidence

glass should have a thickness of 112.7 nm and a refractive index $n = 1.22$. The simulated reflectivity spectrum in this configuration is shown on Fig. 3.2b. The thickness of the layer can be fairly well controlled by adapted liquid deposition techniques (such as dip-coating, spin-coating). As discussed in the first part of the chapter, several chemical strategies can be envisioned to finely modulate the refractive index of the sol-gel layers (down to 1.22). As the lowest index available for dense materials is 1.38 [15, 16, 43], which belongs to magnesium fluoride, additional porosity is necessary to decrease the optical density of thin films. As discussed above, two main strategies can be used to add porosity. Templating processes based on utilization of porogens like surfactants [44], block-copolymers or latex beads [45] which are removed by a thermal post-treatment or by an additional washing step, as shown Fig. 3.3a, b. Alternatively, the deposition of preformed nanoparticles is another useful technique, since the porosity can be localized in the inter-particle space between them as reported in Fig. 3.3c, d.

Taking into account the chemical composition of each layer and subwavelength heterogeneities (such as porosity) or specific morphologies, an approach named *Effective Medium Theory* was proposed to precisely control the optical properties of the final coating [47, 48]. A large number of models exist depending on the approximations used to predict how light can travel through different kinds of heterogeneities in dielectric media. The *Effective Medium Approximation* (EMA) proved to be an essential tool to develop antireflective coatings. Among EMA's models, two are widely used to integrate local variations in a dielectric material as follows:

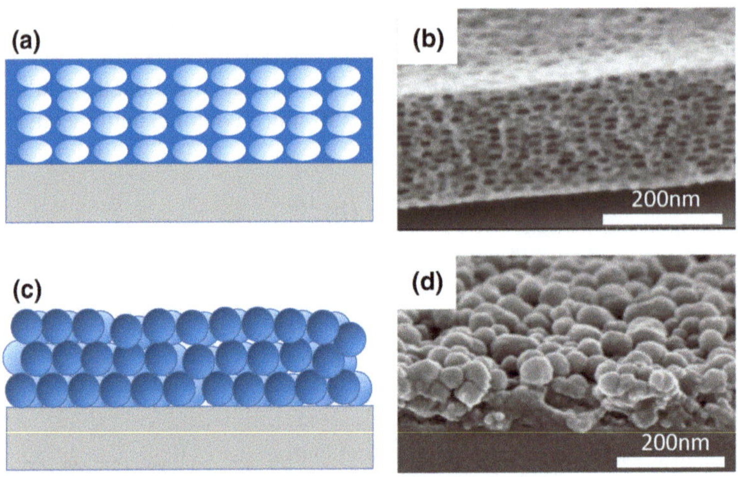

Fig. 3.3 Approaches toward porous film. **a** Schematic representation of a porous thin film after calcination of an organic micelles template, **b** SEM image of a thin TiO_2 layer with 20 nm large pores, **c** schematic representation of a porous thin film composed by a stack of nanoparticles and **d** SEM image of a thin film composed of silica nanoparticles [46]

Maxwell-Garnett [47]:

$$\left[\frac{\tilde{\varepsilon}-\tilde{\varepsilon}_1}{\tilde{\varepsilon}+2\tilde{\varepsilon}_1}\right]=(1-f_1)\left[\frac{\tilde{\varepsilon}_2-\tilde{\varepsilon}_1}{\tilde{\varepsilon}_2+2\tilde{\varepsilon}_1}\right] \quad (2)$$

Bruggeman [48]:

$$f_1\left[\frac{\tilde{\varepsilon}_1-\tilde{\varepsilon}}{\tilde{\varepsilon}_1+2\tilde{\varepsilon}}\right]+f_2\left[\frac{\tilde{\varepsilon}_2-\tilde{\varepsilon}}{\tilde{\varepsilon}_2+2\tilde{\varepsilon}}\right]=0 \quad (3)$$

Each model considers the effective dielectric constant $\tilde{\varepsilon}$ as a mix between two different media (f_1, $\tilde{\varepsilon}_1$) and (f_2, $\tilde{\varepsilon}_2$) with f_1 and f_2 their volume fraction ($f_2 = 1 - f_1$). Corrections must be added in case of nonspherical inclusions, like an eccentricity parameter for ellipsoids. If none of the materials used in the coating absorbs light at the considered wavelengths, the effective dielectric constant of each layer can be reduced to its real part, namely the square of their refractive index as:

$\tilde{n} = n - ik = \sqrt{\tilde{\varepsilon}}$ with $\tilde{n}$ the complex refractive index and k the extinction coefficient at the considered wavelength.

Coming back to the reflectivity spectrum in Fig. 3.2b it can be seen that the antireflective performances are maximized for 550 nm and they are depending on the wavelength. In general this is one of the main drawbacks of the antireflective coatings based on single layers; they are strongly dependent on the incident wavelength and the incident angle, which is a significant restriction to their efficiency if a broadband and wide-angle antireflectivity is needed. Other strategies based on multilayer coatings were consequently developed to reach these goals. The two main strategies based on multilayer antireflective coatings are:

- Gradient-type multilayers: if a strong index contrast is responsible for reflection occurring at interfaces, a decrease of this contrast leads to a lower reflectivity.
- Interference-type multilayers: a stack of layers alternating high and low refractive materials. As explained in case of a monolayer, the reflected rays must be out of phase. This can be achieved on broadband spectra if the number of layers and their optical properties are finely tuned.

In the following sections the two strategies will be discussed more in detail.

3.2.2 *Gradient-Type Antireflective Coatings*

The main goal of the gradient-type antireflective coatings is to decrease the refractive index contrast between the two optical media. Indeed, a continuous variation of refractive index bends the light rays toward the substrate consequently diminishing reflection as showed on Fig. 3.4a Lord Rayleigh was the first to propose this strategy in 1879 [49] and to demonstrate its physical accuracy: a continuous decreasing of the optical density in a material strongly affects its reflectivity in air. A significant point that needs to be underlined is that the gradient-type antireflective coatings are much less sensitive to thickness variations, incident angles or wavelengths limitations than interference-type. However, this advantage is generally balanced by the low mechanical resistance of the coatings made using this process/made in this way. This weakness directly relies on the fact that, at the interface with air, a very porous layer is needed to reach a very low refractive index, typically around 1.10. In the case of silica it represents a material with porous volume fraction of 75–80%vol.

Inducing perpendicular refractive index graded variation has been, and still represents, a technological challenge in terms of materials and deposition techniques. Some strategies can be found in the literature.

Multilayer stacks are commonly used to decrease the refractive index contrast simply by adding successive layers with decreasing refractive index starting from the substrate as schematically represented in Fig. 3.4b. Even if a perfectly continuous variation would require an infinite number of layers, very good anti-reflective properties are obtained only by depositing two or three layers. Figure 3.4c shows a SEM micrograph of a sol-gel coating composed of a graded bilayer [50]. Here, the refractive index gradient results from an increasing porosity

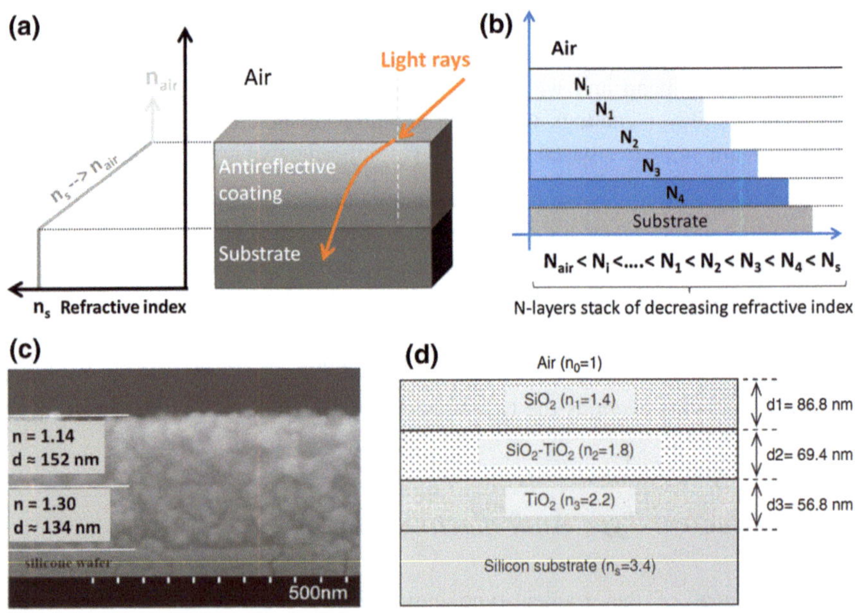

Fig. 3.4 **a** Schematic representation of light bending in a gradient-type antireflective coating, **b** schematic diagram of a multilayer stack designed for gradient-type antireflective coatings, **c** SEM picture of sol-gel coatings composed of two different silica nanoparticles thin films [50], and **d** schematic representation of a tri-layer coating for solar cells based on a composition gradient of SiO_2–TiO_2 [51]

due to cavities between deposited nanoparticles. Figure 3.4d schematically depicts a tri-layer antireflective coating for use in solar cells [51]. In this case, the refractive index gradient is based on a mix between TiO_2 (high refractive index) and SiO_2 (low refractive index).

A second approach to producing gradient-type coatings is based on surface nanostructuration. This strategy can be defined as biomimetic since antireflective structures are often observed in nature. The moth eye surfaces are characterized by complex hierarchical topographies made of micrometric hemispheres covered by nanocones. In the case of moth eyes, the nanopatterns periodicity is smaller than their wavelength; as conceptually represented in Fig. 3.5a, when interacting with the surface, the light is deviated toward the substrate. Along with moths, several insects developed complex nanopatterned surfaces to enhance their vision. For example, *Attacus Atlas* is a butterfly with a very complex eye structure [52] made of a microscopic hexagonal lens entirely covered by nanopillars (Fig. 3.5b). The graded profile of the nanopatterns causes a continuous decrease of the refractive index (Fig. 3.5c) and thus antireflection. However, more complex phenomena are also observed with nonhomogeneous, larger patterns like multiple internal reflections, which also enhanced the transmission of light. The fabrication of biomimetic graded coatings often requires an additional step to shape the sol-gel layer. A simple shaping strategy is the chemical etching of a thin film or of the substrate itself [53, 54]. More sophisticated techniques, like dry etching by a femtosecond laser [55] or by Reactive Ion Etching [56] (RIE), allow to better control the final shape of the remaining patterns, especially if a mask is used during the etching step. A recent example of nanopatterned coating prepared by sol-gel bas been proposed by Kitamura et al. [56]. A mesoporous silica thin film was etched by RIE in presence of fluorine-containing gas and without additional mask to pattern the surface. After the treatment, the remaining layer showed nanopinnacles smaller than 300 nm which were similar to the nanopillars of a moth eye and significantly decreased the reflectance down to 1% for the visible range compared to 4.5% with bare silica glass (Fig. 3.5d). Replication techniques are in general more effective to reproduce biomimetic-graded coatings due to the very specific shape of moth eyes. The successful replication of the hierarchical micro- and nanostructures was achieved by molding UV curable polymers directly on a *Attacus Atlas* eye [52] as shown on Fig. 3.5e. A similar molding approach has also been developed for sol-gel-based coatings suitable for optical applications, especially light guides [57]. Concerning sol-gel antireflective coatings, special molds have also been designed to reproduce a moth eye structure by soft-lithographic routes [58]. The typical protocol is presented on Fig. 3.5f: starting from a nickel master engraved by physical techniques a Polydimethylsiloxane (PDMS) mold is prepared. A sol-gel thin film is then applied on the PDMS mold and transferred on the final glass substrate. Soft-lithographic routes have been used to directly shape hemispherical nanopatterns on sol-gel thin films (Fig. 3.5g) [58]. The challenging step is the master preparation on large scales with a perfect control of the shape and depth of the nano features in order to obtain an optically performing, mechanically stable coating.

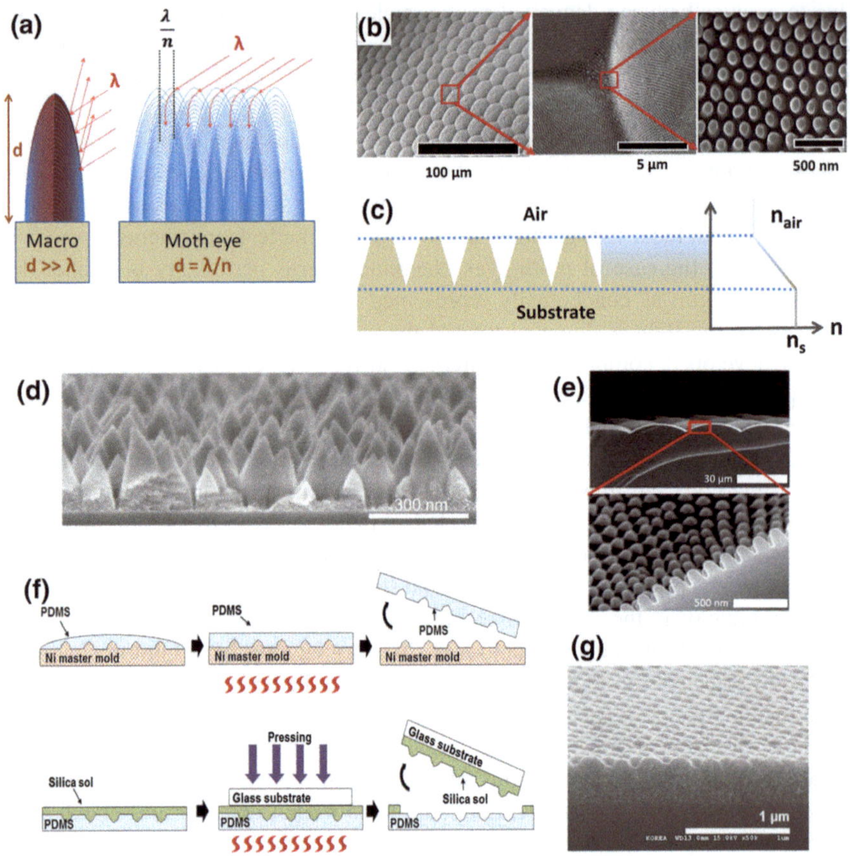

Fig. 3.5 **a** Schematic diagram of the different light travel upon contact with a macroscopic object ($d \gg \lambda$) or with the moth eye surface ($d = \lambda/n$), **b** SEM images of a butterfly eye [52] (Attacus Atlas) at different scales (100, 5 µm and 500 nm), **c** schematic representation of a gradient-type coating made with nanocones with a periodic arrangement closed to λ/n with n a whole number, **d** SEM images of mesoporous silica nanopinnacles prepared by RIE [56], **e** SEM images of a moth eye replica made in UV curable polyurethane at different scales (30 µm and 500 nm) [52], **f** procedure to fabricate silica patterns with moth eye morphology on glass substrate, and **g** cross-sectional SEM image of a silica film embossed with sub-micrometric patterns [58]

3.2.3 *Interference-Type Antireflective Coatings*

Alternating layers of high and low refractive index is the second main strategy used to design antireflective coatings. Interference-type coatings were intensively studied since the first patents in 1939 and 1949 on sol-gel antireflective coatings [59]. As interference-type coatings are easily produced on large-scale substrates, a broad number of commercial applications were successfully developed since the 1960s including contrast enhancement filters for monitors, laser optics, glass or plastic lenses, coatings for glass showcases and more recently coatings designed for silicon solar cells.

In comparison with gradient-type coatings, here the thickness control is a key aspect. Indeed, destructive interferences are very sensitive to the optical path of light rays through the thin films as depicted in Fig. 3.6a.

For a perpendicular incident light ray crossing a thin film, with n_i and h its refractive index and thickness, respectively, the optical path L is defined by:

$$L = 2n_i h \quad (4)$$

In order to obtain destructive interference between R_1 and R_2, the phase difference must be equal to the half of their period λ. This condition is satisfied if:

$$L = 2n_i h = \frac{\lambda}{2} \pm k \quad \text{(With } k \text{ a whole number)} \quad (5)$$

When $k = 0$, one can define the relationship between the film thickness, its refractive index and the wavelength λ of the incident ray:

$$h = \frac{\lambda}{4n_i} \quad (6)$$

The coating architecture is a key aspect to achieve broadband antireflection; however, a simple alternating stack or additional layers do not automatically improve the antireflective properties of the coating. This can be clearly observed in the simulated spectra on Fig. 3.6b in which stacking two, three or four layers lead to slightly different optical response, but similar antireflectivity performances, at least in the visible range. Looking at the spectra in Fig. 3.6b, the stack architectures are characterized by antireflective behavior with the presence of "border effects" (increase of reflectivity) behind and above the wavelength range. The border effects are responsible for the appearance/color of coatings. This is easily visible on commercial tri-layers coating based on sol-gel methods [60], slightly blueish or

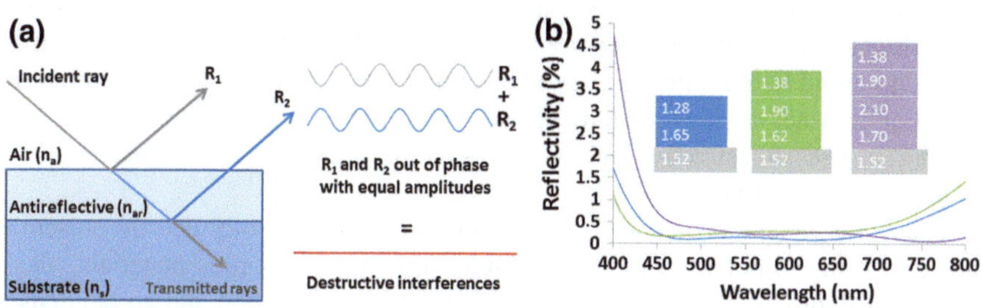

Fig. 3.6 **a** Schematic diagram of an interference-type monolayer coating and **b** simulated reflectivity spectra of a bilayer, a tri-layer and a quadrilayer on a glass substrate at normal incidence (only refractive index are indicated)

greenish like Conturan® and Amiran® (Schott AG) or like ProView green and ProView blue (Prinz Optics).

On the other hand, the main advantage of interference-type coatings is their mechanical stability compared to gradient-type coatings. As very low refractive index layers are not necessarily needed in interference-type coatings, dense layers or slightly porous layers are used, preventing scratch or delamination under mechanical stress.

3.2.4 *Weather-Resistant and Self-cleaning Antireflective Coatings*

Antireflective coatings are strongly sensitive to their environment and several factors can affect their optical performance with time. Scratch-resistant and mechanically robust coatings were developed to improve the durability of antireflective coatings for outdoor applications. Weather conditions could also alter the optical properties of the layers. For example, the refractive index of a nanoporous coating can drastically increase due to fact that water capillary condenses in small pores at relative low humidity. In addition, high surface energy sol-gel materials are particularly susceptible to contamination by volatile organic compounds [61] which is an issue for the optical stability of the coating.

All these issues pushed researchers to add new functionalities to antireflective coatings to add self-cleaning or water-repellent capabilities. Sol-gel methods enabled the development of many multifunctional coatings, specifically tailored to maintain their optical performances unaltered with time. Indeed, in order to tune the coating surface energy, hydrophobic or hydrophilic molecular functions could be added either during the deposition step or with a posttreatment. Depending on the desired protection and on the initial antireflective coating, the architecture has also to be designed to preserve the optical quality of the final coating. In the next paragraphs two types of multifunctional thin films will be presented: weather-resistant antireflective coatings and self-cleaning antireflective coatings.

Most of the sol-gel-based antireflective coatings are chemically hydrophilic at their inner and outer surface. Water-uptake into the pores of antireflective coatings is a critical issue in the case of porous coatings. A small pore size (<50 nm) is needed to avoid light scattering; however, at this scale, the porosity can be filled-up with liquid water at low relative humidity through capillary condensation [62] causing a rise in refractive index followed by a significant decrease of the antireflective performance.

This water-uptake is drastically reduced if hydrophobic molecules are present at the surface of the pores. In the case of micelle templated porous materials, alkylsilanes, or fluoroalkylsilanes can be added at the initial solution and co-condensed with conventional "hydrophilic" monomers or oligomers. In this case the calcination temperature of porogen molecules needs to be carefully chosen to prevent the alkyl-groups from being degraded. Unfortunately a lower temperature also leads to less condensed networks with weaker mechanical properties. Ammonia vapor treatments were also proposed to overcome this condensation problem at low temperature, especially in the case of antireflective coatings on polymeric substrates [63]. After the vapor treatment, porogen molecules can be simply washed away by hydro-alcoholic solution. An alternative is the post-functionalization of the film either in vapor phase (volatile precursors like fluoroalkylsilanes) or in liquid phase.

Water is not the only contaminant modifying the optical properties of thin films. Antireflective coatings prepared by sol-gel are mainly composed of high surface energy oxide or fluoride layers and therefore ambient organic and inorganic pollutants can be adsorbed to the surface with serious consequences on the antireflective efficiency of the device. A first strategy to limit this contamination and/or favor an efficient decontamination is to create "self-cleaning," superhydrophobic surfaces characterized by low surface energies. As shown in the scheme of Fig. 3.7a in presence of highly hydrophobic surfaces, characterized by narrow contact angle hysteresis, all adsorbed contaminants on such surfaces (organic or inorganic) are rinsed way by the rolling of water droplets (such as rain). In nature, such functional superhydrophobic properties are found on Lotus leaves. The surface is based on the combination of hydrophobic micro- and nanopatterns leading to a water-repellent and self-cleaning surface. Many examples of biomimetic coatings reproducing this Lotus Effect can be found in the literature [64]. They are generally based on the generation of roughness combined with functionalization with low surface-energy organosilanes for instance. However, in the case of optical coatings, micropatterns must be avoided because they would lead to light scattering. Good optical performance together with very high static contact angles and low sliding angles were obtained with nanopatterned systems (like nanoparticle films), functionalized with alkyl- or fluoro-silanes [65] as shown in Fig. 3.7b, c. Here again, the main drawback is the mechanical stability. If the coating is submitted to abrasive forces, the surface patterning and functionalization would be slowly damaged causing a loss of the superhydrophobicity. After some time, pollutants will stick again to the surface, clearly diminishing the durability of the multifunctional coating.

Another "self-cleaning" strategy is based on photocatalytic decomposition of the adsorbed organic pollutants by light-induced redox-reactions as represented in Fig. 3.7a. This is an ideal solution for outdoor applications, in which the sun light represents the endless energy source for such a devices. On the other hand, the most common photocatalytic material is the crystalline titanium dioxide which is characterized by a high refractive index (2.2–2.4 at 550 nm by sol-gel routes). The utilization of very high refractive index materials generally leads to a significant increase in reflectivity, or at least, makes the design of an antireflective coating more challenging. The main strategy to overcome this issue is based on the incorporation of TiO_2 nanocrystals into low refractive index silica matrices. A first example of transparent, self-cleaning coatings was obtained by using preformed TiO_2 colloidal particles that were further dispersed within a transparent silica binder with a mesoporous structure. Film porosity in the nanometer range was controlled by achieving the sol-gel silica condensation around self-organized micellar assemblies

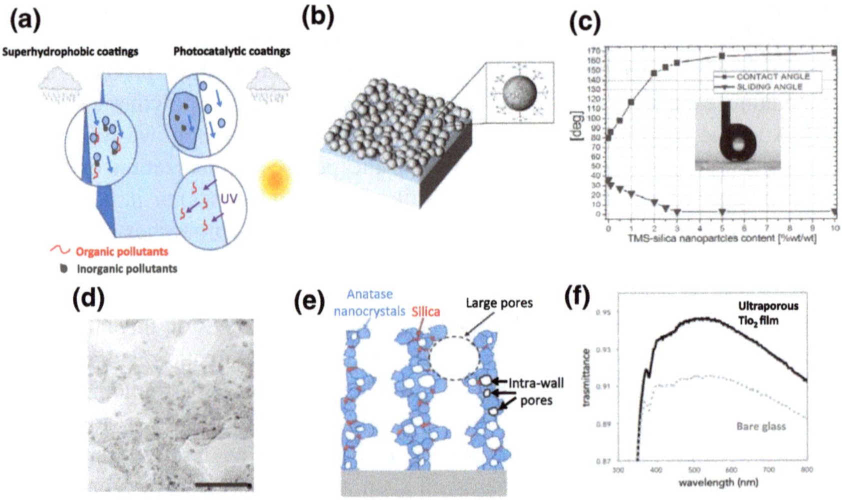

Fig. 3.7 **a** Schematic operating modes of superhydrophobic and photocatalytic self-cleaning surfaces, **b** schematic view of a superhydrophobic coating by depositing silica nanoparticles functionalized with hexamethyldisilazane, **c** static and sliding angles obtained with different amounts of these nanoparticles (20 nm average size) [65], **d** preformed anatase nanoparticles incorporated in a porous silica matrix [66], e schematic illustration of the hierarchical structure of the ultraporous TiO_2-based films and **f** its transmission spectrum compared to bare glass [67]

of a templating copolymer surfactant [68]. Following a very similar approach, more recently, antireflective, photocatalytic coatings were obtained by co-deposition of silica-based precursors, block-copolymer micelles and preformed anatase particles as shown in Fig. 3.7d [66]. Co-deposition of TiO_2 and SiO_2 nanoparticles allowed formation of porous antireflective coatings with a tunable final refractive index depending on the particles size and the TiO_2/SiO_2 ratio [18]. In all these examples, replacing “active” TiO_2 with “inert” SiO_2 generally reduced the photocatalytic activity of the layers. Instead of adding silica, another way to decrease the refractive index of TiO_2-based films is to increase the porous volume ratio. Ultraporous titania films with a refractive index down to 1.27 and 80% of porosity were recently synthetized. The good thermal stability of the films was due to the presence of a very low amount of silica (10%) in the inorganic phase and a hierarchical porous system [67]. The structure and the transmission spectrum of this ultraporous titania layer is shown in Fig. 3.7e, f. Due to its high surface area and to its hierarchical structure, this ultraporous titania layer is an ideal material for photocatalytic decomposition and air depollution while the antireflective properties are preserved.

Other compounds are also suitable for photocatalysis, like zinc oxide, but they are more sensitive to chemical degradation and to weathering conditions. Even if the adsorption band gap of titanium dioxide is below the visible spectrum, its photocatalytic activity is efficient upon natural sunlight activation. Compounds with a lower band gap can absorb light in the visible range, but will consequently add a certain shade, which could be incompatible with transparent displays/windows/

solar cells. Doping strategies to lower TiO_2 bandgap have attracted a lot of interest, but in the case of multifunctional coatings, no simple and cost-effective solution has been found to justify the moderate increase in photoactivity due to additional doping steps.

In addition to its photoactivity, another intrinsic feature of TiO_2 is that under UV radiation it becomes superhydrophilic and its contact angle is very low (<5°). In this configuration, water droplets spread entirely on the surface consequently forming a thin water layer. Due to this property, inorganic pollutants, which are not degraded by photocatalysis, can be rinsed when the water layer flows on the coating (Fig. 3.7a). This phenomenon was adapted for architectural glass for outdoors application because natural sunlight is sufficient to activate TiO_2-coated surfaces. Even if adsorbed pollutants increase the coating contact angle, the self-cleaning cycle is regenerated by alternating photocatalysis and rinsing steps and therefore the durability of such coatings are maintained for several years on commercial windows. Interestingly, non-photo-induced and stable superhydrophilicity was shown for certain porous TiO_2 films without any UV activation and may extend the self-cleaning properties of such coatings [69].

On the other hand, nanoporous layer coatings hydrophilic TiO_2 are still submitted to capillary filling under humid conditions as discussed above. Specific coating architectures were proposed to solve this incompatibility, for example, a bilayer coating made of a hydrophobic porous silica layer and a very thin porous TiO_2 layer on the top [70]. Superhydrophilicity is also known to add antifogging properties to the surface. Indeed, fogging occurs when light is scattered by micrometric water droplets condensing on a (cold) surface from ambient atmosphere. When condensation takes places on superhydrophilic surfaces, these small droplets entirely spread on the substrate to form a thin water layer which does not scatter light anymore [71]. However, under certain circumstances this water layer provokes slight image distortions; in addition superhydrophilic coatings are more frost sensitive which can be a problem, for example, on car windshields.

Due to their promising applications, multifunctional coatings have attracted a lot of interest. But no magic solution exists nowadays, the conflicting requirements for antireflective, self-cleaning or antifogging properties still remain a challenge. Sol-gel chemistry is a powerful method to solve these problems in terms of materials, physical properties, shapes, and porosity. As a strong industry-driven research area, multifunctional coatings will remain a creative domain in the future to enhance the optical properties of displays and smart windows.

3.3 Sol-Gel-Based Photonic Crystals

Among the various applications of sol-gel fabricated thin films photonic crystals are one of the most important. Photonic crystals are periodic structures in dielectric materials whose periodicity allows for strong interaction with light of wavelength of the same scale. In general, the only information required to adequately describe a photonic crystal is a map of the refractive index depending on the position. While photonic crystals can be fabricated from metals, absorbing, and transparent materials the latter type, in particular for sol-gel crystals, are the most relevant and only these will be considered in this section.

Photonic crystals are commonly classified depending on the number of the dimensions of their periodicity. One-dimensional photonic crystals, also known as Bragg reflectors, have been known starting from the second part of the nineteenth century [72]. They are constituted of alternating layers of dielectric materials with different refractive indexes. Two-and three-dimensional photonic crystals are much more complex and their properties have been studied only a century later, starting in the late 80s. Some examples of the possible geometries in 2D are arrays of dielectric pillars in air or holes in dielectric material while for the 3D case one has opals or inverse opals and the so-called "wood piles" structures.

The important geometric and physical quantities in such 1D, 2D, and 3D systems will be discussed. In particular, the geometrically simplest case of 1D. We suppose that the structure is periodic in the z direction and consists of layers that are infinite xy planes. A plane wave propagating in the z direction has to follow the Bloch theorem so the electric and the magnetic fields can be written as the product:

$$e^{ikz}\, u(z) \tag{7}$$

where $u(z)$ is a periodic function with the same periodicity of crystal and e^{ikz} represents the free wave with vector $k = 2\pi/\lambda$ (λ is the light wavelength inside the medium). A wave at extremely low frequency ω has an extremely long, practically infinite wavelength, and it interacts with the structure as if it were a bulk material with an average dielectric constant thus ignoring the finer periodic structure. Thus, the behavior of this low-frequency regime is independent of the refractive indices and respective thicknesses of the layers. For higher frequency regimes, where the wavelength is comparable to the period α of the multilayer structure a form of resonance takes places with subsequent effect on the propagation of light. Two different states with differing frequencies exist at the same wavelength that is precisely double the periodicity. In these two cases, the nodes of the waves stay either in the low or in the high refractive index regions, respectively, for the lower and higher frequencies. Any other frequency between these two cannot propagate in the photonic crystal and only evanescent waves can be found. This interval between the aforementioned upper and lower frequencies is called the "band gap" of the photonic crystal structure. Frequencies above the first band gap propagate through the structure without resonating. Subsequent band gaps happen in correspondence to wavelengths of $\lambda = 2\alpha/N$ with N a natural number higher than 2. The light interacts again strongly with the structure forming other degenerate states. The band gap is wider when the refractive indices of the materials differ more and when the optical lengths differ less, i.e., when $d1{\cdot}n1 \approx d2{\cdot}n2$ with $d1$ and $d2$ the thickness of the layers ($d1 + d2 = \alpha$) and $n1$ and $n2$ the refractive indexes of the two materials, respectively. Some analytical formulas relating the band gap width and associated frequencies with the parameters of the system can be found elsewhere [30]. Fig. 3.8

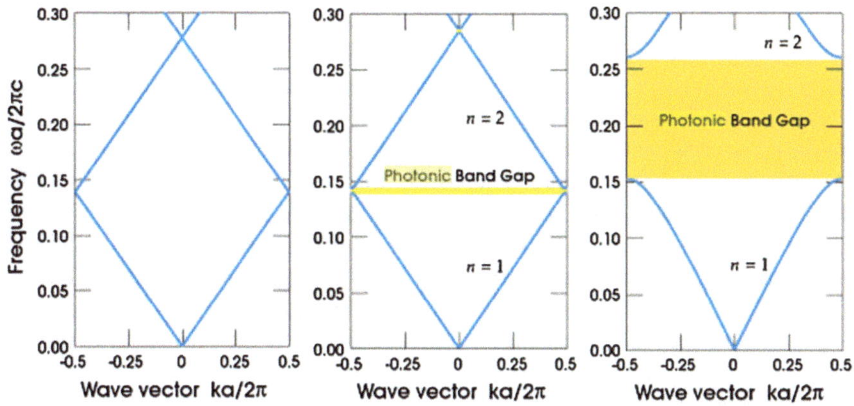

Fig. 3.8 Band structure of a multilayer of 2 materials with $d1 = d2 = \alpha/2$. (*Left*) In the first case the refractive index of both layers is the same $n1 = 3.60$ and $n2 = 3.60$. (*Center*) In the second one the refractive index are, respectively, a value of $n1 = 3.60$ and $n2 = 3.45$. (*Right*) In the third case the refractive index values are $n1 = 3.60$ and $n2 = 1.00$. The forbidden frequency interval increases in size increasing the difference between the refractive index values [30]

reports three common behaviors of the band structures for wave-vectors directed in the normal direction to the layer planes. The values of both frequencies and wave-vectors are normalized with respect to the period α. The left-hand plot represents a homogeneous medium; here there are no resonances and the frequency is linearly dependent on the wave-vector. The center plot corresponds to a photonic crystal with two alternating materials with the values of the refractive indices differing slightly. The right plot, instead, corresponds to a case with a much higher refractive index contrast.

This kind of graph allows to identify which frequencies can propagate in the structure. A plane wave entering the photonic crystal structure with a given normalized frequency $\omega\alpha/2\pi c$ outside the band gap will propagate with the corresponding normalized wave-vector ($2\pi\alpha/\lambda$ ratio) determined by the blue line. If the frequency is in the band gap, the wave is reflected and only an evanescent wave penetrates in the first layers of the photonic crystal. For a deeper understanding of 1D photonic crystal properties we refer to a specific monography [30]. In the next section, some recent examples of functional 1D photonic crystals made by sol-gel chemistry will be presented. In the following Sections 2D and 3D sol-gel photonic crystals will also be discussed.

3.3.1 1D Photonic Crystals

Sol-gel chemistry allows fabrication of layered structures in an extremely controlled way through liquid deposition techniques [31]. The thin film deposition process (by spin-coating or dip-coating) can be repeated with various solution compositions to obtain a stack of thin films with different refractive indices. As discussed in the first section of this chapter, high refractive index materials, such as titania, zirconia, niobia, or zinc oxide can be applied as optical coatings. These materials can be coupled with low-index ones, such as silica or aluminosilicate glasses to form the 1D photonic crystal structure. Moreover, the refractive index can be further decreased by introducing some porosity (as in the case of antireflective coatings). Porous structures can also be obtained by the successive deposition of thin films of nanoparticles or layer-by-layer deposition [73, 74]. A common issue in the synthesis of very thick 1D photonic crystals is the formation of cracks. In general, for thick samples, the mechanical stresses are created during the calcination of the thin films, needed to condense the inorganic network, and in the case of porous film, eliminate the organic porogens. Adequate temperature ramps are often used to avoid excessive stresses while obtaining thick samples necessary to for desired optical properties. Another common issue that can arise when using porous layers is the filling of the empty porosity when applying the successive depositions. A common way to do this is to apply a thermal treatment at average temperature after each deposition in order to consolidate the structure without decomposing the organic template. Once the stack is completed, a final thermal treatment at higher temperature is applied to further condense the network and free the porosity [75].

In some cases, a "defect layer" can be added to the regular 1D photonic crystal structure in order to break the periodicity. This "defect" appears in the band gap as a small range of allowed frequencies. This entails a corresponding diminution in reflectivity of incoming light. Defect layers can be obtained by changing either the thickness or the refractive index of a single layer. The example in Fig. 3.9 introduces a porous TiO_2 film in the middle of a stack of dense SiO_2–TiO_2 layers (a SEM image of the structure is reported in Fig. 3.9a) [76]. The "defect" in the band gap can be clearly visualized in Fig. 3.9b: the black solid line corresponds to

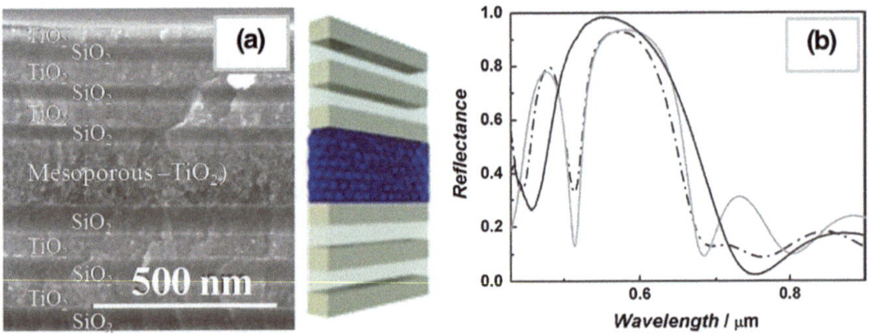

Fig. 3.9 **a** SEM image of a 1D photonic crystal of SiO_2 and TiO_2 layers with a mesoporous TiO_2 layer creating a "defect" in the structure and **b** the presence of this additional layer creates the sharp diminution in reflectivity for wavelengths around 0.52 μm. The reflectivity of a sample without mesoporous layer is reported by the *black solid line*. The reflectivity of the sample in **a** is reported by the *black dotted line*. The *gray line* is the fitting of the last measurement [76]

the regular stack without defects while the black dotted line corresponds to reflectivity of the structure with the defect. The downwards peak in reflectivity around 520 nm makes the 1D photonic structure behave locally as a light filter that selects a narrow window of wavelengths. The position of the defect in the band gap can be tuned by acting on the defect layer thickness and/or refractive index, this will be shown to be extremely useful for applications such as sensing.

The use of sol-gel 1D photonic crystals generally belongs to two main areas; the sensing of volatile/solute molecules and the coupling of the properties of optically active specimens with the resonant structure.

In the first case the typical adsorbing features of porous sol-gel materials are coupled with the optical properties of the photonic crystals in order to detect either vapors of water or other solvents. Alternatively, ions in solution, pH variations, or thermal effects can be detected and evaluated too. When looking for the identification of volatile compounds, one often refers to the "photonic nose" [73]. The general idea is to exploit the adsorption/condensation phenomena of volatile vapors in mesoporous thin films upon exposure to the atmosphere. The parameters affecting the adsorption/condensation in the mesoporosity are the pore size and shape, the surface properties, and the volatility of the specimen. Vapor uptake into the mesoporous layers leads to refractive index increase followed by a shift of the position of the reflected band (if the 1D photonic crystal is composed of mesoporous layers) or of the defect window (if the defect layer is mesoporous). This modification induces a change in the color of the photonic crystal [76]. Colorimetric combinatorial analysis can then be utilized for detection and identification of mixture of volatile compounds. These platforms are composed of arrays of 1D porous photonic crystals in which each component is characterized by a distinct surface energy. This can be done by post functionalization with different molecules (such as different alkoxysilanes). When the system is exposed to vapors or a mix of vapors, each element of the array responds in a distinct manner. Analyzing the various responses, one can identify/quantify different volatile species. Notably, the system is good enough to identify the byproducts of different bacteria growing colonies. This example is shown in Fig. 3.10: the system consists in a matrix of the

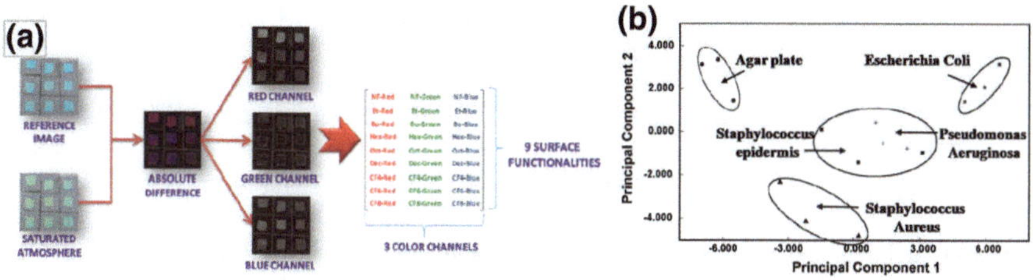

Fig. 3.10 **a** The array of 1D photonic crystals with different functionalizations of the porosity for each element is analyzed subtracting the image in absence of volatiles from the image in saturated atmosphere. The image is then split in the color channels and the data are analyzed through the principal component analysis and **b** identification of the volatile species produced by various bacteria [73]

photonic components, in which the exposure-induced-color variation of each element of the array (each differently functionalized) is analyzed through the principal component analysis (PCA) [73].

In the case of sensing applications a relevant benchmark is the detection limit of a given technique. Concentrations down to 10 ppm (18.8 mg/m$^{3}\approx$1.5% P/P_0 at 20 °C) of ethanol in nitrogen have been shown to be detectable through a complex composite stack device containing an OLED, a PDMS chamber for the injection of the gas, a SiO_2–TiO_2 1D photonic crystal and an organic photodetector (OPD) [77]. Figure 3.11a reports the scheme of the integrated device. Such a high sensitivity is obtained due to the emission spectrum of the OLED being placed at the edge of the reflectivity band of the photonic crystal. When little amounts of ethanol penetrate the porous structure the reflectivity band shifts. This modification induces a decrease of the quantity of photons reflected by the photonic crystals. As a consequence, in presence of ethanol, the OPD absorbs more photons translating it into an electrical signal that is measured to determine the concentration. Figure 3.11b shows the emission and reflectivity band of the OLED and photonic crystal together with the corresponding increase in current from the OPD. This system is also interesting because of its portability.

Sol-gel 1D photonic crystals were also used to detect slight variations in temperature [78] and even reveal the presence of ions/surfactants or pH variations. If clay layers are introduced in the stack, they can be interlaced with large ions, an effect that can be used to modify the thickness of the corresponding layers forming the so-called "photonic clays" [79]. One must mention that for these applications one of the layers or the defect layer can also be composed of materials other than

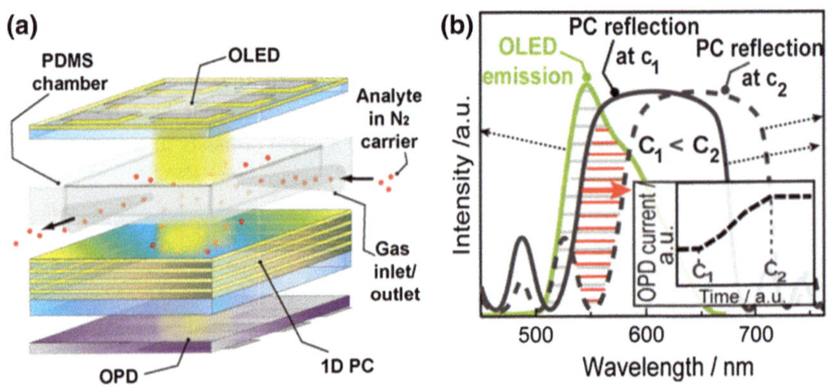

Fig. 3.11 **a** Integrated devices for the detection of ethanol down to 10 ppm in nitrogen. The system is composed of an OLED emitting light passing through a chamber introducing ethanol vapors into the system, the 1D photonic crystals which adsorbs ethanol and an optical photodetector (OPD) which convert the light in electric current. The system works due to the displacement of the reflective spectrum of the photonic crystals by increasing the ethanol concentrations. The *red area* corresponds to the higher quantity of photons reaching the OPD upon exposure of ethanol that can then be identified by an increase of the OPD current in the box [77]

sol-gel ones, such as metal organic frameworks [80] or other materials that can interact more specifically with the target molecules.

The second branch of the 1D photonic crystal applications is strictly related to physical effects. For example the overlap of the photonic band gap of the photonic crystal with the energy levels of fluorescent dyes or quantum dots can have major effect on their photoluminescent properties. In this case, the term “microcavity” is often used when the photoactive specimen is put in the middle of two reflective structure either both composed of photonic crystals or of a photonic crystal and a metallic or other reflective material [81]. If such a microcavity is coupled to a photonic crystal showing a reflectance with a defect and having quantum dots in the defect layer, it is possible, for example, to tune their emission spectrum through a variation of the incidence angle of the excitation. The emission can indeed occur only at the wavelengths allowed by the window in the reflectivity of the photonic crystal which are also angle dependent. In Fig. 3.12a the reflectance spectra of a structure composed of a quartz substrate, a 1D photonic crystal of SiO_2–TiO_2 alternating layers, a ZrO_2 layer containing CdS and CdSe quantum dots, and finally a metallic Ag layer are plotted in function of the angle between the incoming light and the quartz substrate. In this case the sol-gel photonic structure is acting as spectral filter. The reflectance windows that can be seen in Fig. 3.12a corresponds to those observed in photoluminescence spectra of the microcavity as function of the angle for a fixed excitation (Fig. 3.12b). Quantum dots, indeed, only emit light with wavelengths within the window defined by the photonic crystal. It is therefore possible to artificially modify the emission spectrum of the quantum dots through coupling it with the photonic crystal structure. In Fig. 3.12c the emission band originating from the quantum dots layer-photonic crystal system is compared to the emission of the quantum dots layer by itself. It is evident that the photonic crystal generates a blue shift in emission in order to be in accordance with the photonic crystal reflectance window. A further and interesting effect is the increase in absolute photoluminescence of the emitted light intensity due to the presence of the photonic crystal microcavity. In Fig. 3.12d the photoluminescent spectrum of the microcavity (Si-photonic crystal–(ZrO_2–CdS–CdSe)–Ag) excited at 10° has a much higher peak with respect of the reference sample composed only of the silicon substrate, a ZrO_2–CdS–CdSe and an Ag layers.

Similar results have been obtained using fluorescent dyes or rare earth elements instead of quantum dots with intensity enhancements by factors of up to 20 [82, 83]. These effects can be extremely interesting for devices requiring the detection of only a few molecules (such as nanofluidic ones).

Many other applications can be imagined exploiting the reflectivity of photonic crystals and their light filtering properties when the defect window is present. Among others, the so-called “light management” or “light harvesting” applications are important for the growing field of photovoltaics; 1D sol-gel photonic crystals with specific architectures are under development for these purposes [84, 85]. In the case of dye-sensitized solar cell, 1D photonic crystals completely composed of TiO_2 with different degrees of porosity (and then refractive index) have shown to

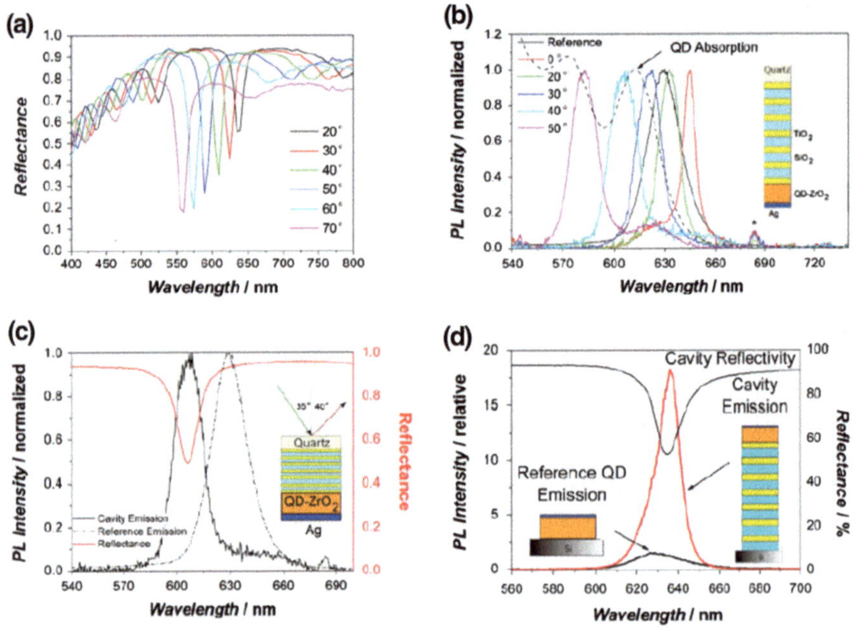

Fig. 3.12 **a** Reflectance spectra of a microcavity composed of a 1D photonic crystal of TiO_2–SiO_2, a ZrO_2 doped with CdS and CdSe quantum dots layer and a metallic silver layer on a quartz substrate in function of the incident angle, **b** photoluminescent spectra in function of the angle of the same structure excited at fixed angle, **c** shift of the luminescent peak due to the resonance with the photonic crystal structure at 40° with excitation at 35°. In *red* the reflectance at that angle, in *black solid* the emission of the microcavity and in *dotted black* the emission of the ZrO_2–CdS–CdSe layer alone, and **d** increase in emitted intensity of the microcavity (in *red*) in respect of the reference ZrO_2–CdS–CdSe on a Si substrate (in *black*) due to the resonance with the cavity reflectivity (in *black*). The light is detected at 10° [81]

increase not only light harvesting but also electron transport and electrolyte infiltration in the electrode [86].

Finally, it is worth mentioning that the development of the original optical devices integrating sol-gel 1D photonic crystals is far to be concluded. The structural complexity of the 1D photonic crystals can be further pushed such as adding gradient of properties. One very recent example of 1D photonic crystals based on TiO_2 and porous SiO_2 was reported thanks to a technique called dip-coating in "acceleration mode" [35]. The graded layers could be obtained by dip-coating at withdrawal speed that is gradually increased during the deposition. The graded 1D photonic crystals, illustrated in Fig. 3.13a exhibits variable band gap as function of the position, as can be seen in the photograph in Fig. 3.13b and from the transmittance spectra of Fig. 3.13c. The approach can be extended to a second dimension with gradients in both the *x* and *y* dimension that allow an even higher complexity and can find application for fast combinatorial screening of the processing parameters (Fig. 3.13d).

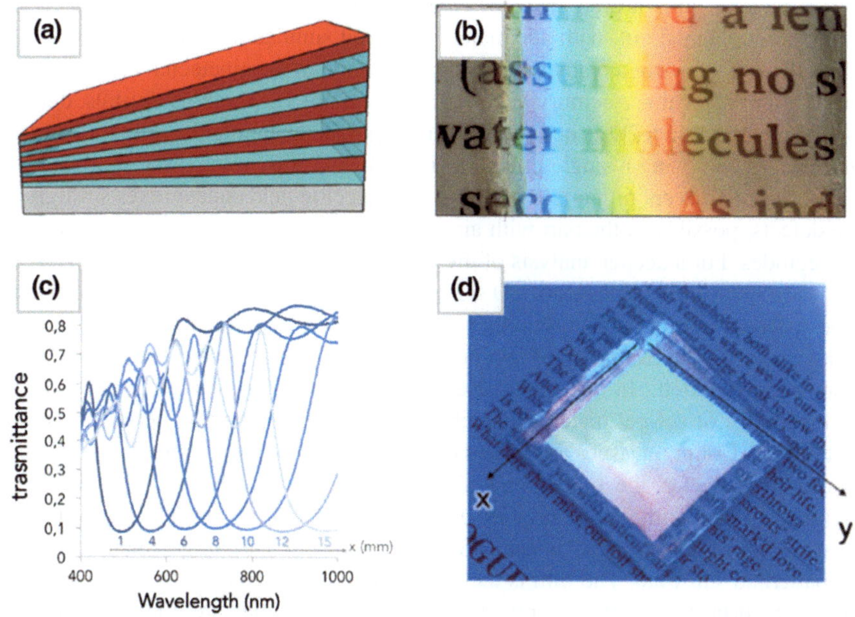

Fig. 3.13 **a** Illustration of the graded 1D photonic crystal, **b** optical photograph of graded 1D photonic crystal on glass substrate, **c** UV–vis transmission spectra of the graded photonic crystal for increasing *x* values, and **d** optical photograph of a photonic crystal with thickness gradients both in the *x* and *y* direction [35]

3.3.2 2D Photonic Crystals

The fabrication of two-dimensional photonic crystals is, in general, a complex operation. Moreover, the finite size of the samples modifies the properties of the photonic structure in a more substantial way than in the case of mono-dimensional photonic crystals. An ideal 2D photonic crystal is a periodic structure, in the *xy* directions and infinite in the *z* one. For example, an infinite and periodic square array of infinite pillars or holes would constitute a 2D photonic crystal. For a 2D photonic crystal, there still is no simple general rule that predicts if a certain structure will have a band gap or not without performing simulations and calculations. A computational modeling approach is thus particularly relevant for 2D photonic crystals, more so than for the 1D case.

Furthermore, real samples are always finite in the *z* direction and this may have important effects and substantially modify the interaction with light. When this becomes relevant for the properties of the sample, one speaks of 2D "photonic crystal slabs." Since finite 2D photonic structures are usually constructed over a substrate, the interaction of light with this underlying structure must also be considered. This adds further complexity to the system and therefore, it is not simple to define general laws as in the case of 1D photonic crystals. The approaches used to obtain systems with desired properties are thus either based on trial and error or are completely planned out, with adequately involved simulations. This is why the 1D photonic crystals were already somewhat understood in the nineteenth century while the 2D (and 3D) came a century later. As in the case of 1D photonic crystals, defects can be generated in a 2D structure from single points (changing the refractive index, diameter or removing a single pillar) or from "lines" (e.g., removal or modification of a series of adjacent pillars). In both cases the electric field concentrates strongly in the defects, possibly in the part with air, which is important for applications such as waveguides. For a deeper analysis of the physics of 2D photonic crystals we refer to [35]. The "real-life" synthesis of 2D slab photonic structures is usually performed as a postprocess of film deposition either with nanoimprint lithography or interference lithography. Some examples of synthesis based on self-assembled structures have also been reported. In the first case, a thin film composed either of sol-gel nanoparticles or sol-gel material is embossed with a mold having periodical structures in 2D. Common issues are related to the fact that the aspect ratio of the structure is usually low. On the other hand, the process is extremely simple and allows creating a large number of surfaces once that the mold has been fabricated [87]. A variation of this method is the direct deposition of the sol-gel materials on a prepatterned substrate. The film is then applied on a surface leaving a 2D patterned photonic surface [88]. When optical lithography techniques are applied to sol-gel materials the initial hybrid thin films are usually obtained mixing a photoactive molecule that interact with light into the sol-gel solution. Patterns created by interference lithography are composed of pillars or other geometries [89, 90]. Self-assembly procedures, instead, involve the deposition of a monolayer of large nanoparticles (with dimensions in the same length scale of the light). It is common to refer to photonic crystals as "colloidal crystals" in both the 2D and 3D case. When sol-gel nanoparticles are applied at an interface (such a solvent air interface) they self-organized by capillary induced lateral mechanical compression. The deposition of the interface particle monolayer films on solid structure is made by the so-called "Langmuir-Blodgett Technique" or, alternatively, via drying the solvent uniformly [91]. Further options involve the evaporation-induced self-assembly during dip-coating [92] or the slow evaporation of a suspension containing nanoparticles without any further action [93]. Some examples of fabrication of 2D photonic crystals obtained by nanoimprint lithography, interference lithography, and self-assembly, respectively are shown in Fig. 3.14.

From the technological point of view, 2D photonic crystals share some applications already shown for their 1D counterparts. Sensing could be performed using porous sol-gel patterned films thanks to the variation of the optical properties of the film upon exposure with the analyte. In the same way light extraction from luminescent films can be enhanced thanks to the scattering properties of the slab structure. In the 1D case, light emission enhancement was obtained by coupling the energy levels of the photonic crystal structures to the ones of the luminescent probe. In this case, the mechanism is based on diffraction. If the light emitted by luminescent specimens is normally confined and trapped in the emitting thin film by total internal reflection, the coupling of the film with a 2D photonic crystal slab structures

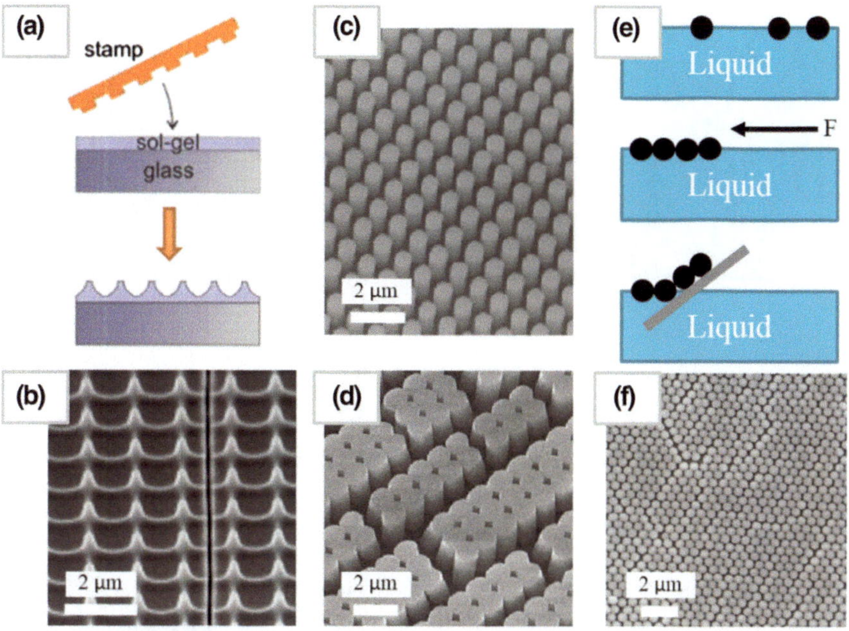

Fig. 3.14 **a** Scheme of fabrication, **b** SEM image of a 2D photonic structure realized by nanoimprint lithography [87], **c** SEM images of two 2D photonic crystals obtained by interference lithography [90], **e** scheme of fabrication and **f** SEM image of a self-assembled monolayer of silica spheres [91]

allows hindering the reflection and obtaining more intense luminescent films [89]. An optical image and a scheme of the system are reported in Fig. 3.15a, b, respectively. Coupling a 2D sol-gel photonic structure with any light-emitting device can therefore be extremely useful to increase its efficiency without increasing substantially its cost. If the light has to follow the inverse path, coming from outside and being absorbed in a device such as solar cells, the surface patterning and the formation of the 2D photonic crystal is useful in increasing performance too. Incident light that is scattered by the photonic structure has indeed a longer path inside the solar cell. Scattered photons are then more likely to be adsorbed giving a higher conversion [94]. In Fig. 3.15d the incident photon-to-current conversion efficiency as function of the wavelength is reported for a niobium oxide dye-sensitized solar cell with non-patterned and patterned surface. An overall increase in efficiency, in particular in the blue part of the spectrum, was observed.

2D photonic crystals are very often utilized to create waveguides and manage light in defects of the structure. However, the presented fabrication techniques, which are commonly utilized with sol-gel materials, are not adapted to prepare structures with completely controlled defects. Self-assembly approaches allow fabrication of large surfaces with a certain number of defects per surface but the

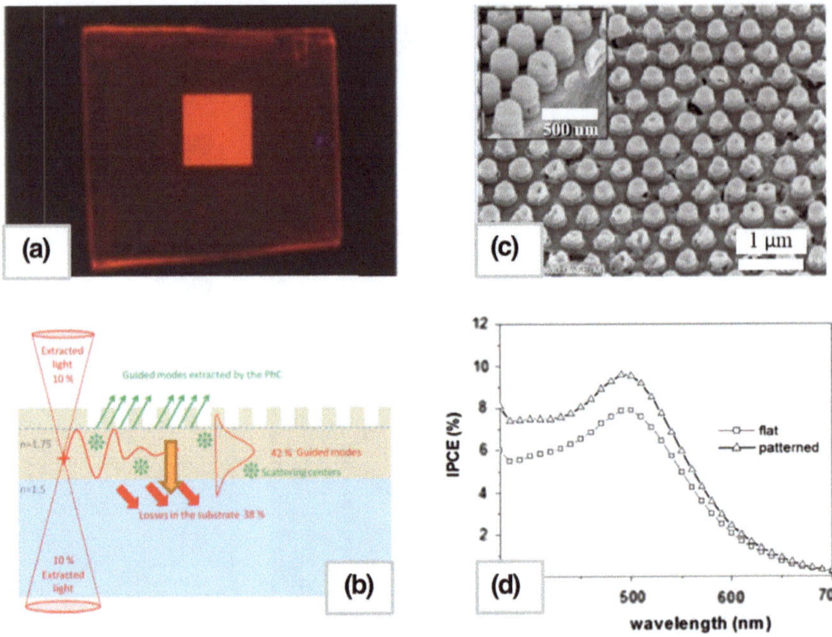

Fig. 3.15 **a** Optical image and a scheme **b** of the light distribution in a nanoimprinted europium doped TiO_2 thin film [89], **c** SEM image and **d** incident photon-to-current efficiency (IPCE) of a nanoimprinted NiO_2-based dye-sensitized solar cell in comparison with a non-patterned one [94]

defects cannot be intentionally introduced and they are an undesired output of the fabrication. Nanoimprint lithography can, in principle, be suitable for controlling defects. Nevertheless, it seems usually more convenient to directly pattern materials (such as polymers, photoresists, or metals) with other top-down lithographic techniques since these niche applications do not require large-scale fabrication.

3.3.3 3D Photonic Crystals

In 3D photonic crystals the periodicity is developed in three dimensions with a photonic band gap in all directions of propagation. As in the case of 2D photonic crystals qualitative evaluations of the band structure based on the system geometry is tricky. Modeling or direct testing, are again needed to obtain the desired properties.

Fabrication of sol-gel 3D photonic crystals was achieved by both bottom-up and top-down methods. As for the 2D case, nanoparticle self-assembly is a common technique to create 3D periodic structures. Self-assembled 3D colloidal structures are commonly called “opals” because the gems having the same name are actually 3D colloidal photonic crystals of silica nanoparticles. When opal structures (commonly obtained with PS or PMMA nanoparticles) are used as the template for the sol-gel replica, a macroporous film is obtained. This structure is commonly called "inverse opal" and, if the refractive index of the sol-gel material is high enough, its band structure has a complete band gap. Examples of opal and inverse opal structures are reported in Fig. 3.16a, b [95]. The photonic properties of the self-assembled colloidal 3D photonic crystals can be tuned varying the particle size, the number of layers and, in some special case, by post-treatment, such as plasma etching, for the inverse opal sol-gel samples [96].

On the other hand top-down approaches allow obtaining more exotic structures such as the so-called "woodpile" photonic crystals showed in Fig. 3.16c (which can exhibit a complete band gap). This structure has been realized by two photon polymerization of a non-shrinking, photosensitive, zirconium sol-gel material [97]. Through the replication of organic structures created by interference lithography (with the help of electrodeposition techniques) inverse opal photonic crystals were created [98]. If the bottom-up approaches are easier and faster, top-down ones allow achievement of structures without defects which can be, in many cases, necessary for the final applications. To conclude, noteworthy examples of sol-gel 3D photonic crystals can also be created by replication of periodic 3D structures found in nature. Butterfly wings [99] or particular beetle cuirasses [100] show iridescent colors due to the periodic structure of their composing living material. These materials can be infiltrated with sol-gel solutions and transformed into complex 3D photonic crystals upon removal of the biological part. In Fig. 3.16d an example of such a structure is reported.

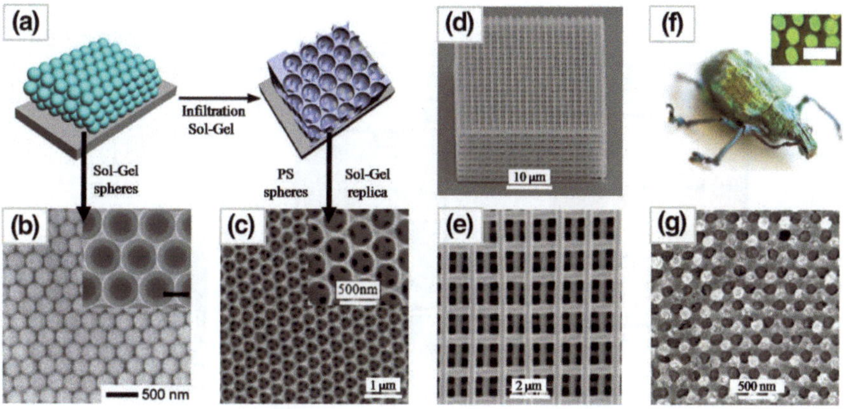

Fig. 3.16 **a** Scheme of the opal (*left*) and inverse opal (*right*) 3D photonic structures [95], **b** SEM image of 3D photonic crystal with opal geometry made of silica nanospheres (with a dense core and a mesoporous shell). In the inset the scale bar is 200 nm [101], **c** SEM image of 3D photonic crystal with inverse opal geometry made of silica replica of polystyrene spheres [102], **d** and **e** "woodpile" 3D photonic crystal realized by holographic lithography of a photoactive zirconium sol-gel material [97], **f** iridescent beetle having photonic structures which is used as template to obtain the photonic crystal silica structure in **g** [100]

Again, as was the case for 1D and 2D photonic crystals, sensing is most common application for 3D photonic crystals. Detection of volatile species through the condensation of vapors in the inner porosity of the material is based on a similar analytical process to their 1D counterparts. In the case of 3D photonic crystals, however, detection of fluids in their liquid state has also been performed. This application is known as "photonic tongue" but words such as "fluid" or "liquid sensor" are more commonly utilized. The principle, from the optical point of view, remains the same: the fluid changes the refractive index of the medium and then the reflected wavelength. The reflection or transmission spectrum can be analyzed in order to identify the nature of the liquid and/or its concentration [38]. An interesting example is the so-called "Watermark-Ink" technology [103]. In this case, the internal surfaces of the 3D photonic structures are functionalized. In this way the porous network can allow, or not, liquid to infiltrate the porosity as function of the internal contact angle of the structure. This can be utilized to detect ethanol concentration in water (Fig. 3.17a) or various solvents (Fig. 3.17b) with a series of different on/off responses. Practically, the pore surfaces are functionalized with trimethylsilyl (TMS), 3,3,3-trifluoropropylsilyl (3FS), *n*-decylsilyl (DEC), or 13FS (1H,1H,2H,2H-perfluorooctyl)silyl; the liquids penetrate into the macropores of the inverse opal structure if their contact angles with the surfaces are lower than 19° (Fig. 3.17c). When the liquid infiltrates the structure, the correspondent letter appears. The blocking mechanism in the pore is schematized in Fig. 3.17d.

Photonic/optical applications of 3D colloidal crystals are also of major importance. As in the case of 1D photonic crystals, light emission of quantum dots or rare earth ions can be increased due to the interactions with the 3D photonic crystals

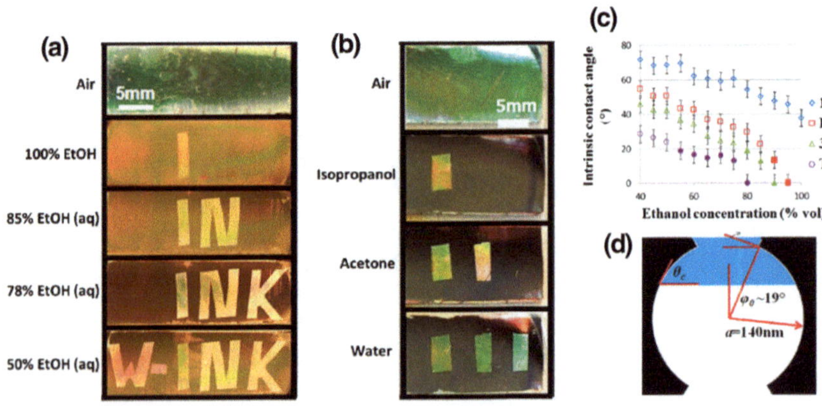

Fig. 3.17 Optical images of the inverse opal functionalized photonic crystals. Each letter of the word "W-INK" is functionalized with different molecules. "W-" 13FS, "I" DEC, "N" 3FS, "K" TMS, background SiOH. In different water-ethanol mixtures different letters appear. **b** Optical images of a film functionalized with three bars (from *left*) 13FS, DEC, 3FS, which are infiltrated differently when immersed in the common solvents, water, acetone, and isopropanol, **c** contact angle of a silicon surface functionalized with the defined groups in function of the ethanol in water concentration and **d** scheme for the blocking mechanism for solvent in filtration [103]

[104]. Effects on the excited state lifetime (with decay time constants increasing up to an order of magnitude), upconversion [105] (the consequential absorption of more than a photon with emission wavelength shorter than the excitation one), energy transfer [106], and amplified spontaneous emission [107] have been investigated and examined for sol-gel, rare earth ion doped, inverse opal structures. These lasts are, in most cases, interesting from a fundamental point of view and they could eventually find applications in the field of physical optics or lasers.

3D sol-gel photonic crystal allows increased performance in solar cells; for example exotic In_2O_3 inverse opal structures can effectively suppress the charge transfer recombination and improve the electron lifetime [108]. Actually many other examples of interesting structures and applications could be found in the literature. However,s a lot of development still needs to be done in the field of 3D photonic structure. For example, great engineering challenges need to be tackled to create controlled defects in 3D structures. However, as in the two-dimensional case, the fabrication methods based on self-assembled sol-gel materials seems not to be adapted as of yet to create structures with controlled defects.

3.4 General Conclusions and Perspectives

As outlined in this chapter, sol-gel chemistry and technology have now reached a quite mature stage in the field of coatings for optics and photonics. The marriage between sol-gel and optics clearly was, and still is, durable and successful; however some challenges still remain to be tackled in the field of sol-gel optical coatings.

With regard to the liquid deposition process, the main progress made over the past number of decades concerns the importance of the evaporation conditions (atmosphere composition and temperature) in order to better control the structure of the coating. However, the effort to develop very large-scale techniques, such as spray-coating, where all coating characteristics (structure and thickness accuracy and homogeneity) would be perfectly adjustable, has become a priority for instance in the case of antireflective coatings.

Integration of sol-gel materials for optics and photonics into nanotechnology devices is also in exponential development. However, much work needs to be done in order to favor the lab-to-market transition of optical nanodevices. In other words, the technology used to demonstrate proofs-of-concept need to evolve toward industrially sustainable production processes. This step requires also a further improvement of specific lithographic techniques for sol-gel and hybrid materials; even more complex 3D architectures could be designed for photonic crystals or even metamaterials.

Finally, future optical devices could be designed to respond in completely new way with to external stimuli (i.e., pH, solvent, light, external fields, temperature, etc.) by mimicking living organism, using light to self-evolve, adapt themselves to the environment, self-regulate or self-repair.

References

1. Dislich, H., Hinz, P.: Proceedings of the international workshop on glasses and glass Ceramics from gels history and principles of the sol-gel process, and some new multicomponent oxide coatings. J. Non-Cryst. Solids **48**(1), 11–16 (1982). doi:10.1016/0022-3093(82)90242-3
2. Livage, J., Henry, M., Sanchez, C.: Sol-gel chemistry of transition metal oxides. Prog. Solid State Chem. **18**(4), 259–341 (1988). doi:10.1016/0079-6786(88)90005-2
3. Joseph, E.J.: Comptes Rendus de l'Academie des Sciences **21**, 502 (1845)
4. Berger G (1939)
5. Schroeder, H.: Thin film formation on glass surfaces in chemical coating processes. Phys. Thin Film **87** (1969)
6. Guizard, C., Lacan, P.: Hybrid organic-inorganic materials applied to membrane preparation. New J. Chem. **18**(10), 1097–1107 (1994)
7. Lu, Y., Ganguli, R., Drewien, C.A., Anderson, M.T., Brinker, C.J., Gong, W., Guo, Y., Soyez, H., Dunn, B., Huang, M.H., Zink, J.I.: Continuous formation of supported cubic and hexagonal mesoporous films by sol-gel dip-coating. Nature **389**(6649), 364–368 (1997)
8. Yang, H., Coombs, N., Sokolov, I., Ozin, G.A.: Free-standing and oriented mesoporous silica films grown at the air-water interface. Nature **381**(6583), 589–592 (1996)
9. Sanchez, C., Belleville, P., Popall, M., Nicole, L.: Applications of advanced hybrid organic-inorganic nanomaterials: from laboratory to market. Chem. Soc. Rev. **40**(2), 696–753 (2011). doi:10.1039/c0cs00136h
10. Lebeau, B., Sanchez, C.: Sol-gel derived hybrid inorganic-organic nanocomposites for optics. Curr. Opin. Solid State Mater. Sci. **4**(1), 11–23 (1999). doi:10.1016/S1359-0286(99)80005-9
11. Sanchez, C., Lebeau, B., Chaput, F., Boilot, J.P.: Optical properties of functional hybrid organic-inorganic nanocomposites. Adv. Mater. **15**(23), 1969–1994 (2003). doi:10.1002/adma.200300389
12. Lebeau, B., Innocenzi, P.: Hybrid materials for optics and photonics. Chem. Soc. Rev. **40**(2), 886–906 (2011). doi:10.1039/c0cs00106f
13. Zayat, M., Almendro, D., Vadillo, V., Levy, D.: Sol-gel optical and electro-optical materials. In: The Sol-Gel Handbook, pp. 1239–1280. Wiley-VCH Verlag GmbH & Co. KGaA (2015). doi:10.1002/9783527670819.ch40
14. Pénard, A.-L., Gacoin, T., Boilot, J.-P.: Functionalized sol-gel coatings for optical applications. Acc. Chem. Res. **40**(9), 895–902 (2007). doi:10.1021/ar600025j
15. Dobrowolski, J.A.: Optical Properties of Films and Coatings in Handbook of Optics, 2nd edn. McGraw-Hill, New York (1995)
16. Stenzel, O.: The physics of thin film optical spectra: an introduction. Springer series in surface sciences, vol. 44. Springer, Berlin (2005)
17. Louis, B., Krins, N., Faustini, M., Grosso, D.: Understanding crystallization of anatase into binary SiO_2/TiO_2 sol-gel optical thin films: an in Situ thermal ellipsometry analysis. J Phys Chem C **115**(7), 3115–3122 (2011). doi:10.1021/jp109653p
18. Lee, D., Rubner, M.F., Cohen, R.E.: All-nanoparticle thin-film coatings. Nano Lett. **6**(10), 2305–2312 (2006). doi:10.1021/nl061776m
19. Sanchez, C., Boissiere, C., Cassaignon, S., Chaneac, C., Durupthy, O., Faustini, M., Grosso, D., Laberty-Robert, C., Nicole, L., Portehault, D., Ribot, F., Rozes, L., Sassoye, C.: Molecular engineering of functional inorganic and hybrid materials. Chem. Mater. **26**(1), 221–238 (2014). doi:10.1021/cm402528b
20. Soler-Illia, G., Azzaroni, O.: Multifunctional hybrids by combining ordered mesoporous materials and macromolecular building blocks. Chem. Soc. Rev. **40**(2), 1107–1150 (2011). doi:10.1039/c0cs00208a

21. Faustini, M., Grosso, D., Boissiere, C., Backov, R., Sanchez, C.: "Integrative sol-gel chemistry": a nanofoundry for materials science. J. Sol-Gel. Sci. Technol. **70**(2), 216–226 (2014). doi:10.1007/s10971-014-3321-9
22. Herrmann-Geppert, I., Bogdanoff, P., Radnik, J., Fengler, S., Dittrich, T., Fiechter, S.: Surface aspects of sol-gel derived hematite films for the photoelectrochemical oxidation of water. Phys. Chem. Chem. Phys. **15**(5), 1389–1398 (2013). doi:10.1039/C2CP42651J
23. Paquez, X., Amiard, G., de Combarieu, G., Boissière, C., Grosso, D.: Resistant RuO_2/SiO_2 absorbing sol-gel coatings for solar energy conversion at high temperature. Chem. Mater. **27**(7), 2711–2717 (2015). doi:10.1021/acs.chemmater.5b00731
24. Avnir, D., Levy, D., Reisfeld, R.: The nature of the silica cage as reflected by spectral changes and enhanced photostability of trapped Rhodamine 6G. J. Phys. Chem. **88**(24), 5956–5959 (1984). doi:10.1021/j150668a042
25. Innocenzi, P., Kozuka, H., Yoko, T.: Fluorescence properties of the Ru(bpy)32 + complex incorporated in sol-gel-derived silica coating films. J. Phys. Chem. B **101**(13), 2285–2291 (1997). doi:10.1021/jp970004z
26. Martucci, A., Fick, J., LeBlanc, S.-É., LoCascio, M., Haché, A.: Optical properties of PbS quantum dot doped sol-gel films. J. Non-Cryst. Solids **345–346**, 639–642 (2004). doi:10.1016/j.jnoncrysol.2004.08.114
27. Epifani, M., Giannini, C., Tapfer, L., Vasanelli, L.: Sol-gel synthesis and characterization of Ag and Au nanoparticles in SiO_2, TiO_2, and ZrO_2 thin films. J. Am. Ceram. Soc. **83**(10), 2385–2393 (2000)
28. Angelome, P.C., Pastoriza-Santos, I., Perez-Juste, J., Rodriguez-Gonzalez, B., Zelcer, A., Soler-Illia, G.J.A.A., Liz-Marzan, L.M.: Growth and branching of gold nanoparticles through mesoporous silica thin films. Nanoscale **4**(3), 931–939 (2012). doi:10.1039/C2NR11547F
29. Levy, D., Serna, C.J., Otón, J.M.: Preparation of electro-optical active liquid crystal microdomains by the sol-gel process. Mater. Lett. **10**(9), 470–476 (1991). doi:10.1016/0167-577X(91)90242-X
30. Joannopoulos, J.D., Johnson, S.G., Winn, J.N., Meade, R.D.: Photonic crystals molding the flow of light Second edition introduction. In: Photonic Crystals: Molding the Flow of Light, 2nd edn, pp. 1–5 (2008)
31. Grosso, D., Boissière, C., Faustini, M.: Thin film deposition techniques. In: The Sol-Gel Handbook. Wiley-VCH Verlag GmbH & Co. KGaA, pp 277–316 (2015). doi:10.1002/9783527670819.ch09
32. Faustini, M., Boissiere, C., Nicole, L., Grosso, D.: From chemical solutions to inorganic nanostructured materials: A journey into evaporation-driven processes. Chem. Mater. **26**(1), 709–723 (2014). doi:10.1021/cm402132y
33. Faustini, M., Louis, B., Albouy, P.A., Kuemmel, M., Grosso, D.: Preparation of sol-gel films by dip-coating in extreme conditions. J. Phys. Chem. C **114**(17), 7637–7645 (2010). doi:10.1021/jp9114755
34. Landau, L., Levich, B.: Dragging of a liquid by a moving plate. Acta Physicochimica Urss **17**, 42–54 (1942)
35. Faustini, M., Ceratti, D.R., Louis, B., Boudot, M., Albouy, P.-A., Boissière, C., Grosso, D.: Engineering functionality gradients by dip coating process in acceleration mode. ACS Appl. Mater. Interfaces **6**(19), 17102–17110 (2014). doi:10.1021/am504770x
36. Ceratti, D.R., Louis, B., Paquez, X., Faustini, M., Grosso, D.: A new dip coating method to obtain large-surface coatings with a minimum of solution. Adv. Mater. **27**(34), 4958–4962 (2015). doi:10.1002/adma.201502518
37. Floch, H.G., Belleville, P.F., Priotton, J.J., Pegon, P.M., Dijonneau, C.S., Guerain, J.: Sol-gel optical coatings for lasers.1. Am. Ceram. Soc. Bull. **74**(10), 60–63 (1995)
38. Flack, W.W., Soong, D.S., Bell, A.T., Hess, D.W.: A mathematical model for spin coating of polymer resists. J. Appl. Phys. **56**(4), 1199–1206 (1984)
39. Britten, J.A., Thomas, I.M.: Sol-gel multilayers applied by a meniscus coating process. MRS Online Proceedings Library Archive **271**, null–null (1992). doi:10.1557/PROC-271-413

40. Belleville, P.F., Floch, H.G., Berger, M.: Sol-gel Optical coatings processed by the laminar-flow coating technique, vol. 1758. Sol-gel Optics Ii. Spie—Int Soc Optical Engineering, Bellingham (1992). doi:10.1117/12.131998
41. Dieudonne, X., Valle, K., Belleville, P.: Towards new levels for stacking of sol-gel functional coatings. Opt. Express **19**(17), 16356–16364 (2011). doi:10.1364/oe.19.016356
42. Walcarius, A., Sibottier, E., Etienne, M., Ghanbaja, J.: Electrochemically assisted self-assembly of mesoporous silica thin films. Nat. Mater. **6**(8), 602–608 (2007). doi:10.1038/nmat1951
43. Raut, H.K., Ganesh, V.A., Nair, A.S., Ramakrishna, S.: Anti-reflective coatings: a critical, in-depth review. Energy Environ. Sci. **4**, 3779 (2011). doi:10.1039/c1ee01297e
44. Pénard, A.-L., Gacoin, T., Boilot, J.-P.: Functionalized sol-gel coatings for optical applications. Acc. Chem. Res. **40**, 895–902 (2007)
45. Guillemot, F., Brunet-Bruneau, A., Bourgeat-Lami, E., Gacoin, T., Barthel, E., Boilot, J.-P.: Latex-templated silica films: tailoring porosity to get a stable low-refractive index. Chem. Mater. **22**, 2822–2828 (2010). doi:10.1021/cm903754k
46. Mizoshita, N., Ishii, M., Kato, N., Tanaka, H.: Hierarchical nanoporous silica films for wear Resistant antireflection coatings. ACS Appl. Mater. Interfaces **7**, 19424–19430 (2015). doi:10.1021/acsami.5b05659
47. Garnett, J.C.M.: Philos Trans R Soc London, Ser A **205**, 237–284 (1906)
48. Bruggeman, D.A.G.: Ann. Phys. **24**, 636–664 (1935)
49. Rayleigh, L.: Proc. London Math. Soc. **s1–11**, 51–56 (1879)
50. Zhang, X.-X., Cai, S., You, D., Yan, L.-H., Lv, H.-B., Yuan, X.-D., Jiang, B.: Template-free sol-gel preparation of superhydrophobic ORMOSIL films for double-wavelength broadband anti-reflective coatings. Adv. Funct. Mater. **23**, 4361–4365 (2013). doi:10.1002/adfm.201203059
51. Lien, S., Wuu, D., Yeh, W., Liu, J.: Tri-layer antireflection coatings ($SiO_2/SiO_2–TiO_2/TiO_2$) for silicon solar cells using a sol-gel technique. Sol. Energy Mater. Sol. Cells **90**, 2710–2719 (2006). doi:10.1016/j.solmat.2006.04.001
52. Ko, D.-H., Tumbleston, J.R., Henderson, K.J., Euliss, L.E., DeSimone, J.M., Lopez, R., Samulski, E.T.: Biomimetic microlens array with anti-reflective "moth-eye" surface. Soft Matter **7**, 6404 (2011). doi:10.1039/c1sm05302g
53. McCollister, H.: Process of making glass articles having anti-reflective coatings and product
54. Liu, L.-Q., Wang, X.-L., Jing, M., Zhang, S.-G., Zhang, G.-Y., Dou, S.-X., Wang, G.: Broadband and omnidirectional, nearly zero reflective photovoltaic glass. Adv. Mater. **24**, 6318–6322 (2012). doi:10.1002/adma.201201740
55. Younkin, R., Carey, J.E., Mazur, E., Levinson, J.A., Friend, C.M.: Infrared absorption by conical silicon microstructures made in a variety of background gases using femtosecond-laser pulses. J. Appl. Phys. **93**, 2626–2629 (2003). doi:10.1063/1.1545159
56. Kitamura, S., Kanno, Y., Watanabe, M., Takahashi, M., Kuroda, K., Miyata, H.: Films with tunable graded refractive index consisting of spontaneously formed mesoporous silica nanopinnacles. ACS Photonics **1**, 47–52 (2014). doi:10.1021/ph400036a
57. Kim, W.-S., Lee, J.-H., Shin, S.-Y., Bae, B.-S., Kim, Y.-C.: Fabrication of ridge waveguides by UV embossing and stamping of sol-gel hybrid materials. IEEE Photonics Technol. Lett. **16**, 1888–1890 (2004). doi:10.1109/LPT.2004.831053
58. Shin, J.-H., Kim, Y.-D., Choi, H.-J., Ryu, S.-W., Lee, H.: Multi-functional SiO_2 moth-eye pattern for photovoltaic applications. Sol. Energy Mater. Sol. Cells **126**, 1–5 (2014). doi:10.1016/j.solmat.2014.03.002
59. Geffcken, W.B.E.: Verfahren zür Anderung des Reflexionsvermögens optisher Gläser (1939)
60. Glaubitt, W., Löbmann, P.: Anti-reflective coatings prepared by sol-gel processing: principles and applications. J. Eur. Ceram. Soc. **32**, 2995–2999 (2012). doi:10.1016/j.jeurceramsoc.2012.02.032
61. Wang, X., Shen, J.: A review of contamination-resistant anti-reflective sol-gel coatings. J. Sol-Gel. Sci. Technol. **61**, 206–212 (2012). doi:10.1007/s10971-011-2615-4

62. Boissiere, C., Grosso, D., Lepoutre, S., Nicole, L., Bruneau, A.B., Sanchez, C.: Porosity and mechanical properties of mesoporous thin films assessed by environmental ellipsometric porosimetry. Langmuir **21**(26), 12362–12371 (2005). doi:10.1021/la050981z
63. Boudot, M., Gaud, V., Louarn, M., Selmane, M., Grosso, D.: Sol-gel based hydrophobic anti-Reflective coatings on organic substrates: a detailed investigation of ammonia vapor treatment (AVT). Chem. Mater. **26**, 1822–1833 (2014). doi:10.1021/cm403787v
64. Glaubitt, W., Löbmann, P.: Anti-soiling effect of porous SiO_2 coatings prepared by sol-gel processing. J. Sol-Gel. Sci. Technol. **59**, 239–244 (2011). doi:10.1007/s10971-011-2489-5
65. Manca, M., Cannavale, A., De Marco, L., Aricò, A.S., Cingolani, R., Gigli, G.: Durable superhydrophobic and anti-reflective surfaces by trimethylsilanized silica nanoparticles-based sol-gel processing. Langmuir **25**, 6357–6362 (2009). doi:10.1021/la804166t
66. Guldin, S., Kohn, P., Stefik, M., Song, J., Divitini, G., Ecarla, F., Ducati, C., Wiesner, U., Steiner, U.: Self-cleaning anti-reflective optical coatings. Nano Lett. **13**, 5329–5335 (2013). doi:10.1021/nl402832u
67. Faustini, M., Grenier, A., Naudin, G., Li, R., Grosso, D.: Ultraporous nanocrystalline TiO_2-based films: synthesis, patterning and application as anti-reflective, self-cleaning, superhydrophilic coatings. Nanoscale **7**, 19419–19425 (2015). doi:10.1039/C5NR06466J
68. Allain, E., Besson, S., Durand, C., Moreau, M., Gacoin, T., Boilot, J.P.: Transparent mesoporous nanocomposite films for self-cleaning applications. Adv. Funct. Mater. **17**(4), 549–554 (2007). doi:10.1002/adfm.200600197
69. Gan, W.Y., Lam, S.W., Chiang, K., Amal, R., Zhao, H., Brungs, M.P.: Novel TiO_2 thin film with non-UV activated superwetting and antifogging behaviours. J. Mater. Chem. **17**, 952–954 (2007). doi:10.1039/b618280a
70. Faustini, M., Nicole, L., Boissière, C., Innocenzi, P., Sanchez, C., Grosso, D.: Hydrophobic, anti-reflective, self-cleaning, and antifogging sol-gel coatings: an example of multifunctional nanostructured materials for photovoltaic cells. Chem. Mater. **22**, 4406–4413 (2010). doi:10.1021/cm100937e
71. Zhang, X., Lan, P., Lu, Y., Li, J., Xu, H., Zhang, J., Lee, Y., Rhee, J.Y., Choy, K.-L., Song, W.: Multifunctional antireflection coatings based on novel hollow silica-silica nanocomposites. Acs Appl. Mater. Interfaces **6**, 1415–1423 (2014). doi:10.1021/am405258d
72. Joannopoulos, J.D., Villeneuve, P.R., Fan, S.H.: Photonic crystals: putting a new twist on light. Nature **386**, 143–149 (1997). doi:10.1038/386143a0
73. Bonifacio, L.D., Puzzo, D.P., Breslav, S., Willey, B.M., McGeer, A., Ozin, G.A.: Towards the photonic nose: a novel platform for molecule and Bacteria identification. Adv. Mater. **22**, 1351-+ (2010). doi:10.1002/adma.200902763
74. Wu, Z., Lee, D., Rubner, M.F., Cohen, R.E.: Structural color in porous, superhydrophilic, and self-cleaning SiO_2/TiO_2 Bragg stacks. Small **3**, 1445–1451 (2007). doi:10.1002/smll.200700084
75. Krins, N., Faustini, M., Louis, B., Grosso, D.: Thick and crack-free nanocrystalline mesoporous TiO_2 films obtained by capillary coating from aqueous solutions. Chem. Mater. **22**(23), 6218–6220 (2010). doi:10.1021/cm102524u
76. Hidalgo, N., Calvo, M.E., Bellino, M.G., Soler-Illia, G.J.A.A., Miguez, H.: Porous supramolecularly templated optical resonators built in 1D photonic crystals. Adv. Funct. Mater. **21**, 2534–2540 (2011). doi:10.1002/adfm.201002486
77. Exner, A.T., Pavlichenko, I., Baierl, D., Schmidt, M., Derondeau, G., Lotsch, B.V., Lugli, P., Scarpa, G.: A step towards the electrophotonic nose: integrating 1D photonic crystals with organic light-emitting diodes and photodetectors. Laser Photonics Rev. **8**, 726–733 (2014). doi:10.1002/lpor.201300220
78. Pavlichenko, I., Exner, A.T., Guehl, M., Lugli, P., Scarpa, G., Lotsch, B.V.: Humidity-enhanced thermally tunable TiO_2/SiO_2 Bragg stacks. J. Phys. Chem. C **116**, 298–305 (2012). doi:10.1021/jp208733t
79. Lotsch, B.V., Ozin, G.A. Clay Bragg stack optical sensors. Adv. Mater. **20**, 4079-+ (2008). doi:10.1002/adma.200800914

80. Hinterholzinger, F.M., Ranft, A., Feckl, J.M., Ruehle, B., Bein, T., Lotsch, B.V.: One-dimensional metal-organic framework photonic crystals used as platforms for vapor sorption. J. Mater. Chem. **22**, 10356–10362 (2012). doi:10.1039/c2jm15685g
81. Jasieniak, J., Sada, C., Chiasera, A., Ferrari, M., Martucci, A., Mulvaney, P.: Sol-gel based vertical optical microcavities with quantum dot defect layers. Adv. Funct. Mater. **18**, 3772–3779 (2008). doi:10.1002/adfm.200800784
82. Ye, J.Y., Ishikawa, M.: Enhancing fluorescence detection with a photonic crystal structure in a total-internal-reflection configuration. Opt. Lett. **33**, 1729–1731 (2008). doi:10.1364/OL.33.001729
83. Das, P., Maiti, R., Ray, S.K., Bhaktha, B.N.S.: Increased photon density of states at defect-mode frequencies led enhancement of tunability of spontaneous emission from Eu^{2+}, 3+ doped SiO_2/SnO_2 one-dimensional photonic crystals. ResearchGate **2**, 036201 (2015). doi:10.1088/2053-1591/2/3/036201
84. Park, J.T., Prosser, J.H., Ahn, S.H., Kim, S.J., Kim, J.H., Lee, D.: Enhancing the performance of solid-state dye-sensitized solar cells using a mesoporous interfacial titania layer with a Bragg stack. Adv. Funct. Mater. **23**, 2193–2200 (2013). doi:10.1002/adfm.201202345
85. Colodrero, S., Mihi, A., Haggman, L., Ocana, M., Boschloo, G., Hagfeldt, A., Miguez, H.: Porous one-Dimensional photonic crystals improve the power-conversion efficiency of dye-sensitized solar cells. Adv. Mater. **21**, 764-+ (2009) doi:10.1002/adma.200703115
86. Park, J.T., Chi, W.S., Kim, S.J., Lee, D., Kim, J.H.: Mesoporous TiO_2 Bragg stack templated by graft copolymer for dye-sensitized solar cells. Sci. Rep. **4**, 5505 (2014). doi:10.1038/srep05505
87. Back, F., Bockmeyer, M., Rudigier-Voigt, E., Loebmann, P.: Periodic nanostructures imprinted on high-temperature stable sol-gel films by ultraviolet-based nanoimprint lithography for photovoltaic and photonic applications. Thin Solid Films **562**, 274–281 (2014). doi:10.1016/j.tsf.2014.04.082
88. Park, C., Kim, H., Park, I.-S., Ko, K.-Y., Kim, K.-K., Lee, B.H., Ahn, J.: The variation of the enhanced PL efficiency of $Y_2O_3:Eu^{3+}$ phosphor films with the height to the ZrO_2 nanoparticle-assisted 2D PCL by reverse nano-imprint lithography. Microelectron. Eng. **136**, 48–50 (2015). doi:10.1016/j.mee.2015.03.035
89. Devys, L., Dantelle, G., Revaux, A., Kubytskyi, V., Paget, D., Benisty, H., Gacoin, T.: Extraction length determination in patterned luminescent sol-gel films. Adv. Opt. Mater. **2**, 81–87 (2014). doi:10.1002/adom.201300304
90. Segawa, H., Misawa, H., Yano, T., Shibata, S.: Fabrication of periodic arrays of top-gathering Titania-organic hybrid pillars derived from multi-beam laser interference—art. no. 61060 N. Proceedings of SPIE—The International Society for Optical Engineering 6106 (2006). doi:10.1117/12.647213
91. Guo, Y., Tang, D., Du, Y., Liu, B.: Controlled fabrication of hexagonally close-packed Langmuir-Blodgett silica particulate mono layers from binary surfactant and solvent systems. Langmuir **29**, 2849–2858 (2013). doi:10.1021/la3049218
92. Armstrong, E., Khunsin, W., Osiak, M., Bloemker, M., Sotomayor Torres, C.M., O'Dwyer, C.: Ordered 2D colloidal photonic crystals on gold substrates by surfactant-assisted fast-rate dip coating. Small **10**, 1895–1901 (2014). doi:10.1002/smll.201303616
93. Zhang, J.-T., Wang, L., Lamont, D.N., Velankar, S.S., Asher, S.A.: Fabrication of large-area two-dimensional colloidal crystals. Angew. Chem.-Int. Ed. **51**, 6117–6120 (2012). doi:10.1002/anie.201105439
94. Cai, Z., Liu, Y.J., Leong, E.S.P., Teng, J., Lu, X.: Highly ordered and gap controllable two-dimensional non-close-packed colloidal crystals and plasmonic-photonic crystals with enhanced optical transmission. J. Mater. Chem. **22**, 24668–24675 (2012). doi:10.1039/c2jm34896a
95. Xia, L., Song, J., Xu, R., Liu, D., Dong, B., Xu, L., Song, H.: Zinc oxide inverse opal electrodes modified by glucose oxidase for electrochemical and photoelectrochemical biosensor. Biosens. Bioelectron. **59**, 350–357 (2014). doi:10.1016/j.bios.2014.03.038

96. Huang, H., Chen, J., Yu, Y., Shi, Z., Moehwald, H., Zhang, G.: Controlled gradient colloidal photonic crystals and their optical properties. Colloids Surf. A-Physicochem. Eng. Aspects **428**, 9–17 (2013). doi:10.1016/j.colsurfa.2013.03.041
97. Ovsianikov, A., Viertl, J., Chichkov, B., Oubaha, M., MacCraith, B., Sakellari, I., Giakoumaki, A., Gray, D., Vamvakaki, M., Farsari, M., Fotakis, C.: Ultra-low shrinkage hybrid photosensitive material for two-photon polymerization microfabrication. ACS Nano **2**, 2257–2262 (2008). doi:10.1021/nn800451w
98. Xu, Y., Zhu, X., Dan, Y., Moon, J.H., Chen, V.W., Johnson, A.T., Perry, J.W., Yang, S.: Electrodeposition of three-dimensional titania photonic crystals from holographically patterned microporous polymer templates. Chem. Mater. **20**, 1816–1823 (2008). doi:10.1021/cm702511k
99. Peng, W., Zhu, S., Wang, W., Zhang, W., Gu, J., Hu, X., Zhang, D., Chen, Z.: 3D network magnetophotonic crystals fabricated on morpho butterfly wing templates. Adv. Funct. Mater. **22**, 2072–2080 (2012). doi:10.1002/adfm.201103032
100. Galusha, J.W., Jorgensen, M.R., Bartl, M.H. Diamond-structured titania photonic-bandgap crystals from biological templates. Adv. Mater. **22**, 107-+ (2010). doi:10.1002/adma.200902852
101. Bai, L., Xie, Z., Cao, K., Zhao, Y., Xu, H., Zhu, C., Mu, Z., Zhong, Q., Gu, Z.: Hybrid mesoporous colloid photonic crystal array for high performance vapor sensing. Nanoscale **6**, 5680–5685 (2014). doi:10.1039/c4nr00361f
102. Guo, W., Wang, M., Xia, W., Dai, L.: Two substrate-confined sol-gel coassembled ordered macroporous silica structures with an open surface. Langmuir **29**, 5944–5951 (2013). doi:10.1021/la304268b
103. Burgess, I.B., Mishchenko, L., Hatton, B.D., Kolle, M., Loncar, M., Aizenberg, J.: Encoding complex wettability patterns in chemically functionalized 3D photonic crystals. J. Am. Chem. Soc. **133**, 12430–12432 (2011). doi:10.1021/ja2053013
104. Lee, K.-T., Park, J.-H., Kwon, S.J., Kwon, H.-K., Kyhm, J., Kwak, K.-W., Jang, H.S., Kim, S.Y., Han, J.S., Lee, S.-H., Shin, D.-H., Ko, H., Han, I.-K., Ju, B.-K., Kwon, S.-H., Ko, D.-H.: Simultaneous enhancement of upconversion and downshifting luminescence via plasmonic structure. Nano Lett. **15**, 2491–2497 (2015). doi:10.1021/nl5049803
105. Tao, L., Xu, W., Zhu, Y., Xu, L., Zhu, H., Liu, Y., Xu, S., Zhou, P., Song, H.: Modulation of upconversion luminescence in Er^{3+} Yb^{3+}-codoped lanthanide oxyfluoride (YOF, GdOF, LaOF) inverse opals. J. Mater. Chem. C **2**, 4186–4195 (2014). doi:10.1039/c4tc00024b
106. Cui, S., Zhu, Y., Xu, W., Zhou, P., Xia, L., Chen, X., Song, H., Han, W.: Self-assembly and modified luminescence properties of NaY(MoO4)(2):Tb^{3+}, Eu^{3+} inverse opals. Dalton Trans. **43**, 13293–13298 (2014). doi:10.1039/c4dt01792g
107. Maskaly, G.R., Petruska, M.A., Nanda, J., Bezel, I.V., Schaller, R.D., Htoon, H., Pietryga, J. M., Klimov, V.I.: Amplified spontaneous emission in semiconductor-nanocrystal/synthetic-opal composites: optical-gain enhancement via a photonic crystal pseudogap. Adv. Mater. **18**, 343-+ (2006). doi:10.1002/adma.200500875
108. Kong, L., Dai, Q., Miao, C., Xu, L., Song, H.: Doped In_2O_3 inverse opals as photoanode for dye sensitized solar cells. J. Colloid Interface Sci. **450**, 196–201 (2015). doi:10.1016/j.jcis.2015.03.024

Chapter 4
The Multiple Roles of Diatoms in Environmental Applications: Prospects for Sol-Gel Modified Diatoms

Yvonne Lang, Francisco del Monte and Peter Dockery

4.1 Overview

Nature presents elaborate patterns and shapes that have fascinated the human mind since the early periods of human history. This fascination has greatly influenced and inspired those working in areas, such as art, design and architecture. Indeed it can be argued that these concepts are fundamental to the worlds of science and engineering, and thus it is no wonder that researchers in disciplines related to material science, such as chemistry, physics and engineering look to nature for creativity. A challenging area of research for material science has been the preparation of materials with complex hierarchical organisations on multiple scales ranging from the nanometre to micrometre. It is not surprising that unravelling the synthesis of complex structures in nature, such as the skeletons of animals or plants has provided direction to devise approaches for the synthesis of artificial complex structures [1]. The skeleton of the unicellular algae, the diatom, is attractive to material scientists due to its highly ornate architecture, and the diversity of structures that exist in this species. Furthermore, the ability of the diatom to synthesise its elaborate cell wall in mild environmental conditions has intrigued many disciples and contributed to the emergence of the fields of biomimetics and biotemplating. The former endeavours to gain an understanding of the structure and function of

Y. Lang (✉) · P. Dockery
School of Medicine, National University of Ireland, Galway, University Road, Galway, Ireland
e-mail: yvonne.lang@nuigalway.ie; lang.yvonne@itsligo.ie

Y. Lang
School of Science, Institute of Technology, Sligo, Ash Lane, Sligo F91 YW50, Ireland

F. del Monte
Instituto de Ciencia de Materiales de Madrid, Consejo Superior de Investigaciones Científicas, Campus de Cantoblanco, 28049 Madrid, Spain

S.C. Pillai and S. Hehir (eds.), *Sol-Gel Materials for Energy, Environment and Electronic Applications*, Advances in Sol-Gel Derived Materials and Technologies, DOI 10.1007/978-3-319-50144-4_4

biological systems for the design and engineering of materials. The latter has been described as follows; “Biotemplating seeks to either replicate the morphological characteristics and the functionality of a biological species or use a biological structure to guide the assembly of inorganic materials” [2].

This chapter will discuss avenues explored to harness the architecture of the diatom and the machinery of the diatom for a broad range of uses including environmental applications, such as catalysis, separation science, filtration, optics, gas detection and nanotechnology [3–10]. A brief overview of diatoms will be followed by a description of approaches to modify the chemistry of the both the living diatom and the diatom frustule for environmental applications. It is important to note that elucidating the molecular life of diatoms is a thriving area of research that will undoubtedly contribute to a greater understanding of biomineralization processes [11, 12], and subsequently the design of hierarchical structures. However, this will be beyond the scope of this chapter.

4.2 Introduction to Diatoms

Diatoms are unicellular algae ubiquitous in seawater and freshwater environments, with the number of species estimated to be 10,000–1,00,000 [13]. The hallmark of the diatom is the intricate architecture of the cell wall, the frustule (Fig. 4.1). The diatom frustule comprises of two valves, the epivalve and the hypovalve that sit together like a Petri dish with the overlapping region surrounded by structures referred to as girdle bands. The frustule is sheathed in an organic casing composed of polysaccharides and proteins [14]. The frustule is composed of amorphous silica conferring certain properties on the diatom cell wall [15–17]. It has been suggested that evolution has optimised the structure of the diatom to play a role in light harvesting [18]. Although the genetic foundations of the species-specific three-dimensional architectures of diatoms are not yet fully elucidated, specific organic molecules are known to guide silica deposition and morphogenesis. The specific transport of silicic acid into the cell is via transport proteins known as silicic acid transporters [19, 20]. It has been suggested that the different types of transporters have differing affinities for silica or differing capacities to transport silica [21, 22]. Once silica is within the parent diatom, it accumulates in a specialised compartment known as the silica deposition vesicle (SDV) [23].

It is within this compartment that the daughter valve is synthesised. The first biomolecules involved in diatom silica formation that have been identified are silaffins [24, 25] and long-chain polyamines [26]. The formation of the valve occurs in a two-dimensional plane first and then expands in the three-dimensional plane [27–31]. The formation of the valve is influenced also by actin and microtubules shaping the mature frustule within the SDV [32, 33]. In addition, environmental conditions such as nutrient availability, light, temperature and salinity can alter the

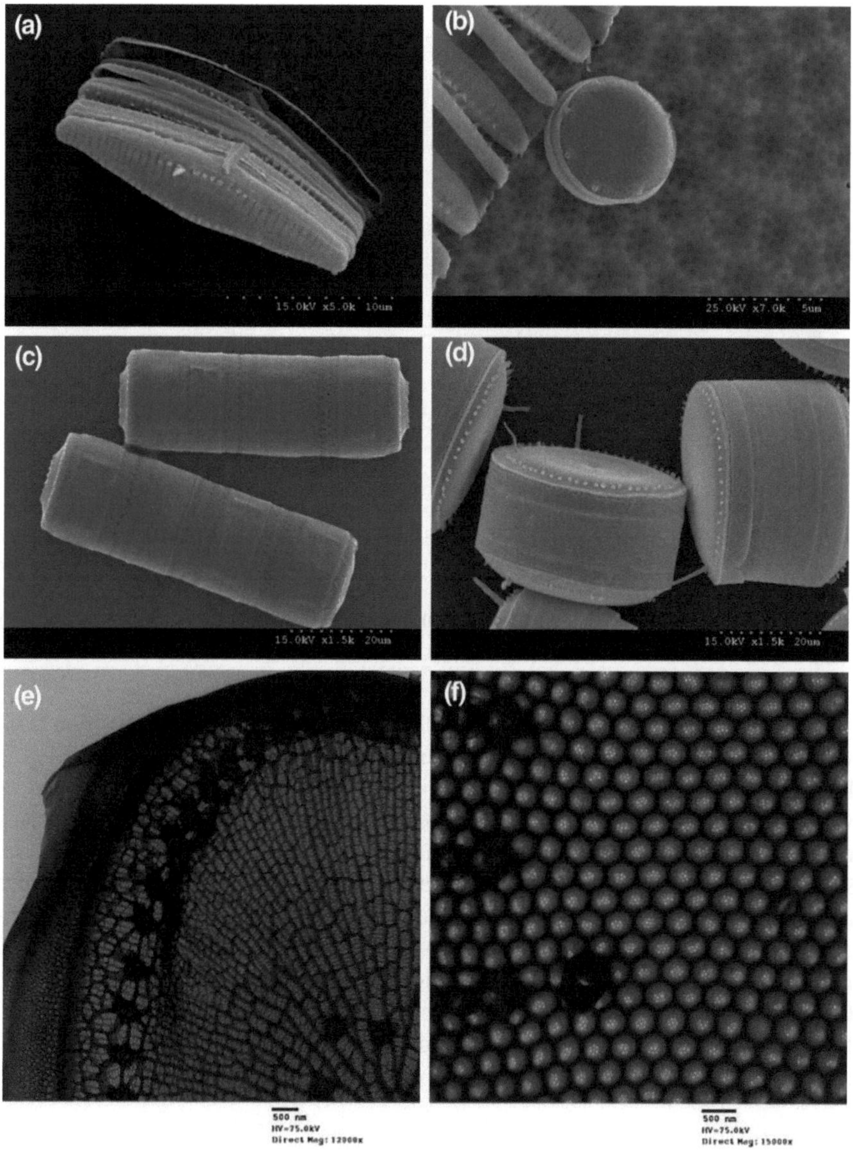

Fig. 4.1 The hallmark of the diatom is the intricate architecture of the cell wall, the frustule. Scanning electron micrographs reveal the unique architecture of different diatom species and the diversity in structures. **a** *Fragilaria* sp. **b** *Coscinodiscus wailesii*, *Stephanodiscus hantzchii*, and *Fragilaria* sp. **c** *Paralia sulcata*, **d** *Thalassiosira* sp. Transmission electron micrographs of *Thalassiosira* sp. **e–f** Illustrate both the order and the porosity on the valve face. Images provided by the Centre for Microscopy and Imaging, National University of Ireland, Galway

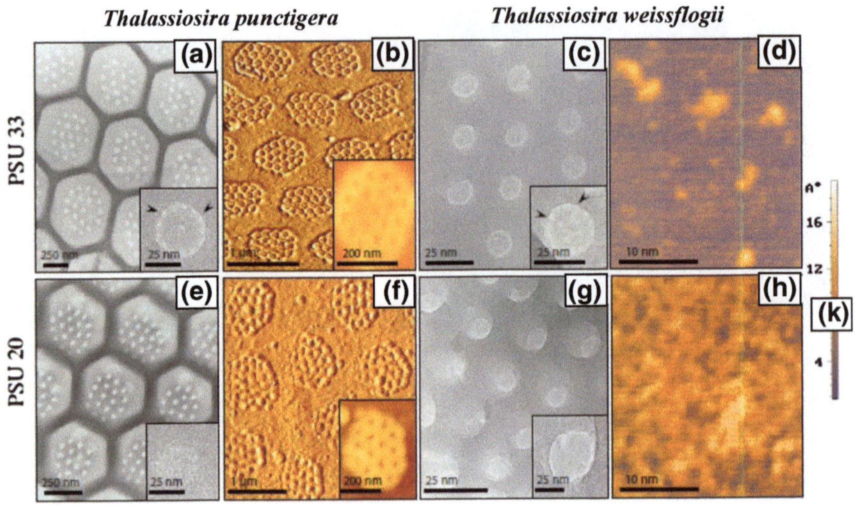

Fig. 4.2 The architectural properties of the diatom frustule are influenced by environmental conditions such as salinity and nutrient availability. High resolution TEM and AFM of biosilica obtained from the diatoms *T. punctigera* and *T. weissflogii* cultured at high salinity (psu 33) and low salinity (psu 20). (Reprinted with permission from [45]. Copyright (2007) National Academy of Sciences, U.S.A.)

growth rate of diatoms by influencing silicic acid uptake, silicic acid deposition rates, valve formation rates, and thus affect the morphology of the diatom [27, 34–45]. Figure 4.2 illustrates the influence of environmental conditions, such as salinity, on valve formation and frustule architecture. The combination of work from many research groups characterising diatom morphology with scanning electron microscopy, transmission electron microscopy and atomic force microscopy has provided an understanding of the complexity of the architecture of the diatom and the interconnected networks within frustules [46–52]. These studies have contributed significantly to illustrating the species-specific architecture of diatom frustule; however, there still remains a vast array of unexplored diatoms. This can be viewed as a gift from nature to material scientists and now all that is required is to mine for the architecture that has been optimised by evolution.

4.3 Modification of Diatom Frustule Chemistry

The chemistry of the diatom biosilica is inert and thus functionalizing the diatom frustule has been reported. The various approaches to alter the chemistry of the frustule include; coating strategies, conversion strategies and more recently in vivo modification of the frustule chemistry.

4.3.1 Coating Strategies to Modify the Diatom

Initial work investigating coating strategies to alter the surface chemistry of the diatom was based on exposure of the cleaned diatom frustule to a precursor solution followed by thermal reaction to generate a coating [53]. Further exploration in this field led to the use of gas/solid displacement reactions [54, 55], atomic layer deposition [56, 57], and sol-gel reactions [8, 58–60]. Diatom composites prepared using a coating approach and their proposed applications are listed in Table 4.1.

The thickness of the layer generated can vary from less than 50 nm [8, 55], to 50–100 nm [61], to 1 μm [53]. The morphology and size of the nanostructures deposited on the diatom frustule are dependent on both the coating material and the type of diatom employed [62–64] (Fig. 4.3).

Table 4.1 Proposed applications of diatom composites prepared via coating strategies

Coating	Proposed application	References
Pt, Au, Ag	Catalysis	[5]
Fe_3O_4	Adsorption	[7]
Eu^{3+}-doped $BaTiO_3$	Optics	[55]
SnO_2	Gas detector	[8]
CdS	Optics	[62]
TiO_2	Photocatalysis	[59]
CdS	Optics	[143]
Au nanoparticle	Photonics	[144]

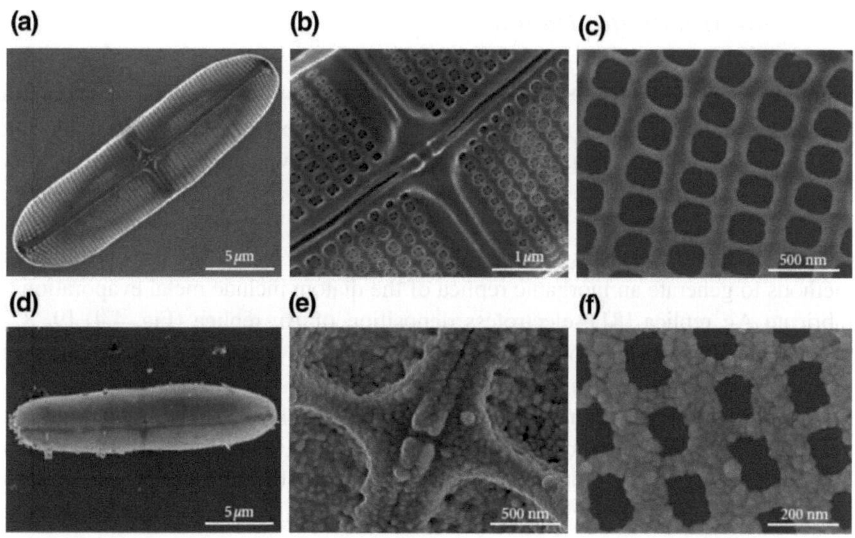

Fig. 4.3 Diatom-cadmium sulphide composite prepared by a coating strategies. SEM images of architectural features of *Pinnularia* sp. **a–c** and CdS coated *Pinnularia* sp. **d–f** The CdS coating on a diatom of spherical particles is clearly visible. (Reprinted with permission from [143])

In the context of sol-gel modified diatoms for environmental applications TiO_2-diatom composites have featured heavily in the literature. The potential to use diatoms as a support for TiO_2 nanoparticles is attractive for photocatalytic applications. It is assumed that the reactive centres of the nanoparticles are more accessible when dispersed on the platform and those nanoparticle properties, such as thermal stability, size, crystallinity and surface area are improved. TiO_2-diatom composites have been prepared by sol-gel methods using tetrabutyl titanate [60] or titanium (IV) isopropoxide [59, 65] as precursors.

The photocatalytic activity of the composites is dependent on crystalline size, TiO_2 loading, and ratio of anatase:rutile TiO_2 [66–68]. Additional approaches to prepare titanium oxide-diatom composites include simple calcinations [69], layer-by-layer assembly [70] hydrolysis-deposition methods [71] and more recently, immobilisation of oxides onto the living diatom via a precipitation process followed by extraction of organic matter to yield frustules coated with oxides [72, 73]. An advance in the development of TiO_2-diatom composite materials for environmental applications involves the addition of graphene to the composite. The preparation of graphene-diatom composites has been described using a combined functionalization and reduction step using direct esterification [74], covalent coupling [75], electrostatic attachment [75] and as aerogels [76]. It is anticipated that the three-component-composite will have enhanced absorption capacities for heavy metals and photocatalytic activity.

4.3.2 Chemical Conversion and Replication Strategies to Modify the Diatom

Methods to modify the chemistry of the complete frustule were pioneered by Sandhage and colleagues reporting the conversion of the diatom frustules into inorganic replicas [77]. A gas/displacement procedure was employed to convert the diatom *Aulacoseira sp.* to a replica composed of MgO. Subsequent work emanating from this seminal research has refined the procedure [78] and resulted in the generation of diatom replicas composed of ZrO_2 [79] and TiO_2 [80]. Additional methods to generate an inorganic replica of the diatom include metal evaporation to fabricate Ag replica [81], electroless deposition of Au replica (Fig. 4.4) [9, 82], sonochemical techniques to generate ZnS replica [83]. Negative replicas have been prepared by vacuum evaporation techniques [84, 85].

The proposed applications of these replicas are detailed in Table 4.2. Attempts to enhance the success of the coating process led to the investigation of a procedure to amplify the number of groups on the surface of the diatom that are available for reaction.

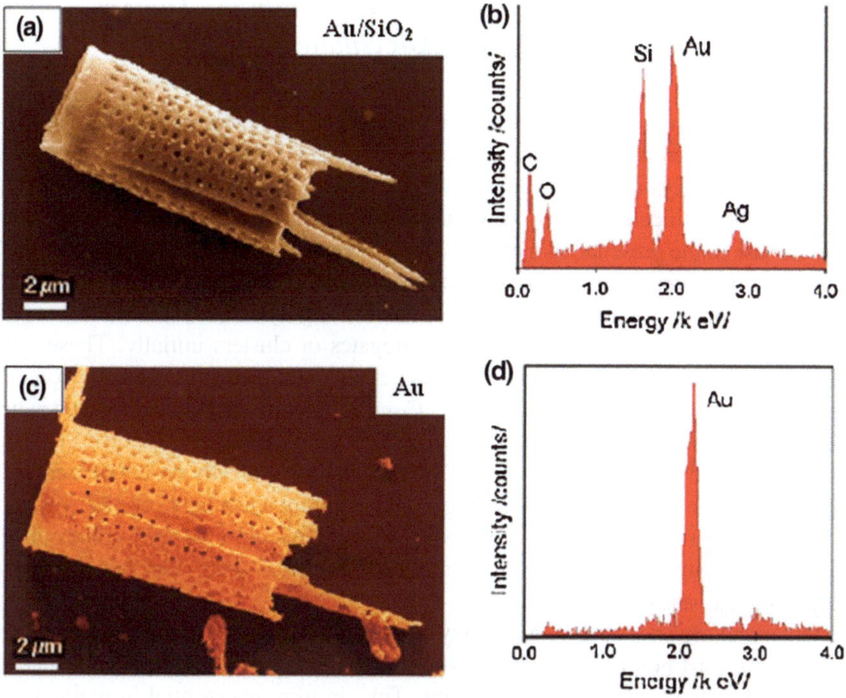

Fig. 4.4 Inorganic three-dimensional replicas of diatom frustules prepared by electroless deposition. SEM images of *Aulacoseira* sp. frustules after electroless deposition of gold, with EDAX spectrum showing gold and silica peaks **a** and **b**. Gold microstructure with frustule morphology generated after removal of silica by HF with EDAX spectrum showing only the gold peak **c** and **d**. (Adapted with permission from [9]. Copyright 2010 American Chemical Society)

Table 4.2 Proposed applications of inorganic diatom replicas

Diatom replica	Proposed application	References
Au	Catalysis	[9]
Ag	Raman substrate	[81]
BN	Reinforcement material	[84]
Cu, Ni	Optics, Energy Storage	[145]
MgO_2	Catalysis, Optics	[77, 78]
TiO_2	Catalysis, Gas sensing	[80]
ZnS	Photonics	[83]
ZrO_2	Reinforcement material	[79]

4.3.3 Sol-Gel Inspired Approaches to Modify the Diatom Frustule Using Organoalkoxysilanes

The sol-gel process is one of the most successful synthetic approaches for the preparation of complex inorganic materials [86]. The sol-gel process involves the sequential hydrolysis and polycondensation of alkoxy silicon derivatives, such as tetraethyl orthosilicate (TEOS) and tetramethyl orthosilicate (TMOS), in aqueous acid or base in the presence of a co-solvent to assist with miscibility of reagents [86]. The sol is the suspension of the first colloids formed during the early stages of the hydrolysis and polycondensation reactions. Further polycondensation between colloids promotes the formation of larger aggregates or clusters initially. These, ultimately, interpenetrate to form a continuous network, and the colloidal sol becomes a gel upon a sudden increase of viscosity. Condensation of organoalkoxysilanes has been reported in sol-gel science because of the wide diversity of properties that the resulting hybrid materials may offer [87, 88]. They are typically co-condensed with alkoxysilanes to tailor the cross-linking of the resulting silica network [89, 90].

Modification of the cleaned non-living diatom frustule has been achieved via silanization with organoalkoxysilanes through reaction with the free hydroxyl groups on the surface [4, 6, 7, 59, 91–95].

Organoalkoxysilanes have an alkoxy group and an organic functional group and provide a robust bond to the silica surface of the diatom through covalent linkage between the alkoxy and hydroxyl groups. This, in turn, is essential in maintaining adhesion to the surface for the attachment of larger molecules. Altering the chemistry of the diatom via silanization can be viewed as a means to prime the surface for further reactions and applications. The applications of organoalkoxysilanes functionalized diatoms are presented in Table 4.3.

Table 4.3 Functionalization of the diatom frustule via silanization and proposed applications

Silanization reagent	Proposed application	References
3-aminopropyltriethoxysilane	Gas detector	[8]
	Molecular sensor	[3]
	Drug delivery	[146]
	Raman substrates	[147]
3-aminopropyltrimethoxysilane	Enzyme support	[91]
	Antibody arrays	[6]
	Adsorbent	[7]
	Biomolecular sensor	[144]
	Photocatalysis	[59]
3-mercaptopropyltrimethoxysilane	Adsorbent	[10]
	Optics	[143]
N-(-3-(trimethoxysilyl)propyl) ethylene diamine	Drug delivery	[146]
	Adsorbent	[10]
Vinyltriethoxysilane	Optics	[143]

There are disadvantages associated with silanization of the surface including a decrease in the pore size and a non-homogeneous distribution of the functional groups on the surface. This is in part due to attachment of silanes at the mouth of the pore restricting access to internal pore sites. These problems can be overcome by incorporating the organic functionality into the diatom frustule during valve synthesis. In addition, introduction of the organic functionality during synthesis may create a more durable linkage between the organic moiety and the siloxane network via non-hydrolyzable Si–C linkages. The interaction of alkoxysilanes with the proteins involved in the biomineralization processes within the diatom has been explored in vitro [96–98]. TMOS has been shown to form SiO_2 in the presence of silaffins in vitro following a typical sol-gel process, whereby the alkoxide undergoes hydrolysis and the resulting sol gradually evolves towards the formation of a continuous colloidal network. Hydrolyzed TMOS and sodium silicate, the natural source of silicic acid for diatoms, have similar chemistry (Scheme 4.1), suggesting that alkoxysilanes and possibly organoalkoxysilanes can serve as precursors capable of forming SiO_2 in the presence of silaffins within the living diatom.

FITC-labelled silanes have been incubated in cultures to monitor valve formation [99]. Mixtures of TMOS and 3-mercaptopropyltrimethoxysilane or TEOS and (chloromethyl)triethoxysilane have been used as a source of silica for diatom growth, the resulting diatoms are functionalized with mercapto- and chloromethyl- moieties, respectively [100, 101]. Further exploration of the use of organoalkoxysilane/alkoxysilane mixtures to modify the chemistry of the living diatom is warranted (Fig. 4.5).

An exciting development in this field may involve translating additional lessons from sol-gel chemistry to modify the living diatom. Soft templating methods can be

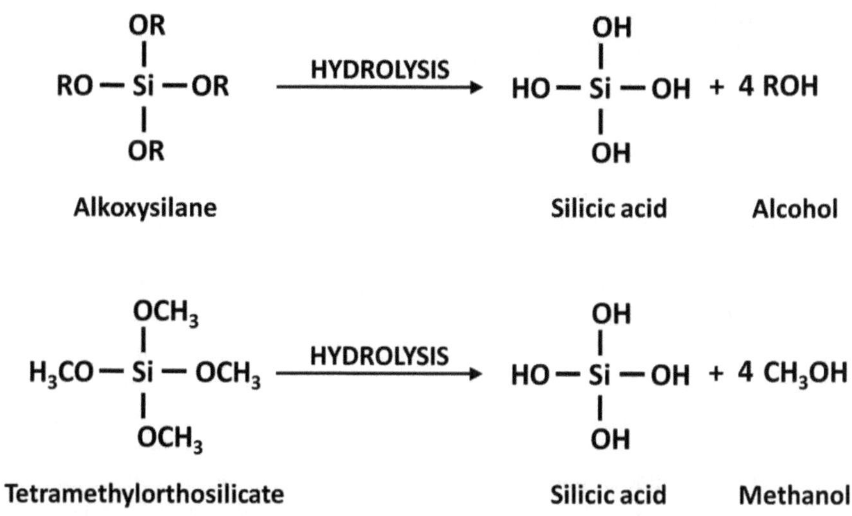

Scheme 4.1 The hydrolysis of alkoxysilanes generates free silicic acid and alcohol The hydrolysis of tetramethylorthosilicate yields silicic acid and methanol

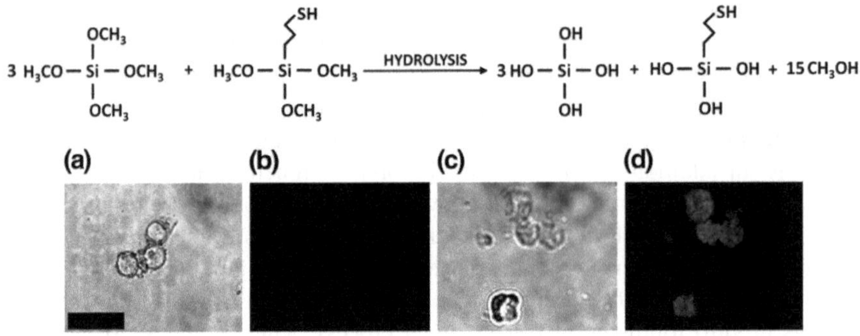

Fig. 4.5 Modification of living diatoms frustule via substitution of silicic acid with alkoxysilanes. Incubation of diatoms with silica precursors tetramethoxysilane (TMOS) and 3-mercaptopropyltrimethoxysilane (MPTMS) enables functionalization of the living diatom with thiol moieties. The incorporation of the thiol moiety was confirmed using a thiol-specific fluorescent probe (7-fluorobenzo-2-oxa-1,3-diazole-4-sulphonic acid ammonium). Brightfield and fluorescent images of the diatom *T. weissflogii* grown in the presence of **a** and **b** silicic acid, or **c** and **d** TMOS and MPTMS. Scale bar 20 μm. (Reprinted with permission from Lang et al. [100])

used to modulate the porous texture of gels prepared by sol-gel approaches. This involves the addition of structure directing agents to the aqueous solution such as surfactants, polymers, or block copolymers, [102–104]. Co-incubation of cultures with silanes and directing agents may present a novel route to tailor both the chemistry and architecture of the frustule in one step.

4.4 The Role of Diatoms as Bioindicators and Bioremediation Agents

The role of diatoms in environmental monitoring is well established with observations of their behaviour becoming a routine measure to evaluate water quality in freshwater and marine environments. The presence of certain diatom species and communities is indicative of the quality of water [105, 106]. Diatoms are sensitive to priority compounds, such as endocrine disrupting chemicals [107–109], polycyclic aromatic hydrocarbons [110, 111], pesticides [112, 113] and heavy metals The sensitivity of diatoms to toxins varies depending on the species and the compound investigated [114–119]. Furthermore, the parameter measured to identify compounds as hazardous may vary depending on the diatom species investigated. It has been reported that exposure to herbicides resulted in morphological abnormalities to the diatom frustule, alterations to the nucleus [113] and alterations to motility [112], without inhibition of growth or alterations to cell physiology. Alterations to gene expression or metabolic activity [120] may be a more useful indicator of contaminates in future as changes to gene expression following exposure to polycyclic aromatic hydrocarbons have been reported and this area is as yet under-explored [111, 121, 122]. In addition, other factors such as salinity [123], presence of bacteria [124], and the exudation of extracellular polymeric substances [125–127] may affect the reactivity of diatoms to compounds of interest. The use of artificially cultured diatoms as bioindicators of water quality collected from freshwater, marine or wastewater sources may benefit from the silane modification approaches described previously. The incorporation of fluorescently labelled silane moieties into the diatom frustule to act as 'switches', with the presence of a toxin of interest activating fluorescence, will provide biosensors to monitor water quality. The composition of such 'switches' may include thiol modified diatoms that emit a signal following adsorption of heavy metals. Advances in this area will rely on the ability to synthesise stable 'switches' and whether the diatom adsorbs or internalises compounds of interest.

The role of diatoms in bioremediation is currently an area of active research with studies investigating the ability of diatoms to remove nitrates, phosphates, pesticides, organic compounds and heavy metals from contaminated water [124, 128–132]. This field of research is in its infancy and many factors need to be addressed before diatoms as bioremediation agents becomes a widespread practice. For example, the ability of diatoms to degrade pesticides is dependent on whether the sample contains multiple pesticides, the diatom species present and diatom-bacteria interactions [132]. The use of artificially prepared diatom 'mats' will assist in overcoming certain issues. The proposed application as an adsorbent material features in all surface functionalized diatoms in Table 4.3. Thus, further exploration of in vivo modifications with the silanes 3-aminopropyltriethoxysilane, 3-aminopropyltrimethoxysilane, *N*-(-3-(trimethoxysilyl)propyl) ethylene diamine and vinyltriethoxysilane could reveal diatoms with enhanced bioremediation properties compared to the native diatom (Scheme 4.2).

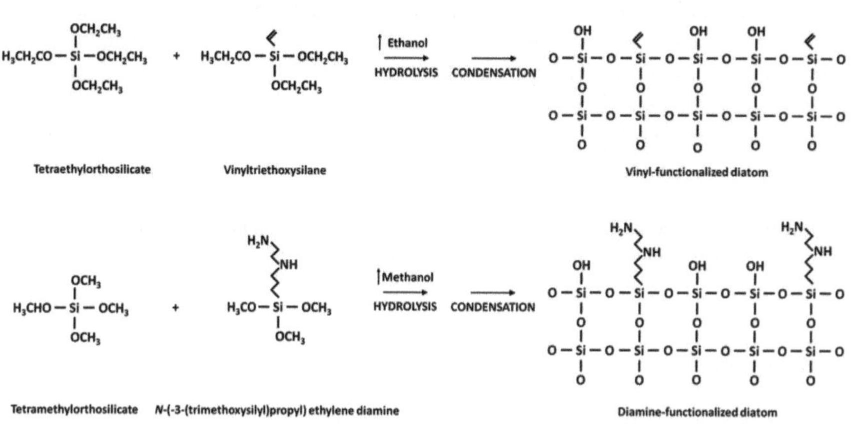

Scheme 4.2 Proposed functionalization of diatom frustule via silanization. Incorporation of vinyl-moieties or diamine moieties into the frustule of diatom may be achieved through a hydrolysis-condensation reaction within the diatom during valve synthesis

Recent studies describing the biosynthesis of gold nanoparticles by diatoms following exposure to gold solution tetrachloroaureate [133] or a solution of potassium gold (III) chloride provided interesting observations regarding the detoxification/decontamination capability of diatoms [134]. The concept of 'diatom factories' is gaining momentum, with potential biotechnological applications of diatoms as bioreactors reported following genetic manipulation [135–137]. Genetic manipulation of the diatom is in its early years, with seminal work reported by Kröger and colleagues [138–140]. Advances in this area may assist in designing diatoms with greater capacities for nanoparticle uptake and concomitant degradation/detoxification abilities.

4.5 Limitations and Perspective

The exploration of diatoms as a biotemplate to generate nanostructure materials for use in medicine, telecommunications, transportation, manufacturing and environmental applications has progressed rapidly since the pioneering work of Sandhage and colleagues [77]. Much of the work reported in this chapter highlights the influence and importance of diatom species on the properties of coated diatoms or diatom replicas. *Aulacoseira* sp. has been the predominant diatom species explored for environmental applications including catalysis, [75], filtration and decontamination [58], and photocatalysis [141]. These studies provide evidence for proof of concept; however, the widespread translation of these applications may require further optimisation with different diatom species. Indeed, continued progress in this field will require a multi-disciplined approach whereby observations from in vitro experiments with silica precursors, identification of critical diatom-microbial interactions and genetic manipulations are viewed together to design new systems. Finally, productivity and cost have remained a limitation to the commercial translation of diatom applications. Huntley and colleagues examined the cost associated with large-scale production and present data that are convincing evidence of the viability of such endeavours [142] giving credibility to the continued research for the multiple roles of diatoms for environmental applications.

Acknowledgments of Funding This work was supported by funding from the Environmental Protection Agency, Ireland (Grant EPA 2014-HW-DS-3).

References

1. Mann, S.: Biomineralization and biomimetic materials chemistry. J. Mater. Chem. **5**, 935–946 (1995)
2. Sotiropoulou, S., Sierra-Sastre, Y., Mark, S.S., Batt, C.A.: Biotemplated nanostructured materials. Chem. Mater. **20**, 821–834 (2008)

3. De Stefano, L., Lamberti, A., Rotiroti, L., De Stefano, M.: Interfacing the nanostructured biosilica microshells of the marine diatom Coscinodiscus wailesii with biological matter. Acta Biomater. **4**, 126–130 (2008)
4. De Stefano, L., Rendina, I., De Stefano, M., Bismuto, A., Maddalena, P.: Marine diatoms as optical chemical sensors. Appl. Phys. Lett. **87**, 233902 (2005)
5. Jantschke, A., Herrmann, A.-K., Lesnyak, V., Eychmüller, A., Brunner, E.: Decoration of diatom biosilica with noble metal and semiconductor nanoparticles (<10 nm): assembly, characterization, and applications. Chem. Asian J. **7**, 85–90 (2012)
6. Townley, H.E., Parker, A.R., White-Cooper, H.: Exploitation of diatom frustules for nanotechnology: tethering active biomolecules. Adv. Funct. Mater. **18**, 369–374 (2008)
7. Wang, G., Fang, Y., Kim, P., Hayek, A., Weatherspoon, M.R., Perry, J.W., Sandhage, K.H., Marder, S.R., Jones, S.C.: Layer-by-layer dendritic growth of hyperbranched thin films for surface sol-gel syntheses of conformal, functional, nanocrystalline oxide coatings on complex 3D (bio)silica templates. Adv. Funct. Mater. **19**, 2768–2776 (2009)
8. Weatherspoon, M.R., Dickerson, M.B., Wang, G., Cai, Y., Shian, S., Jones, S.C., Marder, S.R., Sandhage, K.H.: Thin, conformal, and continuous SnO_2 coatings on three-dimensional biosilica templates through hydroxy-group amplification and layer-by-layer alkoxide deposition. Angew. Chem. Int. Ed. Engl. **46**, 5724–5727 (2007)
9. Yu, Y., Addai-Mensah, J., Losic, D.: Synthesis of self-supporting gold microstructures with three-dimensional morphologies by direct replication of diatom templates. Langmuir **26**, 14068–14072 (2010)
10. Yu, Y., Addai-Mensah, J., Losic, D.: Chemical functionalization of diatom silica microparticles for adsorption of gold (III) ions. J. Nanosci. Nanotechnol. **11**, 10349–10356 (2011)
11. Brunner, E., Groger, C., Lutz, K., Richthammer, P., Spinde, K., Sumper, M.: Analytical studies of silica biomineralization: towards an understanding of silica processing by diatoms. Appl. Microbiol. Biotechnol. **84**, 607–616 (2009)
12. Kroger, N., Poulsen, N.: Diatoms-from cell wall biogenesis to nanotechnology. Annu. Rev. Genet. **42**, 83–107 (2008)
13. Round, F.E., Crawford, R.M., Mann, D.G.: The diatoms: biology and morphology of the genera. Cambridge University Press, New York (1990)
14. Hecky, R.E., Mopper, K., Kilham, P., Degens, E.T.: The amino acid and sugar composition of diatom cell-walls. Mar. Biol. **19**, 323–331 (1973)
15. Davidson, A.T., Bramich, D., Marchant, H.J., McMinn, A.: Effects of UV-B irradiation on growth and survival of Antarctic marine diatoms. Mar. Biol. **119**, 507–515 (1994)
16. Milligan, A.J., Varela, D.E., Brzezinski, M.A., Morel, F.O.M.M.: Dynamics of silicon metabolism and silicon isotopic discrimination in a marine diatom as a function of pCO(2). Limnol. Oceanogr. **49**, 322–329 (2004)
17. Morel, F.M.M., Cox, E.H., Kraepiel, A.M.L., Lane, T.W., Milligan, A.J., Schaperdoth, I., Reinfelder, J.R., Tortell, P.D.: Acquisition of inorganic carbon by the marine diatom *Thalassiosira weissflogii*. Funct. Plant Biol. **29**, 301–308 (2002)
18. Fuhrmann, T., Landwehr, S., El Rharbi-Kucki, M., Sumper, M.: Diatoms as living photonic crystals. Appl. Phy. B **78**, 257–260 (2004)
19. Hildebrand, M., Volcani, B.E., Gassmann, W., Schroeder, J.I.: A gene family of silicon transporters. Nature **385**, 688–689 (1997)
20. Thamatrakoln, K., Hildebrand, M.: Analysis of *Thalassiosira pseudonana* silicon transporters indicates distinct regulatory levels and transport activity through the cell cycle. Eukaryot. Cell **6**, 271–279 (2007)
21. Conway, H.L., Harrison, P.J.: Marine diatoms grown in chemostats under silicate or ammonium limitation. IV. Transient response of *Chaetoceros debilis*, *Skeletonema costatum*, and *Thalassiosira gravida* to a single addition of the limiting nutrient. Mar. Biol. **43**, 33–43 (1977)

22. Conway, H.L., Harrison, P.J., Davis, C.O.: Marine diatoms grown in chemostats under silicate or ammonium limitation. II. Transient response of *Skeletonema costatum* to a single addition of the limiting nutrient. Mar. Biol. **35**, 187–199 (1976)
23. Drum, R.W., Pankratz, H.S.: Post mitotic fine structure of *Gomphonema parvulum*. J. Ultrastruct. Res. **10**, 217–223 (1964)
24. Kroger, N., Deutzmann, R., Sumper, M.: Polycationic peptides from diatom biosilica that direct silica nanosphere formation. Science **286**, 1129–1132 (1999)
25. Kroger, N., Lorenz, S., Brunner, E., Sumper, M.: Self-assembly of highly phosphorylated silaffins and their function in biosilica morphogenesis. Science **298**, 584–586 (2002)
26. Kroger, N., Deutzmann, R., Bergsdorf, C., Sumper, M.: Species-specific polyamines from diatoms control silica morphology. Proc Natl Acad Sci USA **97**, 14133–14138 (2000)
27. Chiappino, M., Volcani, B.E.: Studies on the biochemistry and fine structure of silicia shell formation in diatoms VII. Sequential cell wall development in the pennate *Navicula pelliculosa*. Protoplasma **93**, 205–221 (1977)
28. Gordon, R., Drum, R.W.: The chemical basis of diatom morphogenesis. In: Richard, G. (ed.) International Review of Cytology. Academic Press (1994)
29. Hazelaar, S., Van Der Strate, H.J., Gieskes, W.W.C., Vrieling, E.G.: Monitoring rapid valve formation in the pennate diatom *Navicula salinarum* (*Bacillariophyceae*). J. Phycol. **41**, 354–358 (2005)
30. Heredia, A., van der Strate, H.J., Delgadillo, I., Basiuk, V.A., Vrieling, E.G.: Analysis of organo–silica interactions during valve formation in synchronously growing cells of the diatom *Navicula pelliculosa*. ChemBioChem **9**, 573–584 (2008)
31. Rogerson, A., Defreitas, A.S.W., Mcinnes, A.G.: Observations on wall morphogenesis in *Coscinodiscus asteromphalus* (Bacillariophyceae). Trans. Am. Microsc. Soc. **105**, 59–67 (1986)
32. Tesson, B., Hildebrand, M.: Dynamics of silica cell wall morphogenesis in the diatom *Cyclotella cryptica*: substructure formation and the role of microfilaments. J. Struct. Biol. **169**, 62–74 (2010)
33. Tesson, B., Hildebrand, M.: Extensive and intimate association of the cytoskeleton with forming silica in diatoms: control over patterning on the meso- and micro-scale. PLoS ONE **5**, e14300 (2010)
34. Brzezinski, M.A.: Cell-cycle effects on the kinetics of silicic-acid uptake and resource competition among diatoms. J. Plankton Res. **14**, 1511–1539 (1992)
35. Brzezinski, M.A., Olson, R.J., Chisholm, S.W.: Silicon availability and cell-cycle progression in marine diatoms. Mar. Ecol. Prog. Ser. **67**, 83–96 (1990)
36. Claquin, P., Martin-Jezequel, V.: Regulation of the Si and C uptake and of the soluble free-silicon pool in a synchronised culture of *Cylindrotheca fusiformis* (Bacillariophyceae): effects on the Si/C ratio. Mar. Biol. **146**, 877–886 (2005)
37. De La Rocha, C.L., Passow, U.: Recovery of *Thalassiosira weissflogii* from nitrogen and silicon starvation. Limnol. Oceanogr. **49**, 245–255 (2004)
38. De La Rocha, C.L., Terbruggen, A., Volker, C., Hohn, S.: Response to and recovery from nitrogen and silicon starvation in *Thalassiosira weissflogii*: growth rates, nutrient uptake and C, Si and N content per cell. Mar. Ecol. Prog. Ser. **412**, 57–68 (2010)
39. Finkel, Z.V., Matheson, K.A., Regan, K.S., Irwin, A.J.: Genotypic and phenotypic variation in diatom silicification under paleo-oceanographic conditions. Geobiology **8**, 433–445 (2010)
40. Garcia, N., Lopez-Elias, J.A., Miranda, A., Martinez-Porchas, M., Huerta, N., Garcia, A.: Effect of salinity on growth and chemical composition of the diatom *Thalassiosira weissflogii* at three culture phases. Lat. Am. J. Aquat. Res. **40**, 435–440 (2012)
41. Olson, R.J., Watras, C., Chisholm, S.W.: Patterns of individual cell-growth in marine centric diatoms. J. Gen. Microbiol. **132**, 1197–1204 (1986)
42. Su, Y., Lundholm, N., Friis, S.r.M, Ellegaard, M.: Implications for photonic applications of diatom growth and frustule nanostructure changes in response to different light wavelengths. Nano Res. 1–10

43. Vaulot, D., Olson, R.J., Merkel, S., Chisholm, S.W.: Cell-cycle response to nutrient starvation in 2 phytoplankton species, *Thalassiosira-weissflogii* and *Hymenomonas-carterae*. Mar. Biol. **95**, 625–630 (1987)
44. Vrieling, E.G., Poort, L., Beelen, T.P.M., Gieskes, W.W.C.: Growth and silica content of the diatoms *Thalassiosira weissflogii* and *Navicula salinarum* at different salinities and enrichments with aluminium. Eur. J. Phycol. **34**, 307–316 (1999)
45. Vrieling, E.G., Sun, Q.Y., Tian, M., Kooyman, P.J., Gieskes, W.W.C., van Santen, R.A., Sommerdijk, N.A.J.M.: Salinity-dependent diatom biosilicification implies an important role of external ionic strength. Proc Natl Acad Sci U S A **104**, 10441–10446 (2007)
46. Crawford, S.A., Higgins, M.J., Mulvaney, P., Wetherbee, R.: Nanostructure of the diatom frustule as revealed by atomic force and scanning electron microscopy. J. Phycol. **37**, 543–554 (2001)
47. Hildebrand, M., Doktycz, M.J., Allison, D.P.: Application of AFM in understanding biomineral formation in diatoms. Pflugers Arch. **456**, 127–137 (2008)
48. Hildebrand, M., Holton, G., Joy, D.C., Doktycz, M.J., Allison, D.P.: Diverse and conserved nano- and mesoscale structures of diatom silica revealed by atomic force microscopy. J. Microsc. **235**, 172–187 (2009)
49. Hildebrand, M., Kim, S., Shi, D., Scott, K., Subramaniam, S.: 3D imaging of diatoms with ion-abrasion scanning electron microscopy. J. Struct. Biol. **166**, 316–328 (2009)
50. Losic, D., Rosengarten, G., Mitchell, J.G., Voelcker, N.H.: Pore architecture of diatom frustules: potential nanostructured membranes for molecular and particle separations. J. Nanosci. Nanotechnol. **6**, 982–989 (2006)
51. Pletikapić, G., Berquand, A., Radić, T.M., Svetličić, V.: Quantitative nanomechanical mapping of marine diatom in seawater using peak force tapping atomic force microscopy. J. Phycol. **48**, 174–185 (2012)
52. Volcani, B.E.: Cell wall formation in diatoms: morphogenesis and biochemistry. In: Simpson, T., Volcani, B. (eds.) Silicon and siliceous structures in biological systems. Springer, New York (1981)
53. Cai, Y., Sandhage, K.H.: Zn_2SiO_4-coated microparticles with biologically-controlled 3D shapes. Phys. Status Solidi (A) Appl. Mater. Sci. **202**, R105–R107 (2005)
54. Weatherspoon, M.R., Allan, S.M., Hunt, E., Cai, Y., Sandhage, K.H.: Sol-gel synthesis on self-replicating single-cell scaffolds: applying complex chemistries to nature's 3-D nanostructured templates. Chem. Commun., 651–653 (2005)
55. Weatherspoon, M.R., Haluska, M.S., Cai, Y., King, J.S., Summers, C.J., Snyder, R.L., Sandhage, K.H.: Phosphor microparticles of controlled three-dimensional shape from phytoplankton. J. Electrochem. Soc. **153**, H34–H37 (2006)
56. Losic, D., Triani, G., Evans, P.J., Atanacio, A., Mitchell, J.G., Voelcker, N.H.: Controlled pore structure modification of diatoms by atomic layer deposition of TiO_2. J. Mater. Chem. **16**, 4029–4034 (2006)
57. Mao, L., Liu, J., Zhu, S., Zhang, D., Chen, Z., Chen, C.: Sonochemical fabrication of mesoporous TiO_2 inside diatom frustules for photocatalyst. Ultrason. Sonochem. **21**, 527–534 (2014)
58. Toster, J., Kusumawardani, I., Eroglu, E., Iyer, K.S., Rosei, F., Raston, C.L.: Superparamagnetic imposed diatom frustules for the effective removal of phosphates. Green Chem. **16**, 82–85 (2014)
59. Van Eynde, E., Tytgat, T., Smits, M., Verbruggen, S.W., Hauchecorne, B., Lenaerts, S.: Biotemplated diatom silica-titania materials for air purification. Photochem. Photobiol. Sci. **12**, 690–695 (2013)
60. Wang, B., de Godoi, F.C., Sun, Z., Zeng, Q., Zheng, S., Frost, R.L.: Synthesis, characterization and activity of an immobilized photocatalyst: natural porous diatomite supported titania nanoparticles. J. Coll Interface Sci. **438**, 204–211 (2015)
61. Shi, J.Y., Yao, Q.Z., Li, X.M., Zhou, G.T, Fu, S.Q.: Formation of asymmetrical structured silica controlled by a phase separation process and implication for biosilicification. PLoS ONE **8** (2013)

62. Gutu, T., Gale, D.K., Jeffryes, C., Wang, W., Chang, C.H., Rorrer, G.L., Jiao, J.: Electron microscopy and optical characterization of cadmium sulphide nanocrystals deposited on the patterned surface of diatombiosilica. J. Nanomater. **9** (2009). http://dx.doi.org/10.1155/2009/860536
63. Lee, D.H., Gutu, T., Jeffryes, C., Rorrer, G.L., Jiao, J., Chang, C.H.: Nanofabrication of green luminescent Zn_2SiO_4: Mn using biogenic silica. Electrochem. Solid State Lett. **10**, K13–K16 (2007)
64. Liu, Z., Fan, T., Zhou, H., Zhang, D., Gong, X., Guo, Q., Ogawa, H.: Synthesis of $ZnFe_2O_4$/SiO_2 composites derived from a diatomite template. Bioinspiration Biomim **2**, 30 (2007)
65. Toster, J., Harnagea, C., Swaminathan Iyer, K., Rosei, F., Raston, C.L.: Controlling anatase coating of diatom frustules by varying the binding layer. CrystEngComm **14**, 3446–3450 (2012)
66. Etacheri, V., Di Valentin, C., Schneider, J., Bahnemann, D., Pillai, S.C.: Visible-light activation of TiO_2 photocatalysts: advances in theory and experiments. J. Photochem. Photobiol. C **25**, 1–29 (2015)
67. Nolan, N.T., Seery, M.K., Pillai, S.C.: Spectroscopic investigation of the Anatase-to-Rutile transformation of Sol-Gel-synthesized TiO_2 photocatalysts. J. Phys. Chem C. **113**, 16151–16157 (2009)
68. Podporska-Carroll, J., Panaitescu, E., Quilty, B., Wang, L., Menon, L., Pillai, S.C.: Antimicrobial properties of highly efficient photocatalytic TiO_2 nanotubes. Appl. Catal. B: Environ. **176–177**, 70–75 (2015)
69. Hsien, K.J., Tsai, W.T., Su, T.Y.: Preparation of diatomite-TiO_2 composite for photodegradation of bisphenol-A in water. J. Sol-Gel. Sci. Technol. **51**, 63–69 (2009)
70. Jia, Y.X., Han, W., Xiong, G.X., Yang, W.S.: Layer-by-layer assembly of TiO_2 colloids onto diatomite to build hierarchical porous materials. J. Colloid Interface Sci. **323**, 326–331 (2008)
71. Sun, Z.M., Hu, Z.B., Yan, Y., Zheng, S.L.: Effect of preparation conditions on the characteristics and photocatalytic activity of TiO_2/purified diatomite composite photocatalysts. Appl. Surf. Sci. **314**, 251–259 (2014)
72. Thakkar, M., Randhawa, V., Mitra, S., Wei, L.: Synthesis of diatom-FeOx composite for removing trace arsenic to meet drinking water standards. J. Coll. Interface Sci. **457**, 169–173 (2015)
73. Thakkar, M., Wu, Z., Wei, L., Mitra, S.: Water defluoridation using a nanostructured diatom-ZrO_2 composite synthesized from algal biomass. J. Coll. Interface Sci. **450**, 239–245 (2015)
74. Dalagan, J.Q., Enriquez, E.P., Li, L.J.: Simultaneous functionalization and reduction of graphene oxide with diatom silica. J. Mater. Sci. **48**, 3415–3421 (2013)
75. Kumeria, T., Bariana, M., Altalhi, T., Kurkuri, M., Gibson, C.T., Yang, W., Losic, D.: Graphene oxide decorated diatom silica particles as new nano-hybrids: towards smart natural drug microcarriers. J. Mater. Chem. B **1**, 6302–6311 (2013)
76. Kabiri, S., Tran, D.N.H., Azari, S., Losic, D.: Graphene-diatom silica aerogels for efficient removal of mercury ions from water. ACS Appl. Mater. Interfaces **7**, 11815–11823 (2015)
77. Sandhage, K.H., Dickerson, M.B., Huseman, P.M., Caranna, M.A., Clifton, J.D., Bull, T.A., Heibel, T.J., Overton, W.R., Schoenwaelder, M.E.A.: Novel, bioclastic route to self-assembled, 3D, chemically tailored meso/nanostructures: shape-preserving reactive conversion of Biosilica (Diatom) microshells. Adv. Mater. **14**, 429–433 (2002)
78. Cai, Y., Allan, S.M., Sandhage, K.H.: Three-dimensional magnesia-based nanocrystal assemblies via low-temperature magnesiothermic reaction of diatom microshells. J. Am. Ceram. Soc. **88**, 2005–2010 (2005)
79. Shian, S., Cai, Y., Weatherspoon, M.R., Allan, S.M., Sandhage, K.H.: Three-dimensional assemblies of zirconia nanocrystals via shape-preserving reactive conversion of diatom microshells. J. Am. Ceram. Soc. **89**, 694–698 (2006)

80. Unocic, R.R., Zalar, F.M., Sarosi, P.M., Cai, Y. & Sandhage, K. H. 2004. Anatase assemblies from algae: coupling biological self-assembly of 3-D nanoparticle structures with synthetic reaction chemistry. Chem. Commun. (Camb.), 796–797
81. Payne, E.K., Rosi, N.L., Xue, C., Mirkin, C.A.: Sacrificial biological templates for the formation of nanostructured metallic microshells. Angew. Chem-Int. Ed. **44**, 5064–5067 (2005)
82. Fang, Y.N., Chen, V.W., Cai, Y., Berrigan, J.D., Marder, S.R., Perry, J.W., Sandhage, K.H.: Biologically enabled syntheses of freestanding metallic structures possessing subwavelength pore arrays for extraordinary (surface plasmon-mediated) infrared transmission. Adv. Funct. Mater. **22**, 2550–2559 (2012)
83. Zhou, H., Fan, T., Li, X., Ding, J., Zhang, D., Li, X., Gao, Y.: Bio-Inspired bottom-up assembly of diatom-templated ordered porous metal chalcogenide meso/nanostructures. Eur. J. Inorg. Chem. **2009**, 211–215 (2009)
84. Kusari, U., Bao, Z., Cai, Y., Ahmad, G., Sandhage, K.H., Sneddon, L.G.: Formation of nanostructured, nanocrystalline boron nitride microparticles with diatom-derived 3-D shapes. Chem. Commun. (Camb.), 1177–1179 (2007)
85. Losic, D., Mitchell, J.G., Voelcker, N.H. Complex gold nanostructures derived by templating from diatom frustules. Chem. Commun. 4905–4907 (2005)
86. Hench, L.L., West, J.K.: The sol-gel Process. Chem. Rev. **90**, 33–72 (1990)
87. Avnir, D., Coradin, T., Lev, O., Livage, J.: Recent bio-applications of sol-gel materials. J. Mater. Chem. **16**, 1013–1030 (2006)
88. Sanchez, C., Lebeau, B., Chaput, F., Boilot, J.P.: Optical properties of functional hybrid organic-inorganic nanocomposites. Adv. Mater. **15**, 1969–1994 (2003)
89. Hoffmann, F., Cornelius, M., Morell, J., Froba, M.: Silica-based mesoporous organic-inorganic hybrid materials. Angew. Chemi-Int. Ed. **45**, 3216–3251 (2006)
90. Wen, J.Y., Wilkes, G.L.: Organic/inorganic hybrid network materials by the sol-gel approach. Chem. Mater. **8**, 1667–1681 (1996)
91. Bayramoglu, G., Akbulut, A., Yakup Arica, M.: Immobilization of tyrosinase on modified diatom biosilica: enzymatic removal of phenolic compounds from aqueous solution. J. Hazard. Mater. **244–245**, 528–536 (2013)
92. Fowler, C.E., Buchber, C., Lebeau, B., Patarin, J., Delacote, C., Walcarius, A.: An aqueous route to organically functionalized silica diatom skeletons. Appl. Surf. Sci. **253**, 5485–5493 (2007)
93. Gale, D.K., Gutu, T., Jiao, J., Chang, C.H., Rorrer, G.L.: Photoluminescence detection of biomolecules by antibody-functionalized diatom biosilica. Adv. Funct. Mater. **19**, 926–933 (2009)
94. Losic, D., Yu, Y., Aw, M.S., Simovic, S., Thierry, B., Addai-Mensah, J.: Surface functionalisation of diatoms with dopamine modified iron-oxide nanoparticles: toward magnetically guided drug microcarriers with biologically derived morphologies. Chem. Commun. (Camb.) **46**, 6323–6325 (2010)
95. Rosi, N.L., Thaxton, C.S., Mirkin, C.A.: Control of nanoparticle assembly by using DNA-modified diatom templates. Angew Chem-Int. Ed. **43**, 5500–5503 (2004)
96. Marner, W.D., Shaikh, A.S., Muller, S.J., Keasling, J.D.: Morphology of artificial silica matrices formed via autosilification of a silaffin/protein polymer chimera. Biomacromolecules **9**, 1–5 (2008)
97. Wieneke, R., Bernecker, A., Riedel, R., Sumper, M., Steinem, C., Geyer, A.: Silica precipitation with synthetic silaffin peptides. Org. Biomol. Chem. **9**, 5482–5486 (2011)
98. Patwardhan, S.V., Mukherjee, N., Steinitz-Kannan, M., Clarson, S.J.: Bioinspired synthesis of new silica structures. Chem. Commun. (Camb), 1122–1123 (2003)
99. Descles, J., Vartanian, M., El Harrak, A., Quinet, M., Bremond, N., Sapriel, G., Bibette, J., Lopez, P.J.: New tools for labeling silica in living diatoms. New Phytol. **177**, 822–829 (2008)

100. Lang, Y., del Monte, F., Collins, L., Rodriguez, B.J., Thompson, K., Dockery, P., Finn, D.P., Pandit, A.: Functionalization of the living diatom *Thalassiosira weissflogii* with thiol moieties. Nat. Commun. **4** (2013)
101. Lang, Y., del Monte, F., Finn, D.P., Wang, W., Pandit, A.: Synthesis of polymer-silica hybrid microparticles with defined geometry using surface initiated atom transfer radical polymerization. Polym. Chem. **6**, 3014–3017 (2015)
102. Kresge, C.T., Leonowicz, M.E., Roth, W.J., Vartuli, J.C., Beck, J.S.: Ordered mesoporous molecular-sieves synthesized by a liquid-crystal template mechanism. Nature **359**, 710–712 (1992)
103. Tanaka, N., Kobayashi, H., Ishizuka, N., Minakuchi, H., Nakanishi, K., Hosoya, K., Ikegami, T.: Monolithic silica columns for high-efficiency chromatographic separations. J. Chromatogr. A **965**, 35–49 (2002)
104. Wan, Y., Zhao, D.Y.: On the controllable soft-templating approach to mesoporous silicates. Chem. Rev. **107**, 2821–2860 (2007)
105. Kireta, A.R., Reavie, E.D., Sgro, G.V., Angradi, T.R., Bolgrien, D.W., Hill, B.H., Jicha, T.M.: Planktonic and periphytic diatoms as indicators of stress on great rivers of the United States: testing water quality and disturbance models. Ecol. Ind. **13**, 222–231 (2012)
106. Oeding, S, Taffs, K.: Are diatoms a reliable and valuable bio-indicator to assess sub-tropical river ecosystem health? Hydrobiologia, 1–19
107. Li, R., Chen, G.-Z., Tam, N.F.Y., Luan, T.-G., Shin, P.K.S., Cheung, S.G., Liu, Y.: Toxicity of bisphenol A and its bioaccumulation and removal by a marine microalga *Stephanodiscus hantzschii*. Ecotoxicol. Environ. Saf. **72**, 321–328 (2009)
108. Liu, Y., Guan, Y.T., Gao, Q.T., Tam, N.F.Y., Zhu, W.P.: Cellular responses, biodegradation and bioaccumulation of endocrine disrupting chemicals in marine diatom *Navicula incerta*. Chemosphere **80**, 592–599 (2010)
109. Luo, L.J., Wang, P., Lin, L., Luan, T.G., Ke, L., Tam, N.F.Y.: Removal and transformation of high molecular weight polycyclic aromatic hydrocarbons in water by live and dead microalgae. Process Biochem. **49**, 1723–1732 (2014)
110. Bopp, S.K., Lettieri, T.: Gene regulation in the marine diatom *Thalassiosira pseudonana* upon exposure to polycyclic aromatic hydrocarbons (PAHs). Gene **396**, 293–302 (2007)
111. Carvalho, R.N., Bopp, S.K., Lettieri, T.: Transcriptomics responses in marine diatom *Thalassiosira pseudonana* exposed to the polycyclic aromatic hydrocarbon benzo[a]pyrene. PLoS ONE **6** (2011)
112. Coquillé, N., Jan, G., Moreira, A., Morin, S.: Use of diatom motility features as endpoints of metolachlor toxicity. Aquat. Toxicol. **158**, 202–210 (2015)
113. Debenest, T., Sivestre, J., Coste, M., Delmas, F., Pinelli, E.: Herbicide effects on freshwater benthic diatoms: induction of nucleus alterations and silica cell wall abnormalities. Aquat. Toxicol. **88**, 88–94 (2008)
114. Basharina, T.N., Danilovtseva, E.N., Zelinskiy, S.N., Klimenkov, I.V., Likhoshway, Y.V., Annenkov, V.V.: The effect of titanium, zirconium and tin on the growth of diatom *Synedra acus* and morphology of its silica valves. Silicon **4**, 239–249 (2012)
115. Bour, A., Mouchet, F., Verneuil, L., Evariste, L., Silvestre, J., Pinelli, E., Gauthier, L.: Toxicity of CeO_2 nanoparticles at different trophic levels—effects on diatoms, chironomids and amphibians. Chemosphere **120**, 230–236 (2015)
116. Choi, C.J., Berges, J.A., Young, E.B.: Rapid effects of diverse toxic water pollutants on chlorophyll a fluorescence: variable responses among freshwater microalgae. Water Res. **46**, 2615–2626 (2012)
117. Libralato, G., Gentile, E., Volpi Ghirardini, A.: Wastewater effects on *Phaeodactylum tricornutum* (Bohlin): setting up a classification system. Ecol. Ind. **60**, 31–37 (2016)
118. Othman, H.B., Leboulanger, C., Le Floc'h, E., Hadj Mabrouk, H., Sakka Hlaili, A.: Toxicity of benz(a)anthracene and fluoranthene to marine phytoplankton in culture: does cell size really matter? J. Hazard. Mater. **243**, 204–211 (2012)

119. Wong, S.W.Y., Leung, P.T.Y., Djurisic, A.B., Leung, K.M.Y.: Toxicities of nano zinc oxide to five marine organisms: influences of aggregate size and ion solubility. Anal. Bioanal. Chem. **396**, 609–618 (2010)
120. Durrieu, C., Guedri, H., Fremion, F., Volatier, L.: Unicellular algae used as biosensors for chemical detection in Mediterranean lagoon and coastal waters. Res. Microbiol. **162**, 910–914 (2011)
121. Carvalho, R.N., Burchardt, A.D., Sena, F., Mariani, G., Mueller, A., Bopp, S.K., Umlauf, G., Lettieri, T.: Gene biomarkers in diatom *Thalassiosira pseudonana* exposed to polycyclic aromatic hydrocarbons from contaminated marine surface sediments. Aquat. Toxicol. **101**, 244–253 (2011)
122. Carvalho, R.N., Lettieri, T.: Proteomic analysis of the marine diatom *Thalassiosira pseudonana* upon exposure to benzo(a)pyrene. BMC Genomics **12** (2011)
123. Yung, M.M.N., Wong, S.W.Y., Kwok, K.W.H., Liu, F.Z., Leung, Y.H., Chan, W.T., Li, X.Y., Djurišić, A.B., Leung, K.M.Y.: Salinity-dependent toxicities of zinc oxide nanoparticles to the marine diatom *Thalassiosira pseudonana*. Aquat. Toxicol. **165**, 31–40 (2015)
124. Li, Y., Gao, J., Meng, F., Chi, J.: Enhanced biodegradation of phthalate acid esters in marine sediments by benthic diatom *Cylindrotheca closterium*. Sci. Total Environ. **508**, 251–257 (2015)
125. Miao, A.J., Schwehr, K.A., Xu, C., Zhang, S.J., Luo, Z., Quigg, A., Santschi, P.H.: The algal toxicity of silver engineered nanoparticles and detoxification by exopolymeric substances. Environ. Pollut. **157**, 3034–3041 (2009)
126. Pletikapic, G., Zutic, V., Vrcek, I.V., Svetlicic, V.: Atomic force microscopy characterization of silver nanoparticles interactions with marine diatom cells and extracellular polymeric substance. J. Mol. Recognit. **25**, 309–317 (2012)
127. Zhang, S., Jiang, Y., Chen, C.S., Creeley, D., Schwehr, K.A., Quigg, A., Chin, W.C., Santschi, P.H.: Ameliorating effects of extracellular polymeric substances excreted by *Thalassiosira pseudonana* on algal toxicity of CdSe quantum dots. Aquat. Toxicol. **126**, 214–223 (2013)
128. Abdelaziz, A.E.M., Leite, G.B., Belhaj, M.A., Hallenbeck, P.C.: Screening microalgae native to Quebec for wastewater treatment and biodiesel production. Bioresour. Technol. **157**, 140–148 (2014)
129. Arini, A., Feurtet-Mazel, A., Maury-Brachet, R., Coste, M., Delmas, F.: Field translocation of diatom biofilms impacted by Cd and Zn to assess decontamination and community restructuring capacities. Ecol. Ind. **18**, 520–531 (2012)
130. Arini, A., Feurtet-Mazel, A., Maury-Brachet, R., Pokrovsky, O.S., Coste, M., Delmas, F.: Recovery potential of periphytic biofilms translocated in artificial streams after industrial contamination (Cd and Zn). Ecotoxicology **21**, 1403–1414 (2012)
131. Arini, A., Feurtet-Mazel, A., Morin, S., Maury-Brachet, R., Coste, M., Delmas, F.: Remediation of a watershed contaminated by heavy metals: a 2-year field biomonitoring of periphytic biofilms. Sci. Total Environ. **425**, 242–253 (2012)
132. Tien, C.-J., Lin, M.-C., Chiu, W.-H., Chen, C.S.: Biodegradation of carbamate pesticides by natural river biofilms in different seasons and their effects on biofilm community structure. Environ. Pollut. **179**, 95–104 (2013)
133. Schröfel, A., Kratošová, G., Bohunická, M., Dobročka, E., Vávra, I.: Biosynthesis of gold nanoparticles using diatoms-silica-gold and EPS-gold bionanocomposite formation. J. Nanopart. Res. **13**, 3207–3216 (2011)
134. Feurtet-Mazel, A., Mornet, S., Charron, L., Mesmer-Dudons, N., Maury-Brachet, R., Baudrimont, M.: Biosynthesis of gold nanoparticles by the living freshwater diatom *Eolimna minima*, a species developed in river biofilms. Environ. Sci. Pollut. Res. (2015)
135. Hempel, F., Bozarth, A.S., Lindenkamp, N., Klingl, A., Zauner, S., Linne, U., Steinbuchel, A., Maier, U.G.: Microalgae as bioreactors for bioplastic production. Microb. Cell Fact. **10**, 81 (2011)

136. Hempel, F., Lau, J., Klingl, A., Maier, U.G.: Algae as protein factories: expression of a human antibody and the respective antigen in the diatom *Phaeodactylum tricornutum*. PLoS ONE **6**, e28424 (2011)
137. Hempel, F., Maier, U.G.: An engineered diatom acting like a plasma cell secreting human IgG antibodies with high efficiency. Microb. Cell Fact. **11**, 126 (2012)
138. Poulsen, N., Berne, C., Spain, J., Kroger, N.: Silica immobilization of an enzyme through genetic engineering of the diatom *Thalassiosira pseudonana*. Angew. Chem. Int. Ed. Engl. **46**, 1843–1846 (2007)
139. Poulsen, N., Kroger, N.: A new molecular tool for transgenic diatoms: control of mRNA and protein biosynthesis by an inducible promoter-terminator cassette. FEBS J. **272**, 3413–3423 (2005)
140. Sheppard, V.C., Scheffel, A., Poulsen, N., Kroger, N.: Live diatom silica immobilization of multimeric and redox-active enzymes. Appl. Environ. Microbiol. **78**, 211–218 (2012)
141. Chandrasekaran, S., Sweetman, M.J., Kant, K., Skinner, W., Losic, D., Nann, T., Voelcker, N.H.: Silicon diatom frustules as nanostructured photoelectrodes. Chem. Commun. **50**, 10441–10444 (2014)
142. Huntley, M.E., Johnson, Z.I., Brown, S.L., Sills, D.L., Gerber, L., Archibald, I., Machesky, S.C., Granados, J., Beal, C., Greene, C.H.: Demonstrated large-scale production of marine microalgae for fuels and feed. Algal Res. **10**, 249–265 (2015)
143. Gutu, T., Gale, D.K., Wang, W., Chang, C.H., Rorrer, G.L., Jiao, J.: Electron microscopy and Raman spectroscopy characterization of nanoparticle coated diatom biosilica. Microsc. Microanal. **15**, 1278–1279 (2009)
144. Toster, J., Iyer, K.S., Burtovyy, R., Burgess, S.S.O., Luzinov, I.A., Raston, C.L.: Regiospecific assembly of gold nanoparticles around the pores of diatoms: toward three-dimensional nanoarrays. J. Am. Chem. Soc. **131**, 8356–8357 (2009)
145. Fang, Y.N., Berrigan, J.D., Cai, Y., Marder, S.R., Sandhage, K.H.: Syntheses of nanostructured Cu- and Ni-based micro-assemblies with selectable 3-D hierarchical biogenic morphologies. J. Mater. Chem. **22**, 1305–1312 (2012)
146. Aw, M.S., Simovic, S., Yu, Y., Addai-Mensah, J., Losic, D.: Porous silica microshells from diatoms as biocarrier for drug delivery applications. Powder Technol. **223**, 52–58 (2012)
147. Chen, J., Qin, G.W., Chen, Q., Yu, J.Y., Li, S., Cao, F., Yang, B., Ren, Y.P.: A synergistic combination of diatomaceous earth with Au nanoparticles as a periodically ordered, button-like substrate for SERS analysis of the chemical composition of eccrine sweat in latent fingerprints. J. Mater. Chem. C **3**, 4933–4944 (2015)

Chapter 5
Sol-Gel Sensors

Aine M. Whelan

5.1 Introduction

Recent advances in sol-gel processing have greatly impacted sensor development. Broadly speaking, sol-gel chemistry has been applied to sensor fabrication in two ways. First, the material synthesised by the sol-gel process can serve as the sensing element. This is particularly seen in gas sensors where metal oxides synthesised by sol-gel techniques are used for detection of gases via changes in their conductivity upon gas adsorption. The mechanism for this will be discussed in greater detail in the section on gas sensing in this chapter.

Alternatively, the sol-gel material can be used as the matrix for encapsulation of a sensing molecule. Sol-gel materials offer a number of advantages in this respect. As outlined in the review of Monton [1], silica-based sol-gel materials are transparent to UV and visible radiation. Thus, they are ideally suited to the fabrication of optical sensors, which rely on changes in absorption, reflection or fluorescence of the sensing molecule for detection of an analyte. They are chemically stable and mechanically robust materials. Properties of the material such as porosity and surface charge can be readily tuned by adjusting parameters of the synthetic process such as the precursor molecules used, the pH and the catalyst used. This allows the microstructure to be optimised for the relevant sensor application. Furthermore, in the case of biosensors, immobilisation of biomolecules such as enzymes, antibodies and aptamers within the sol-gel matrix allows retention of their biological activity and provides resistance to degradation [2]. Finally, these materials can be prepared in a diverse range of geometric structures including thin films, monoliths, powders and fibres, thus permitting a wide choice of sensor configurations.

A.M. Whelan (✉)
School of Chemical and Pharmaceutical Sciences,
Dublin Institute of Technology, Kevin Street, Dublin 2, Ireland
e-mail: 450391@dit.ie

S.C. Pillai and S. Hehir (eds.), *Sol-Gel Materials for Energy, Environment and Electronic Applications*, Advances in Sol-Gel Derived Materials and Technologies,
DOI 10.1007/978-3-319-50144-4_5

This chapter will focus primarily on developments in sol-gel-based sensors published in the past five years. The use of sol-gel-based sensors for the detection of gases, monitoring of humidity and pH, cation and anion detection will be reviewed. Furthermore, the use of sol-gel techniques for preparation of biosensors for monitoring of biologically relevant species will be discussed. Finally, the development of sol-gel-based sensors for forensic analysis will be outlined.

5.2 Gas Sensors

5.2.1 Oxygen

The majority of sol-gel-based sensors for oxygen are optical sensors. In these systems, the optically active transducer molecule is immobilised within a sol-gel matrix. The presence of oxygen results in a change in the optical properties of the transducer, be it a change in its absorption, fluorescence or reflectance. Typically, silica matrices are used since they are transparent. A variety of chemical reagents have been incorporated into sol-gel matrices for this purpose, including Ru (II) polypyridyl complexes [3] as well as metalloporphyrins [4]. In these sensors, the oxygen mediated quenching of the metal complex fluorescence can be monitored to determine oxygen concentration (Fig. 5.1).

The tailoring of sol-gel films for optical sensing of oxygen was discussed by McDonagh et al. [5]. In their work, the presence of oxygen was indicated by the quenching of the fluorescence of a ruthenium complex, $Ru(Ph_2phen)_3^{2+}$. The authors argued that the immobilisation of the ruthenium complex within a sol-gel matrix allowed the development of a variety of different sensor configurations, namely evanescent-wave-excited fluorescence of sol-gel-coated optical fibres and direct excitation of sol-gel-coated planar glass substrates. They described the effect of changes in the molar water/precursor ratio on the microstructure of the sol-gel

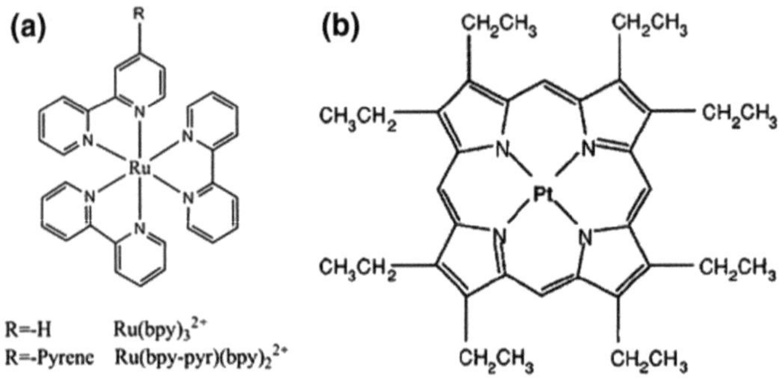

Fig. 5.1 **a** Ruthenium polypyridyl complexes used for oxygen sensing. **b** Platinum porphyrin complex used for oxygen sensing

film and the consequent effects for sensor performance. They also noted that tetraethoxysilane (TEOS) derived films gave poor results for dissolved oxygen sensing due to the hydrophilicity of the matrix. By addition of modified precursors such as methyltriethoxysilane (MTEOS) and ethyltriethoxysilane (ETEOS) it was possible to prepare films that were more hydrophobic in nature and showed enhanced fluorescence quenching.

Sol-gel chemistry has also been employed for the development of a sensor to measure oxygen concentration in exhaled breath [6]. Such sensors are useful for non-invasive monitoring of patient health. Fluorescent iridium complexes were used as the oxygen sensitive fluorophores in this system. These complexes were doped into hybrid organically modified silicas (ORMOSILs) composed of n-propyltrimethoxysilane (n-propyl-TriMOS) and perfluoroalkyl ORMOSIL 3,3,3-trifluoropropyltrimethoxysilane (TFP-TriMOS). An optical fibre was then dipcoated with the doped ORMOSIL. The luminescence of the iridium complex can be excited by the evanescent wave that can be produced on the optical fibre surface. This luminescence can be quenched by O_2, thus allowing detection of oxygen. Fluorinated ORMOSIL was selected as the support material for the sensor because this material has a high permeability to oxygen. Furthermore, the material is very hydrophobic and thus, insensitive to relative humidity. The performance of the sensor was tested at temperatures of 25, 35 and 45 °C and found to be stable over this temperature range. The sensor gave a linear response over 0–100% O_2 concentration range with a limit of detection of 0.009% O_2

The sol-gel process has also been used to develop sensors for monitoring oxygen concentrations within living cells. The Kopelman group prepared silica nanoparticles which contained an oxygen sensitive fluorescent indicator, Ru(II) tris (4,7-diphenyl-1,10-phenanthroline) chloride ($[Ru(dpp)_3]^{2+}$) and an oxygen-insensitive fluorescent dye, Oregon Green 488-dextran (Spiro[isobenzofuran-1 (3H),9′-[9H]xanthen]-3-one, 2′,7′-difluoro-3′,6′-dihydroxy) [7]. These nanoparticles were inserted into rat glioma cells and the intracellular oxygen concentration was monitored by a ratiometric method. The ratio of the fluorescence of the oxygen sensitive indicator to the fluorescence of the oxygen-insensitive dye was measured and used to determine oxygen concentration. The stability of the nanoparticles was monitored over a three-day period and showed that there was no detectable leaching of the dye molecules out of the sensor matrix during this time.

$[Ru(dpp)_3]^{2+}$ doped silica nanoparticles were also used by Korzeniowska et al. [8] for monitoring intracellular levels of oxygen. However, rather than using a ratiometric method, they used fluorescence-lifetime imaging microscopy (FLIM) to measure the luminescence lifetime of the ruthenium complex in the presence of oxygen. Since luminescence-lifetime measurements are insensitive to interferents such as dye concentration, variation in excitation source intensity and signal loss within the optical path, they are considered more reliable for intracellular sensing. This method was used to determine the dissolved oxygen in human epithelial cells at 7.8 ppm, which was found to be in agreement with previously published data [9] (Fig. 5.2).

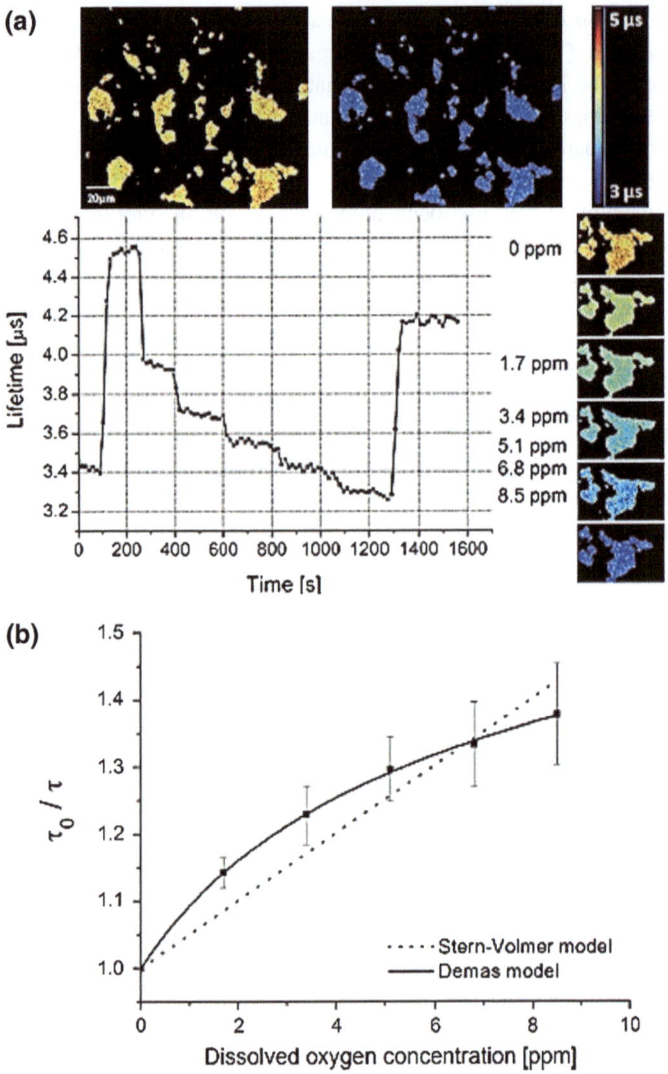

Fig. 5.2 Fluorescence lifetime imaging microscopy-based calibration of $[Ru(dpp_3)]^{2+}$–doped organically modified silica nanoparticles performed inside respiring human osteosarcoma epithelial cells **a** lifetime images and time course data and **b** corresponding calibration curve ($n = 3$). *Colour scale* corresponds to the lifetime of $[Ru(dpp)_3]^{2+}$ figure reproduced with permission from Ref. [8]. Copyright Royal Society of Chemistry 2015

5.2.2 *Carbon Dioxide*

Many food products are packaged under a protective atmosphere of carbon dioxide in order to decrease the growth of bacteria within the food. The monitoring of the package for leaking of the protective carbon dioxide atmosphere has traditionally been done by extraction of the atmosphere with a needle followed by infrared absorption spectrophotometry. This is a destructive process, leading to loss of product. An alternative non-destructive approach to this was developed by McCraith et al. [10]. They prepared sensor strips which could be incorporated into the packaging. These strips comprised an organically modified silica (ormosil) matrix which contained a fluorescent pH indicator HPTS (1-hydroxypyrene trisulfonate). Carbon dioxide levels could be detected via a dual luminophore referencing sensing scheme. In this mode of detection, the analyte-sensitive fluorescence intensity signal is converted into the phase domain by immobilisation with an inert long-lifetime reference luminophore. The luminophores must show significant overlap between their excitation and emission spectra. Therefore, in this instance, a ruthenium complex, ruthenium (II) tris (4,7-diphenyl 1,10-phenanthroline) (Ru $(dpp)_3)^{2+}$ and the HTPS dye were used as candidates for the DLR type CO_2 sensor because their excitation and emission wavelengths are well matched. A handheld scanner could be used for external interrogation of the sensor strip (Fig. 5.3).

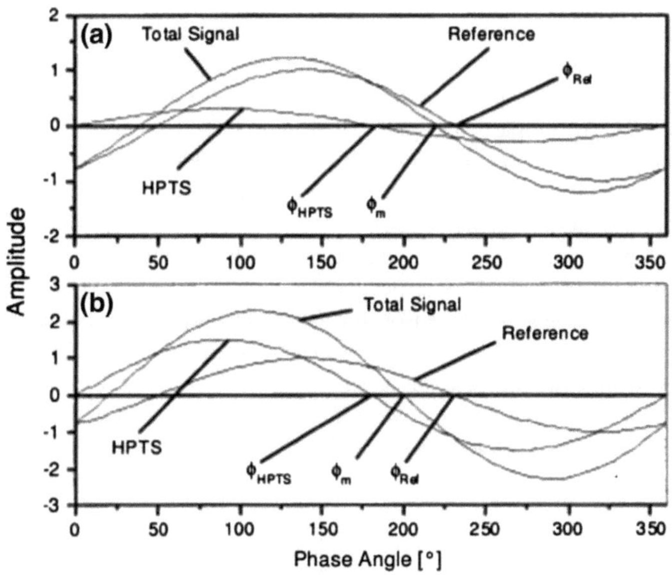

Fig. 5.3 Illustration of the principle of dual luminophore referencing system. Single sine wave signals are produced by the reference luminophore (reference) and the analyte-sensitive luminophore (HPTS). The superposition of these two signals yields the detected signal (total signal). The reference phase angle (ϕRef) and the HPTS phase angle ($\phi HPTS = 0$) remain constant. The total luminescence phase angle (ϕm) is a function of the amplitude of the HPTS signal in the presence (**a**) and absence (**b**) of carbon dioxide. Figure reproduced with permission from Ref. [10]. Copyright Royal Society of Chemistry 2002

A similar system for CO_2 sensing was devised by Xue et al. [11]. However, they also added titanium dioxide nanoparticles into the sol-gel sensor films to enhance the interaction of incident light with the film, via Mie scattering. This led to increased sensitivity of the system to CO_2, with detection in the 0.03–30% range for CO_2 with an LOD of 0.08% (80 ppm CO_2).

Metal oxide semiconductors are known to demonstrate changes in conductivity upon adsorption of molecules from the gas phase [12]. When the target gas interacts with the surface of the metal oxide film (typically via oxygen ions adsorbed on the surface), this leads to a change in the concentration of charge carriers in the material. This results in a change in the conductivity (or resistivity,) of the material. The nature of the charge carriers present depends on whether the material is an n-type semiconductor or a p-type semiconductor. In n-type semiconductors, the majority charge carriers are electrons. Therefore, interaction with a reducing gas results in an increase in conductivity. However, interaction of an oxidising gas with an n-type semiconductor serves to decrease number of charge carrying electrons in the sensing layer, thus yielding a decrease in conductivity.

In p-type semiconductors, positive holes are the majority charge carriers. These materials demonstrate an increase in conductivity in the presence of an oxidising gas (since the gas has increased the number of positive holes present). A resistance increase with a reducing gas is observed, where the negative charge introduced into the material reduces the positive (hole) charge carrier concentration.

Titania (TiO_2) nanoparticles synthesised by a sol-gel process were used in the fabrication of a sensor for carbon dioxide [13]. The nanoparticles were incorporated into thin film devices and the performance of these devices was compared to those of thin films which did not contain nanoparticles. The devices containing TiO_2 were found to be more sensitive to CO_2 than the other films and had response time and recovery time of 95 and 110 s respectively.

A different sol-gel-based sensor for CO_2 was reported by Fan et al. [14] $La_{1-x}Sr_xFeO_3$ nanocrystalline powders were prepared by a sol-gel method. These materials showed an increase in resistance with increasing CO_2 concentration. However, the response and recovery time were considerably slower than the titania-based devices reported above, at 11 and 5 min, respectively for 2000 ppm CO_2 in dry air.

5.2.3 Ammonia

Long-term exposure (greater than 8 h) or inhalation of ammonia concentrations greater than 5000 ppm can lead to serious health problems [15]. Therefore, the development of sensors for ammonia is of importance. Zinc oxide nanofibers were prepared by a sol-gel electrospinning technique and then used to detect gaseous ammonia [16]. The precursor fibres were spun directly onto a quartz substrate and

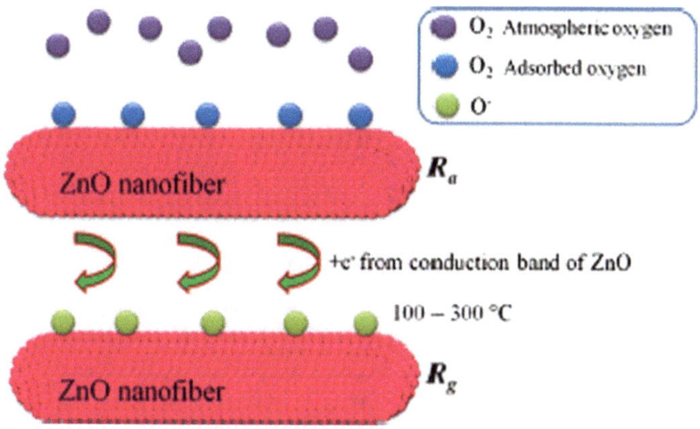

Fig. 5.4 Ammonia sensing mechanism of zinc oxide nanofibers. Figure reproduced with permission from Ref. [16]. Copyright Elsevier B.V. 2014

then calcined at different temperatures (773, 973, 1173 K) for 2 h. Then silver electrodes were attached to the nanofiber mat such that electrical resistance measurements could be carried out. Increasing the calcination temperature was found to increase the crystallinity and crystallite size within the fibres. ZnO fibres calcined at 1173 K showed the greatest sensitivity towards ammonia. This was attributed to the higher surface area and porosity of this material compared to the ZnO nanofibers calcined at lower temperatures. In all cases, however, the sensor response saturated at 100 ppm NH_3 concentration (Fig. 5.4).

MoO_3 thin film sensors for the detection of ammonia were prepared by two methods—sol-gel technique and by ion beam deposition [17]. The presence of ammonia was signalled by changes in the electrical resistance of the film. This change was proportional to the concentration of ammonia present. The performances of the two sensors were compared. The response time was found to be faster for the sensors prepared by ion beam deposition (a few seconds vs. 15–20 s for sol-gel prepared thin film). It was suggested that differences in gas sensitivity of the two films could be due to differences in the microstructure of MoO_3.

Markovics and Kovacs [18] compared the use of two different support materials in the fabrication of optical sensors for ammonia. In one sensor, a sol-gel film was used for immobilisation of bromocresol green indicator; in the other the dye was adsorbed onto anodized alumina plates. Their results showed that the sensitivities, response times and recovery times of the sensors were highly dependent on the nature of the support material used. The response and recovery time was found to be faster for the alumina-based sensors compared to the sol-gel-based sensors. However the dynamic range for ammonia detection was found to be greater for the sol-gel-based sensors (Fig. 5.5).

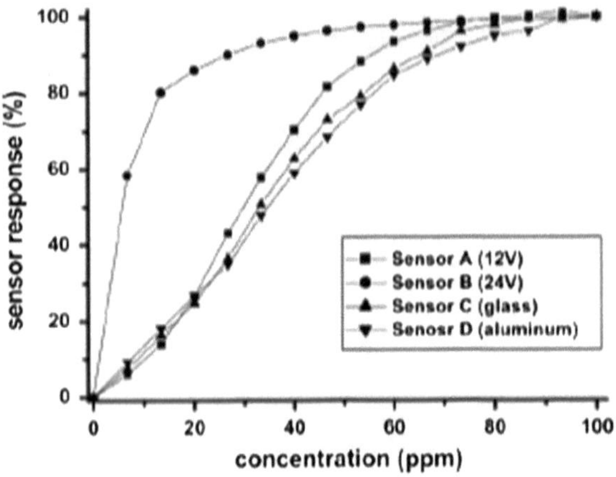

Fig. 5.5 Normalized calibration curves of sensors for gaseous ammonia in the 1–100 ppm range. Sensors A and B were prepared by direct adsorption of ammonia sensitive indicator dye onto anodized alumina support, sensor C was prepared by immobilisation of the dye within a sol-gel film, sensor D was made with a combined alumina support-sol-gel layer composition. Figure reproduced with permission from Ref. [18]. Copyright Elsevier B.V. 2013

5.2.4 Hydrogen

A nanocrystalline TiO_2 thin film was prepared by sol-gel method and used for the detection of hydrogen [19]. The film showed p-type conductivity. When hydrogen was adsorbed on the film, it reacted with the adsorbed oxygen on the surface. This led to the release of free electrons. These partially annihilated the holes and thus, the conductivity of the film was reduced. This change in conductivity was found to be proportional to the concentration of hydrogen present. The optimum operating temperature range for the sensor was found to be between 75 and 225 °C. Response times of 2 s were obtained while the recovery time was found to vary depending on whether the hydrogen was present in nitrogen carrier gas or in air. In nitrogen carrier gas, the recovery time was ~473 s while in air it was ~35 s.

An alternative sensor for hydrogen gas was prepared by Hou et al. [20]. They used Al-doped ZnO nanocrystalline films as the sensing element. They studied the effect of using different zinc precursors (zinc nitrate and zinc acetate) on the performance of the sensor. They reported that films prepared using the zinc acetate precursor showed higher sensitivity and faster response and recovery times than those prepared using zinc nitrate precursor. They attributed the superior response of the zinc acetate derived film to the wrinkly skeletal network structure of the film.

5.2.5 Nitric Oxide

The determination of gaseous nitric oxide is relevant as it is has a variety of physiological functions [21]. Sol-gel encapsulated manganese myoglobin, (a heme

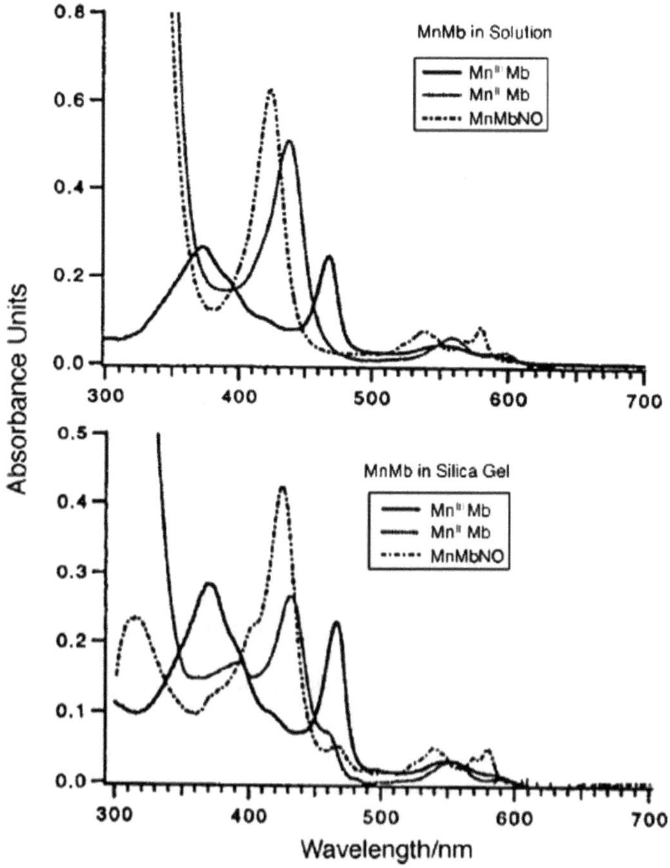

Fig. 5.6 Detection of nitric oxide was successfully achieved in silica-encapsulated manganese myoglobin (MnMb) as the same characteristic absorption changes were observed upon metal reduction and NO binding as compared to MnMb in aqueous buffer. Figure reproduced with permission from Ref. [22]. Copyright Royal Society of Chemistry 1999

protein) was used for the detection of this species [22]. Immobilisation of the heme biomolecule within a silica matrix allowed it to be stabilised and protected from biodegradation. The binding of NO to the manganese myoglobin triggers changes in the visible absorption spectrum of the protein, thus signalling the formation of the MnMbNO adduct (Fig. 5.6).

A hybrid material of carbon nanotubes and silica was used to fabricate a NO sensor [23]. The material was prepared by a microwave assisted sol-gel process. Adsorption of NO onto the sensor surface led to a reduction in sensor resistance, indicating that the hybrid material was an n-type semiconductor. The sensor was reported to detect NO in the range of 30–90 ppb, while operating at room temperature.

A tungsten oxide based nanoprobe for measurement of nitric oxide in exhaled human breath was synthesised by Gouma and Kalyanasundaram [24]. The synthesis was controlled such that a stoichiometric and single phase nanostructure monoclinic phase of WO_3 was produced. This structure, which is a distorted rhenium oxide cubic structure, exhibits specific affinity toward NO in the presence of interfering gases commonly found in exhaled human breath, such as ethanol, methanol, isoprene, acetone and CO. The sensor gave a response in the order of seconds to NO concentrations of 1 ppm, 500 and 300 ppb. Thus, it is particularly suited to monitoring NO concentrations in human breath.

5.2.6 *Nitrogen Dioxide*

Tungsten oxide nanoparticles were synthesised by a sol-gel method, then drop cast as a thin film in order to prepare a sensor for detection of NO_2 [25]. The changes in the resistance of the film were measured upon exposure to fixed concentrations of gases such as NO_2, NH_3, Cl_2, H_2S and CH_3OH. The film showed the greatest response to NO_2 with a very low response towards the other gases. The response of the sensor to a range of NO_2 concentrations between 5 and 160 ppm was then analysed. The highest response was obtained at 100 ppm. At this concentration and at an operating temperature of 200 °C, a response time of 24 s and recovery time of 5 min were obtained.

Wang et al. utilised a different approach for detection of nitrogen dioxide [26]. They prepared a fluorescent carbon quantum dot-silica aerogel hybrid material—the fluorescence of this material was found to be selectively quenched by NO_2. The limit of detection of this sensor was found to be 250 ppb.

5.2.7 *Other Gases*

An ORMOSIL formulation was used for the preparation of a colorimetric sensor array which functioned as an optoelectronic nose for detection of toxic gases [27]. Four different types of chemically responsive dyes were incorporated into the array: metal containing dyes that respond to changes in Lewis basicity, pH indicators, dyes with large permanent dipoles that react to changes in polarity and metal salts that respond to redox reactions. The incorporation of these dyes into the ORMOSIL matrix provided good stability, facilitated the processing of the sensor into a format that could be used with a handheld scanner and also led to improved sensitivity. Nineteen different toxic industrial chemicals (TICs) could be discriminated within two minutes of exposure at IDLH (immediately dangerous to life or health) concentration. The array response to each TIC produces a unique pattern, as illustrated in Fig. 5.7.

A sensor for detection of formaldehyde was prepared by incorporation of acetylacetone within a silica matrix prepared by the sol-gel process [28]. Upon adsorption of formaldehyde from the air, the formaldehyde reacts with the beta

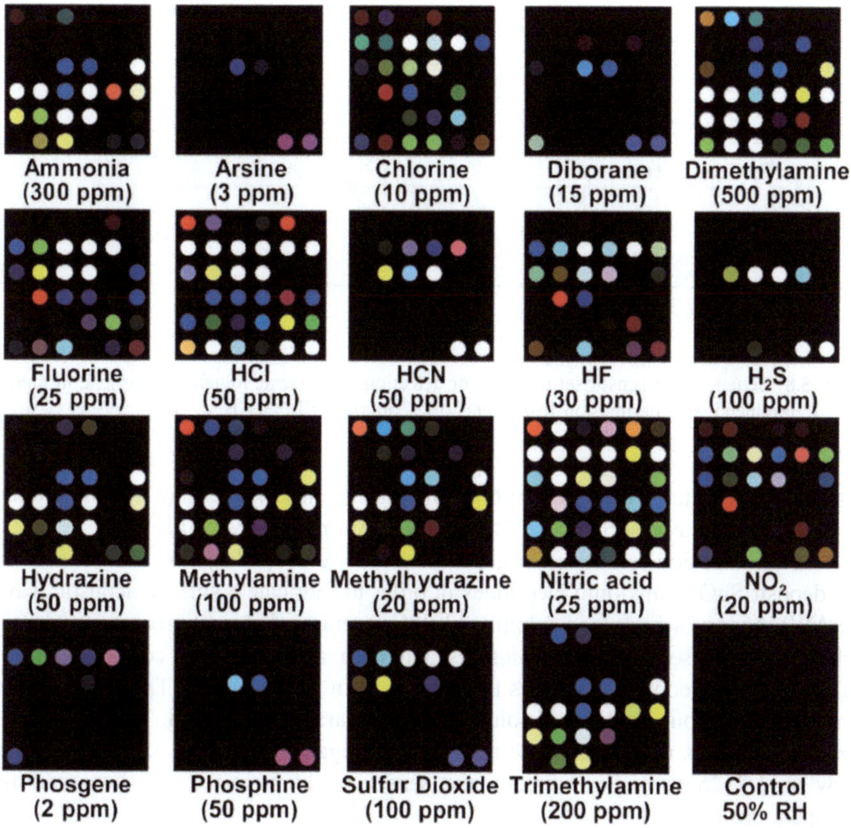

Fig. 5.7 Distinctive colour change patterns produced by optoelectronic nose upon 2 min exposure to representative toxic industrial chemicals (TICs) at their IDLH (immediately dangerous to life or health). Reproduced with permission from Ref. [27]. Copyright Nature Publishing Group, 2009

diketone to give a yellow coloured product, lutidine. Thus, this provides a simple, facile and rapid means of determining the presence of formaldehyde. Use of a sol-gel matrix also had the advantage of allowing a solvent free detection system, thereby avoiding sample preparation steps such as pre-concentration or extraction. The limit of detection of the sensor was found to be 0.03 parts per million by volume. The selectivity of the sensor was tested by measuring the sensor response towards species such as benzaldehyde, acetaldehyde, acetone and butanone. The system was reported to be highly selective for formaldehyde.

Sol-gel chemistry was employed to develop a sensor for the detection of benzene, toluene and xylene (BTX) at ppb concentrations in indoor air [29]. At present, there is no simple and direct method for detection of BTX in the ppb range. Nanoporous concave organically modified silica disks were synthesised, which served to trap and concentrate the volatile organic hydrocarbon species. These disks were incorporated into an experimental chamber which also held a

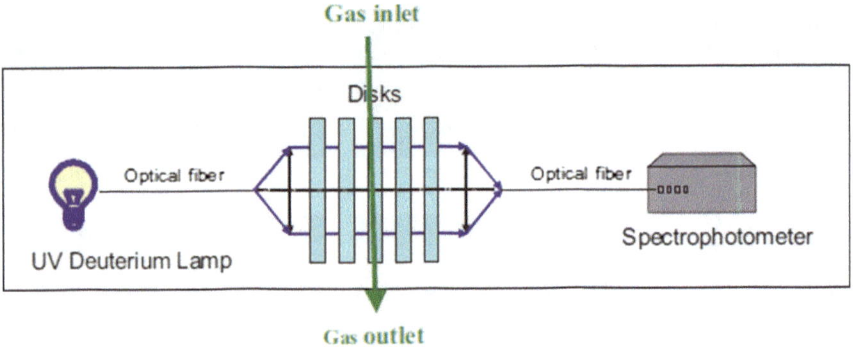

Fig. 5.8 Schematic of sensor set-up used for detection of benzene, toluene and xylene in indoor air. Reproduced with permission from Ref. [29]. Copyright Elsevier B.V. 2013

spectrophotometer. The concentrations of benzene and xylene were then monitored by measuring the absorbance at 254 and 271 nm respectively (Fig. 5.8).

An aqueous non corrosive sol-gel method was developed by Luo and co-workers to deposit SnO_2 films onto electrodes in order to fabricate a surface acoustic wave (SAW) sensor for hydrogen sulphide [30]. Conventional sol-gel processes using tin chloride precursors, ethanol solutions and high acidity led to corrosion of the aluminium and copper electrodes typically used in these devices. This new method employed aqueous oxalic acid solutions for preparation of the sol. The SnO_2 films prepared by this route were able to detect hydrogen sulphide at concentrations as low as 1.7125 ppm. Furthermore, the sensor could be used at a low working temperature of 30 °C.

5.3 Humidity Sensors

The monitoring and control of environmental humidity is an important issue for semiconductor and pharmaceutical manufacture [31]. Therefore, the development of humidity sensors that are sensitive, small in size, flexible in design and low in cost is of interest. The use of fibre optics in the sensor configuration offers a solution to some of these challenges. Zhao and Duan [32] designed a humidity sensor in which an optical fibre was coated in a silica sol-gel film doped with Methylene Blue dye. They reported that this device had a limit of detection of 0.062% relative humidity. The sensor had a good linear relationship between the optical absorbance of the fibre and the corresponding relative humidity value, in the range of 1.1–4.1%. The authors suggested that this sensor could be used as a supplement to traditional fibre optic relative humidity sensors which are insensitive below 30% R.H. (Fig. 5.9).

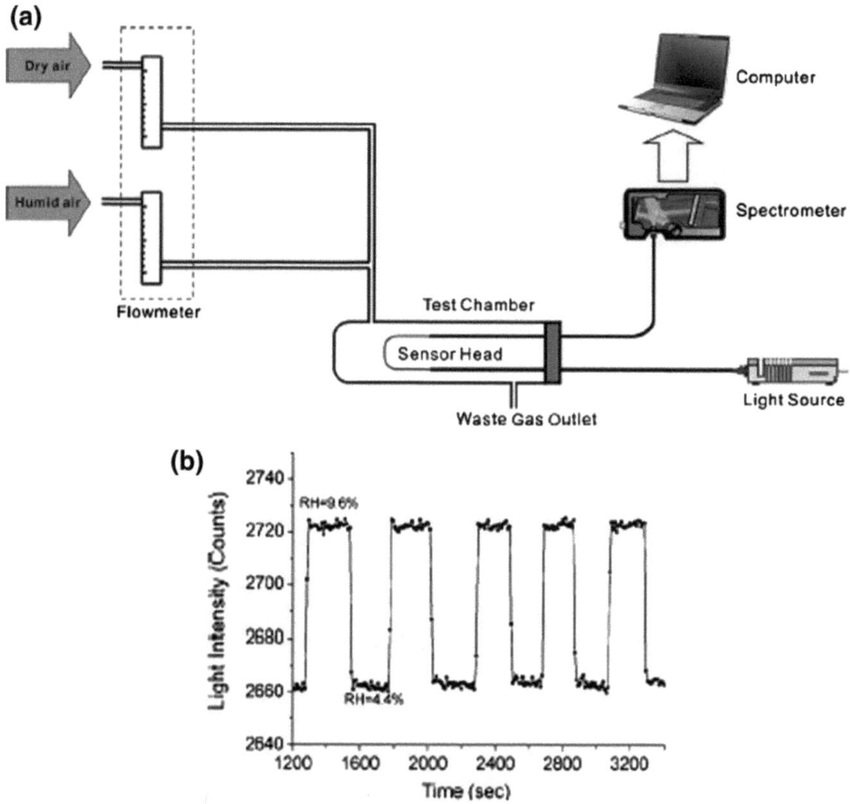

Fig. 5.9 **a** Schematic diagram of sol-gel fibre optic sensor used by Zhao and Duan to measure relative humidity, **b** Reversibility of the sensor between relative humidity of 9.6 and 4.4% (291 K, ambient pressure). Figure adapted with permission from Ref. [32]. Copyright Elsevier B.V. 2011

Rather than using an optical sensor, Islam and co-workers developed a capacitive sensor for measuring relative humidity [33]. Gold electrodes were dip coated with a thin film of nanoporous alumina, which had been prepared by the sol-gel method. In order to optimise the sensor performance, various parameters of the sol-gel process and the deposition process were varied such that a uniform crack free thin film was obtained. The capacitance of the sensor was shown to vary with variation in moisture from 2.5 to 25 ppm in a nitrogen atmosphere. The sensor output was found to be highly repeatable with response and recovery times of 47 and 118 s respectively.

A mesoporous silica thin film was used to enhance the sensitivity of a resonant chemical sensor, used for the detection of relative humidity and carbon dioxide [34]. An aminopropyl functionalised mesoporous silica thin film was grown directly on the surface of a capacitative micromachined ultrasonic transducer (CMUT) to

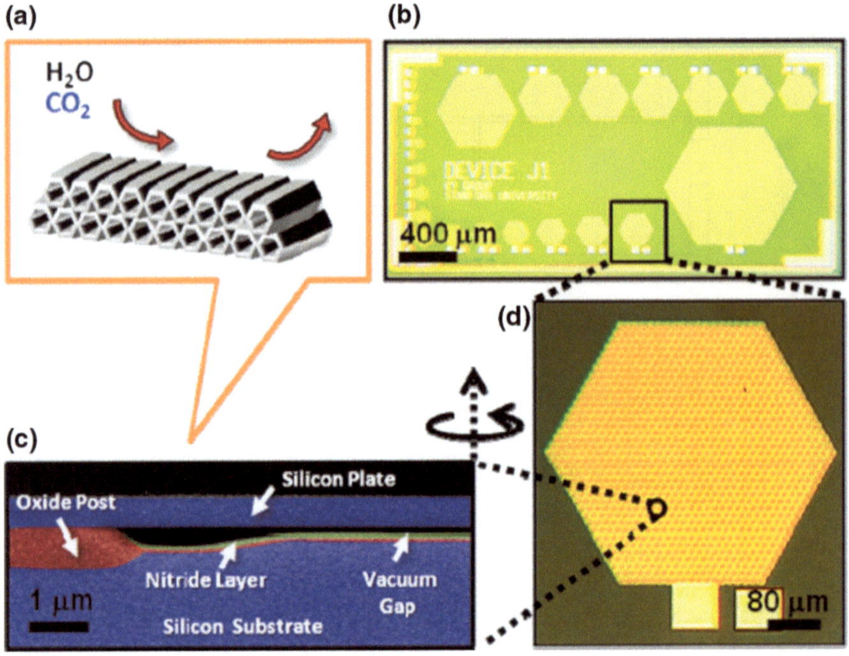

Fig. 5.10 **a** Illustration of the mesoporous silica thin film used for detection of CO_2 and humidity. **b** Optical picture of a single die, showing an array of multiple elements. **c** Cross-section of a single CMUT resonator, imaged by scanning electron microscopy. **d** Optical picture of a single element with 1027 cells. Figure reproduced with permission from Ref. [34]. Copyright American Chemical Society 2012

induce chemisorption between the amine groups and the CO_2. It was found that the silica coated CMUT showed an enhanced volume sensitivity to relative humidity compared to CMUTs coated with sputtered oxide and non-templated oxide (7.5 and 70 times greater, respectively). For CO_2 detection, use of an APTES functionalised silica layer yielded sensitivity approximately 38 times larger than that obtained using only mesoporous silica coating (Fig. 5.10).

Undoped and Sn-doped zinc oxide nanopowders were prepared by the sol-gel method and their humidity sensing properties were investigated by Ates et al. [35] The powders were coated onto the surface of a clean quartz crystal which formed part of a quartz crystal microbalance (QCM). The change in the QCM frequency with changes in relative humidity was recorded. The frequency was found to decrease with adsorption of humidity molecules and increase upon desorption of the molecules. Nanopowders were used as it was expected that the sensing ability would be greater due to the greater surface area achieved using these materials. It was found that the frequency change during this adsorption desorption cycle was greater for the undoped ZnO than the Sn-doped ZnO. This was attributed to the fact that the Sn doping causes the zinc oxide particle size to increase, thus reducing the surface to volume ratio. A summary of the key features of all the gas sensors discussed above can be found in Table 5.1 at the end of this chapter.

5.4 Cation Detection

An electrochemical sensor for the determination of Al(III) was prepared by Mashhadizadeh and Khani [36]. Gold nanoparticles were assembled onto a thiolated sol-gel silica network. This sol-gel nanoparticle hybrid was mixed with graphite and an ionophore, DNMA, to yield a carbon paste, which in turn was used for the preparation of electrodes. This sensor was shown to have a lower limit of detection (0.054 ppb) and a wider working range than previously reported Al^{3+} selective electrodes [37].

Feng et al. [38] reported the detection of trace levels of eight different heavy metal ions using colorimetric indicators immobilised within an organically modified silica matrix (ormosil). This matrix was chosen as the host material for the indicators as it had a high surface area, was optically transparent and was stable over a wide pH range. This material was used to fabricate a sensor array, based on the assembly of five flow cells, each containing sol-gel modified cellulose acetate/nitrate membranes. It was suggested that it should be possible to adapt this system to construct a portable handheld device for heavy metal detection.

Urease enzyme was incorporated into a silicate sol-gel matrix in order to construct a sensor for heavy metal ions [39]. Urease converts urea into ammonium and bicarbonate ions. However, the enzyme activity is known to be inhibited by heavy metal ions. Therefore, the decreased activity of the urease enzyme can be used to signal the presence of heavy metal ions such as mercury and cadmium. This sensor was used to detect mercury in the dynamic range of 0.005–0.5 mM, copper ion in the range 0.02–1.0 mM, cadmium ion in the range 1.0–10 mM and lead ions in the range 0.9–10 mM.

Another sol-gel-based sensor for detection of cadmium and lead ions was reported by Yantasee et al. [40]. A composite of self-assembled thiol monolayers on a mesoporous silica substrate and a Nafion binder was used for the construction of electrodes. The thiol groups served to pre-concentrate the metal ions of interest in the sample solutions. This was followed by detection by an anodic stripping voltammetry technique. This detection was found to be sensitive to solution pH. The optimal pH for Cd^{2+} measurement was in the range of 6–8. Fortunately, this corresponds well to the pH of natural waters (river water, ground water, seawater) and urine, thus, demonstrating the usefulness of this system for monitoring of such matrices. Limits of detection of 2.5 ppb for Cadmium and 0.5 ppb for lead were achieved with this sensor.

A sol-gel based sensor for detection of copper in urine was developed by Jeronimo et al. [41]. High levels of copper in urine (i.e. 500–100 μg per day) are associated with abnormalities in copper metabolism, such as Wilson's disease [42]. Therefore, it was considered of interest to develop a simple, sensitive and cheap system for monitoring urine copper levels. PAR (4-(2-pyridylazo)resorcinol), an azo dye, was incorporated into a sol-gel matrix. This material was then cast as a thin film by spin coating onto glass substrates. The PAR dye forms a coloured complex with copper ions. Thus, changes in the absorption spectrum of the film could be used for detection of copper ions, as the formation of the complex results in a decreased absorption at 410 nm and appearance of a new peak at 500 nm. The sensor was shown to give a linear response for Cu(II) concentrations between 5 and 80 μg per litre. The accuracy of the sensor was assessed by comparison of analytical results obtained using this method with analysis of the same samples by ICP-MS. The results showed very good agreement with the values obtained by ICP-MS.

An alternative approach to the preparation of a sol-gel-based copper sensor was described by Carrington and Xue [43]. They prepared amine grafted monoliths which formed blue Cu^{2+} amine complexes upon exposure to copper ions in solution. The authors argued that the use of monoliths offered a number of advantages over sol-gel thin films. They provided a longer sensor path length and they avoided reagent leaching as the functional groups are bound to the gel interior, rather than being encapsulated in the gel. Carrington and Xue [44] also used sol-gel monolith based sensors for the detection of Cr(VI). Pyridine functionalised monoliths were prepared which demonstrated a colour change to yellow upon exposure to ppm level Cr(VI) solutions. In order to detect Cr(VI) at ppb concentrations, it was necessary to add diphenylcarbazide to monoliths previously exposed to Cr(VI) solutions. This yielded a colour change to magenta, which could be measured by visible spectroscopy. Use of this system allowed detection of Cr(VI) at concentration of 10 ppb.

Sol-gel processes were used to prepare a sensor for colorimetric detection of mercury ions [45]. Azobenzene coupled receptors were synthesised and immobilised to mesoporous silica via sol-gel grafting method. Upon addition of mercury ion, a colour change from yellow to red was observed. No such colour change was observed for other ions tested (K^+, Ca^{2+}, Sr^{2+}, Co^{2+}, Cd^{2+}, Pb^{2+}, Zn^{2+}, Fe^{3+}, Cu^{2+}). This sensor gave a linear response in the range 1–10 μM.

5.5 Anion Detection

Dhanya et al. [46] prepared a sensor for nitrite by immobilisation of the fluorescent dye, Rhodamine 6G in a silica sol-gel matrix. This mixture was then cast as a film onto test strips. A colour change from pink to orange was observed on exposure to nitrite. Additionally, changes in the emission spectrum of the dye were observed. The authors tested the selectivity of their sensor towards nitrite by addition of other species (nitrate, bromide, phosphate, iodide, perchlorate ions) to the nitrite test solution. They reported that the sensor gave superior specificity to test strips based on the Greiss reaction for nitrite and commercial electrochemical sensors. The test strips were also used for determination of nitrite in natural waters such as tap, well and sea waters.

A different approach to nitrite sensing was reported by Wang et al. [47]. They prepared silicomolybdate-methylsilicate modified graphene electrodes by a sol-gel technique. These electrodes were then used for the amperometric sensing of nitrite since it is known that the $SiMo_{12}$ anion can catalyse the electroreduction of nitrite. The electrodes were found to be selective for nitrite given that a 1200-fold excess of ions such as fluoride, chloride, bromide, phosphate and carbonate did not interfere with the nitrite detection. The electrode response was found to be linear in the range of 5×10^{-6} to 1.5×10^{-2} mol dm^{-3} for nitrite.

A sensor for fluoride ions was prepared by covalent bonding of a luminescent terbium complex to a mesoporous silica matrix [48]. This complex showed significant changes in its fluorescence upon hydrogen bonding to fluoride guest ions, thus allowing detection of fluoride at an LOD of 1 μM.

An interesting sol-gel based approach to iodide sensing was demonstrated by Shen et al. [49]. They took advantage of the fact that supramolecular hydrogels prepared using low molecular weight gelators are known to undergo a sol-gel state transition when triggered by stimuli such as heat, light, pH. Supramolecular hydrogels of Ag(I)-Glutathione co-ordination polymers were prepared. These gels were found to undergo a sol-gel transition upon addition of iodide ion. Thus, this provided a means for rapid naked eye detection of iodide ion. The selectivity of the system for iodide was investigated and it was found that the addition of other anions such as fluoride, chloride, bromide and phosphate did not lead to any significant change in the state of the gel.

A 'turn on' fluoride ion sensor was prepared by using the sol-gel process to dope aminosilane functionalised silica nanoparticles with silyl-ether protected fluorescein isothiocyanate [50]. The sensitivity of the sensor was found to depend on the type of aminosilane used to functionalise the nanoparticles since this affected the loading of the fluorescent dye. Thus, the authors suggested that this allows the detection range of the sensor to be easily varied by simply changing the aminosilane. The selectivity of the sensor was tested by measuring its response in the presence of ions such as Cl^-, Br^-, I^- and NO_3^-. The results showed that only F-induced a significant change in emission intensity. A summary of the key features of all ion sensors discussed above can be found in Table 5.2 at the end of this chapter.

5.6 PH Sensors

Dye doped sol-gels have also been employed for pH sensing. Wang et al. prepared one such sensor by addition of phenol red to a sol-gel formed by co-condensation of TEOS and Phenyl-triethoxysilane (Ph-triEOS) [51]. Use of the Ph-triEOS precursor provided a solution to the problems of dye leaching and small response signal associated with previously reported organically modified sol-gel pH sensors [52]. The doped sol-gel showed changes in the optical absorbance at 560 nm as the pH was changed. The sensor was shown to be stable for up to a year, with continuous usage of the sensor over a three-month period having no deleterious impact on the sensor performance (Fig. 5.11).

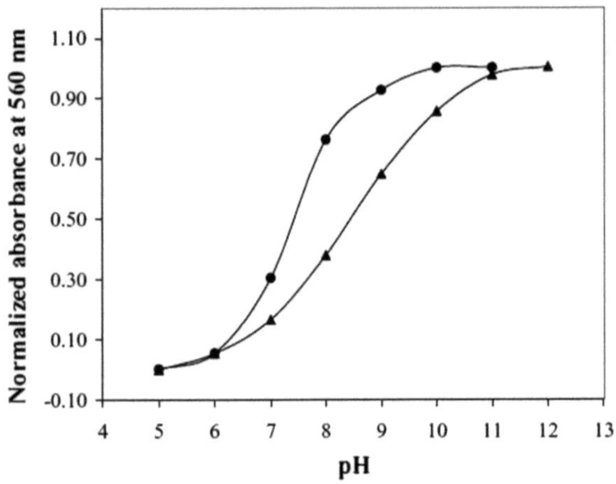

Fig. 5.11 Response curve of the (*filled circle*) phenol red in solution (concentration: 1.41×10^{-4} M) and (*filled triangle*) the phenol red/TEOS/Ph-TriEOS (5% v/v/v) sol-gel glass film at 558 nm as a function of buffer pH. Figure reproduced with permission from Ref. [52]. Copyright Elsevier B.V. 2003

Similarly, sol-gel methods were used to fabricate a fibre optic pH sensor based on a sol-gel film doped with neutral red indicator [53]. This system gave a response time of less than 20 s in the pH range of 6–9.

A different approach to sol-gel-based pH sensing was favoured by Lobnik and et al. [54]. A luminescent europium complex was synthesised and entrapped into a variety of sol-gel matrices. The luminescence of the complex was independent of pH in the range 0–10. Bromothymol blue, a non-fluorescent pH indicator with an alkaline absorption maximum close to the emission maximum of the Eu complex was also incorporated into the sol-gel matrix. The response of this indicator to pH changes in the pH range 5–10.8 was monitored. The results showed that the emission of the europium complex was absorbed by the indicator at alkaline pH and restored to original level at acidic pH (Fig. 5.12).

While the previous studies described above relate to optical pH sensors, electrochemical pH sensors based on sol-gel matrices have also been developed. For example, in the work of Huang et al. an iridium oxide thin film was deposited onto

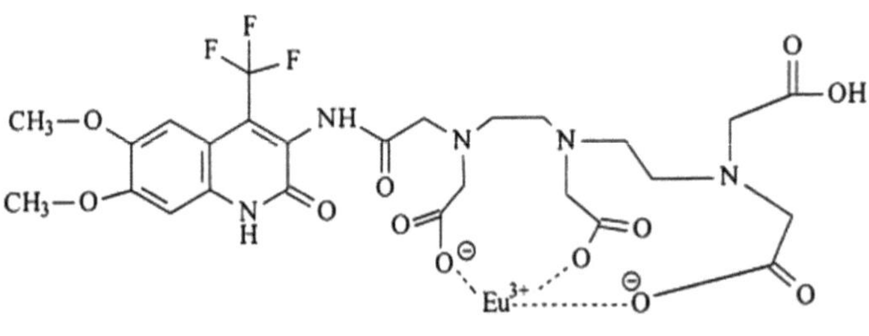

Fig. 5.12 Luminescent europium complex which was incorporated into sol-gel matrix and used for pH sensing. Figure reproduced with permission from Ref. [54]. Copyright Elsevier 2001

gold electrodes via sol-gel process [55]. This was used for the electrochemical detection of pH. The researchers argued that this sensor used less power than optical pH sensors. They compared the performance of this sensor with commercially available glass-based pH electrodes and demonstrated that the iridium oxide based sensor gave a similar performance but had the advantages of being smaller in size and flexible and deformable. They were able to roll up the pH sensor array and insert it into a long tube with diameter of 2 cm. The response time of the sensor was 2 s which was an improvement over the response time of other metal oxide pH electrodes [56].

5.7 Biosensing

Sol-gel technology also offers advantages for the development of biosensors, given that sol-gels are optically transparent, compatible with many organic compounds and stable in harsh environments. However, problems can be encountered with gel shrinkage, fragility of the gel structure or poor adhesion of the sol-gel to other substrates [57]. Park and Clark reported the fabrication of a sol-gel encapsulated enzyme array which was stable enough to endure repeated cycles of bioscreening [58]. In order to overcome the problem of poor adhesion of the enzyme–silica mixture to a glass slide, the slide was coated with polyvinyl acetate. Polyvinyl alcohol was added to the enzyme solution to reduce sol-gel shrinkage. A variety of different enzymes were immobilised in the sol-gel, including glucose oxidase, horseradish peroxidase, lipases and hydrolases. The activity of all encapsulated enzymes was equivalent to that of the soluble enzymes. However, the immobilised enzymes showed increased thermostability relative to the soluble enzymes.

Glucose oxidase was immobilised on a nanostructured cerium oxide film deposited on a gold electrode in order to yield an amperometric glucose biosensor [59]. The cerium oxide film was prepared by sol-gel technique. The limit of detection of this sensor was found to be 12.0 μM with linearity in the range 50–400 mg/dL. This work was further developed by preparation of cerium oxide doped nanoporous titania film for glucose sensing [60].

A different approach was taken by Xu and co-workers in that the detection of glucose was visually signalled by changes in sol-gel structure triggered by hydrogen peroxide [61]. They designed a sensor whereby the presence of glucose is signalled by hydrogen peroxide induced sol-gel transition of an acrylic acid-gold nanoparticle based hydrogel. When glucose oxidase enzyme is incubated with glucose in a solution, hydrogen peroxide is produced. The addition of Fe^{2+} yields hydroxyl radicals ·OH which trigger the polymerization of the acrylic acid and subsequent sol-gel transition. The formation of the gel leads to the aggregation of the gold nanoparticles present providing a colorimetric response to the presence of glucose. A limit of detection of 1 μM for glucose was obtained using this system.

A contact lens with integrated glucose sensor was developed for the in situ monitoring of glucose levels in tear fluid [62]. There is reasonable correlation between tear glucose levels and blood glucose levels [63], therefore, this was considered a good non-invasive approach to the issue of continuous blood glucose monitoring. Glucose oxidase (GOD) was immobilised in a TiO_2 sol-gel membrane which was applied to coat the microfabricated electrode components integrated onto the contact lens. The results showed that the presence of the GOD/titania sol-gel film greatly increased the glucose signal obtained, relative to use of GOD alone (more than 30 times increased). This was attributed to the ability of the sol-gel film to hold the glucose oxidase enzyme in place. In the absence of interferents, the sensor could detect glucose at an LOD of 0.01 mM, which is one-tenth of the minimum concentration of glucose in tear fluid (Fig. 5.13).

A sol-gel-based sensor for the simultaneous electrochemical determination of ascorbic acid and uric acid in urine was prepared by immobilising methylene blue in a methyltrisiloxane sol-gel film which was coated onto a glassy carbon electrode [64]. Limits of detection of 5 nM for ascorbic acid and 1 nM for uric acid were achieved using this sensor. It was also noted that the system had good stability with no leaching of the methylene blue during 2 months immersion of the electrode in aqueous media.

The sol-gel process was used to produce enzyme doped inks which could be inkjet printed onto paper strips to produce a colorimetric sensor for neurotoxins [65]. Incorporation of the enzymes within the sol-gel matrix allowed retention of their biological activity, while at the same time providing them in a format suitable for inkjet printing. The advantage of using inkjet printing for deposition of the sol-gel mixture, rather than a method such as spin coating, is that it is readily scalable for large scale production. Acetylcholinesterase enzyme was immobilised in silica. Upon binding of organophosphates (e.g. paraoxon) the enzyme activity was reduced and the residual activity of the enzyme was monitored using the Ellman colorimetric assay. A limit of detection of 100 nM was achieved for paraoxon.

Another sensor for detection of organophosphates was prepared by encapsulation of genetically engineered bacterial cells within a porous silica matrix [66]. The cells were engineered to express organophosphorous hydrolase enzyme on their surfaces. The authors of this work argued that this approach avoided the leaching that could be associated with encapsulation of the enzyme alone.

The combination of live microorganisms with silica matrices was also exploited by Ponamerova et al. to produce a biosensor for methanol [67]. A suspension of yeast cells was added to a mixture of TEOS and MTES. Addition of a basic catalyst triggered the formation of silica capsules around the yeast cells. This biomatrix was then deposited onto the surface of a Clark oxygen electrode. The resultant system could be used to determine methanol concentration since the immobilised cells oxidise the methanol during their respiration. This results in a decrease in oxygen concentration, which can be detected as a change in electric current with time. The sensor was used to determine methanol content in process waste from a chemical plant. The methanol concentrations were independently verified by gas chromatography. The results showed good agreement between both analytical methods, however, the biosensor gave more rapid results without the need for extensive sample preparation (Fig. 5.14).

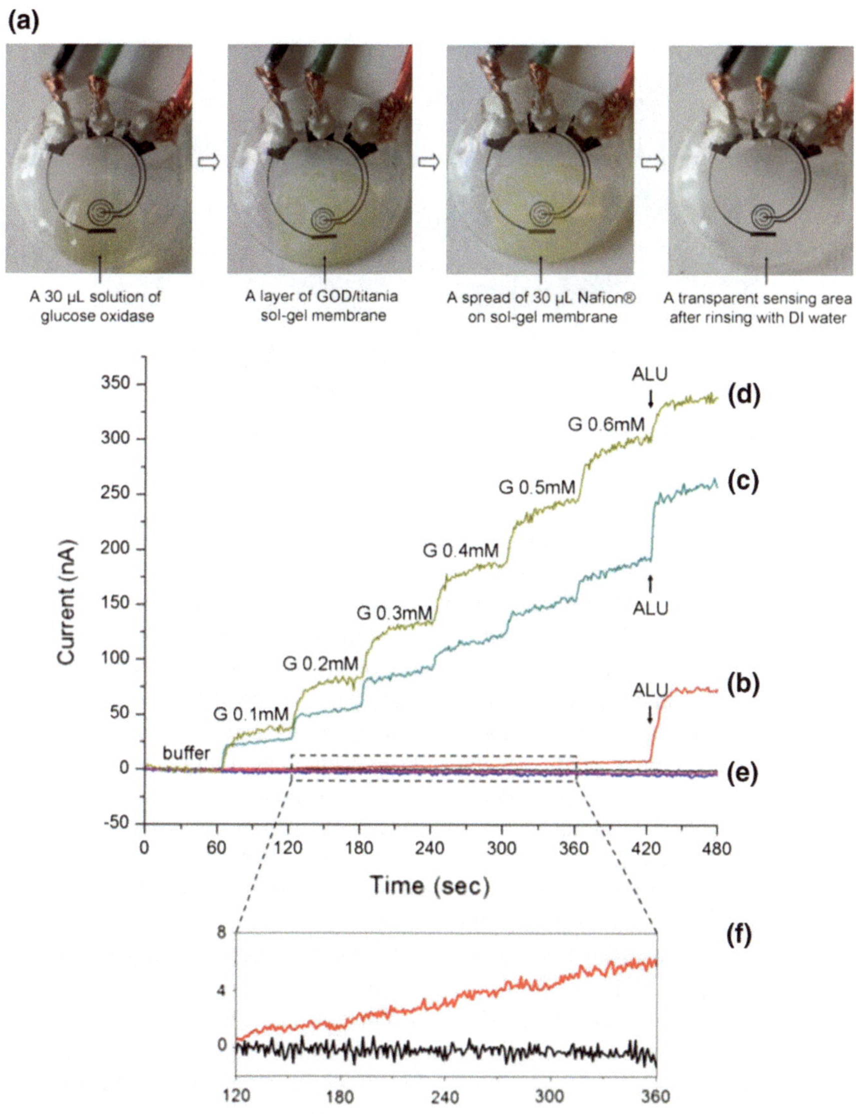

Fig. 5.13 Integration of glucose sensor with contact lens and sensor response to varying glucose concentrations. Reproduced with permission from Ref. [62]. Copyright Elsevier 2011

Aptamers are single stranded nucleic acids that can fold into distinct three-dimensional structures that can function as receptors for certain proteins and cells. They can be derived from DNA and RNA, however, there is a greater variety of aptamers known from RNA. While these molecules show great promise as

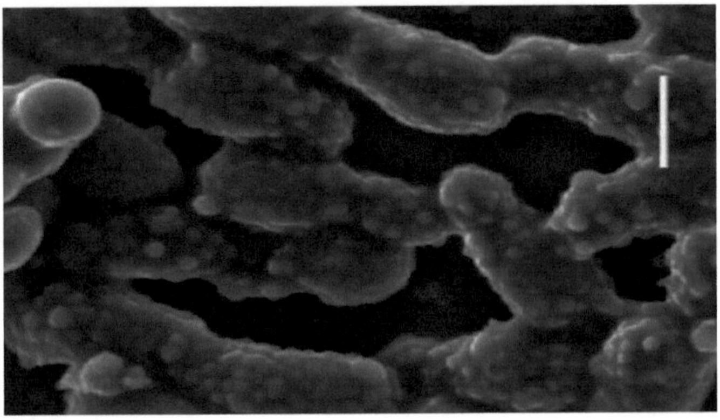

Fig. 5.14 Yeast encapsulation in silica matrix. Reproduced with permission from Ref. [67]. Copyright Elsevier 2015

recognition components in diagnostics, the application of these materials to sensors has been hindered by their chemical instability. Furthermore, it is necessary to incorporate these molecules onto or into a suitable solid surface in order to construct an aptamer RNA-based bioarray. Carrasquilla et al. reported the use of a low temperature sol-gel method for encapsulation of structure switching RNA aptamers in porous silica and organosilane matrices [68]. They studied the chemical stability, resistance to nuclease activity, leaching and signalling capabilities of the encapsulated RNA aptamers relative to the free molecules in solution. The degree of leaching was found to be dependent on the type of precursor used to prepare the sol-gel with greater leaching of the aptamers derived from the more hydrophobic matrices. The optimum material for entrapment of the RNA aptamers was found to be a sol-gel derived from a mixture of TMOS (tetramethylorthosilicate) (60%) and MTMS (methyltrimethoxysilane) (40%) (v/v).

A sol-gel-based sensor for detection of *V. cholerae* bacteria was developed by Solanki et al. [69]. This bacterium is responsible for the production of the cholera toxin, which can cause outbreaks of diarrheal disease in the human population [70]. Hence, it is of interest to develop a rapid, reliable and cheap sensor for detection of this species. Existing immunosensors for these bacteria have a lengthy and expensive fabrication process. In this work, the researchers prepared a nanostructured ZrO_2 thin film by a low temperature sol-gel process. This had the advantage of yielding a large area of substrate at relatively low cost. This film was then functionalised with a single strand DNA sequence based on the O1 gene (O1 is an antigen associated with the cholera bacteria). The composite was then used to fabricate a bioelectrode. Changes in the magnitude of current upon binding of genomic DNA were monitored by differential pulse voltammetry.

Choline is produced in the body for the regulation of insulin, transportation of fats in the liver and for the synthesis of cell membranes [71]. It can also be provided by consumption of certain foods. Low levels of choline intake in women have been associated with a greater risk of giving birth to a child with neural tube defects [72]. Therefore, it is of great importance to have a reliable sensor available for monitoring the concentration of this essential nutrient. Mazurenko et al. [73] prepared an amperometric biosensor for choline by modification of gold electrodes with a silica film containing choline oxidase. The silica film was prepared by the sol-gel method. It was reported that the addition of CTAB (Cetyl trimethylammonium bromide) to the film increased the stability of the sensor. The addition of a certain concentration of CTAB was shown to permit retention of 50% of sensor activity after 3 weeks of continuous measurements. In contrast, use of higher or lower concentrations of CTAB showed losses of up to 80% of sensor activity after 8 days. The researchers speculated that the addition of CTAB led to a more uniform distribution of the choline oxidase enzyme within the silica network. They used this sensor for the determination of choline in baby food samples.

The determination of total polyphenolic content of food has received much attention of late due to the antioxidant properties of these compounds [74]. Laccase enzymes are copper containing enzymes which oxidise phenols to phenones with attendant reduction of oxygen. Therefore, these enzymes can be used for the electrochemical determination of polyphenol content by monitoring the reduction current of quinines generated as products of the enzyme reaction. Casero et al. [75] investigated three different approaches for the immobilisation of laccase enzymes onto gold electrode surfaces for preparation of polyphenol biosensors. They noted that the sensors developed by immobilisation of the enzyme in a sol-gel network showed higher sensitivity and a lower limit of detection than those sensors prepared using adsorption or covalent attachment of the enzyme to the gold electrode surface.

An interesting approach to the detection of the herbicide, simazine, was reported by Pena-Vazquez et al. [76]. Traditional analytical methods for detection of herbicides involve sample clean up followed by LC-MS or GC-MS. A more rapid means of detection was proposed wherein a biosensor was prepared by immobilisation of freshwater green algae in a sodium silicate matrix. This matrix was chosen as it is biocompatible and chemically stable. Herbicide induced changes in the amount of chlorophyll fluorescence were used to quantify the amount of simazine present. Addition of herbicide inhibits photosynthetic activity at PS11, resulting in increases in chlorophyll fluorescence intensity. Three different species of algae were tested and the Dictiosphaerium chlorelloides biosensor was found to give the lowest limit of detection and the broadest calibration range. However, the sensor was not sufficiently sensitive to detect herbicide at the level specified by the European Community directive for drinking water (0.1 μg per litre) so more work was needed to improve the sensitivity of the system [77].

In recent years, the detection of melamine has become a topic of research interest after it was illegally added to dairy products to increase their nitrogen concentration, thereby falsely indicating increased protein concentration [78]. In 2008, over 50,000 infants fell ill in China after drinking formula milk deliberately adulterated

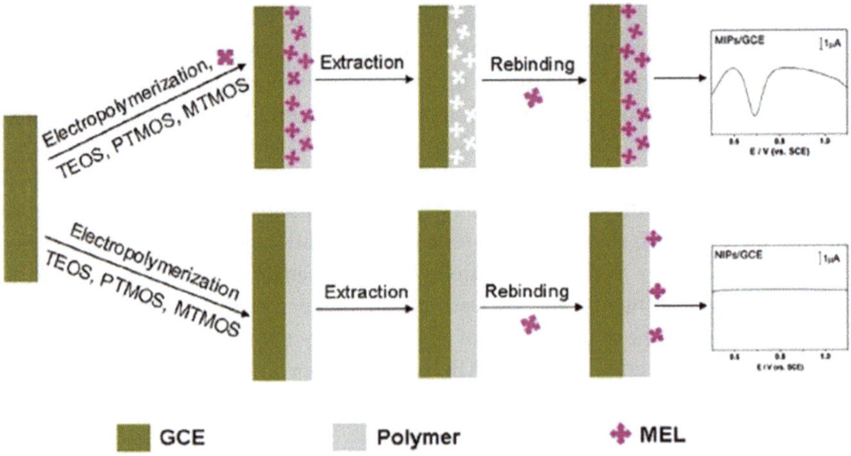

Fig. 5.15 Preparation of molecularly imprinted polymer electrode for detection of melamine (*GCE* glassy carbon electrode, *MEL* melamine molecule). Reproduced with permission from Ref. [80]. Copyright Elsevier B.V. 2012

with melamine [79]. The presence of levels of melamine above the safety limit (2.5 ppm in US and Europe) can cause kidney failure and death and hence, it is of particular importance to rapidly and accurately determine melamine concentrations in food samples [80]. A molecularly imprinted polymer–silica sol-gel hybrid was prepared using melamine and this material was then coated onto glassy carbon electrodes in order to fabricate an electrochemical sensor [81]. The sensor was used to analyse milk powder samples that had been spiked with melamine and recoveries of 90.8–109.2% were obtained. A limit of detection of 6.8×10^{-8} mol L^{-1} was achieved (Fig. 5.15).

Anatase titanium dioxide was synthesised via sol-gel process and deposited onto indium tin oxide glass. Cholesterol oxidase (ChOx) enzyme was then immobilised on the titania surface via electrostatic interactions in order to fabricate an electrochemical sensor for cholesterol [82]. This bioelectrode was incorporated into a microfluidics system. The sensitivity of the ChOx/ant-TiO_2 based microsystem was reported to be 94.65 $\mu A/mM/cm^2$, which compared favourably with other cholesterol biosensors.

A diagnostic assay for rheumatoid arthritis was prepared by synthesis of a three-dimensional structure of ZnO nanorods [83]. This system allowed for detection of anti-CCP (cyclic citrullinated peptide) antibodies which are present in rheumatoid arthritis patients several years before the onset of clinical symptoms. Since the CCP sequence contains a high content of basic amino acids (histidine and arginine) it is reported to have a strong affinity for the ZnO surface and can be readily immobilised there without need for surface functionalization. Binding of the captured anti-CCP antibodies is then signalled by a secondary antibody conjugated with fluorescent dye.

5.8 Forensic Analysis

A sol-gel based colorimetric sensor for methamphetamine was developed by Choodum et al. [84]. Simon's reagent was entrapped within a sol-gel matrix. Addition of standard solutions of methamphetamine of varying concentrations led to a colour change from yellow to purple-blue. Digital image colorimetry was used to assess the colour change results in order to quantitatively analyse for methamphetamine. The intensity of the RGB values obtained from the sensor could be correlated with methamphetamine concentration. The sensor could be stored for up to 3 months in a freezer prior to use without significant deterioration of sensor performance. The sensor was used for the analysis of methamphetamine in casework tablet samples and in urine samples.

Sol-gel chemistry has also been applied to the detection of latent fingerprints. Liu and co-workers prepared silicon dioxide based xerogels which were doped with Eu^{3+}/sensitizer complexes [85]. The doped xerogels were finely ground to a powder, mixed with magnetic iron powders, then used for the visualisation of fresh and aged fingerprints on a variety of porous and non-porous surfaces. (The magnetic iron powder was added to facilitate application of the xerogel powder to the surface). Latent fingerprints were successfully developed on all substrate types. Interestingly, the best results were obtained with xerogels prepared using TEOS precursor rather than TMOS or TKIS. This was due to the higher level of fluorescence intensity of this doped xerogel compared to the other xerogels.

Mesoporous ORMOSIL thin films were prepared by a sol-gel process and then doped with pyrene in order to fabricate a sensor for nitro explosives [86]. It is known that pyrene excimer emission can be quenched by exposure to nitroaromatic explosives such as TNT (trinitrotoluene). Prior attempts to incorporate pyrene into thin films prepared from polymers such as polystyrene for sensing purposes were hindered by the resultant poor formation of the excimer. However, the incorporation of pyrene into the porous ORMOSIL thin films yielded strong excimer emission. Exposure of the film to TNT vapour resulted in quenching of the pyrene emission. This quenching effect could be observed by naked eye when the films were illuminated with UV light. Exposure of the sensor to toluene, ammonium nitrate, benzene and methanol did not result in significant fluorescence quenching, thus, the sensor was deemed to have good selectivity to nitroaromatic explosives. Furthermore, the sensor was shown to be reusable up to five cycles.

5.9 Conclusions

The significant number of sol-gel-based sensors outlined herein clearly demonstrates the continued popularity of this technique for sensor production. The ease of preparation, the biocompatibility of many of the materials and the low cost of

Table 5.1 Gas sensors prepared using sol-gel methods

Analyte	Reagent	Sol-gel precursor/additives	Linear range/detection limit	Principle of detection	Refs.
Oxygen	$Ru(bpy)_3$, $Ru(dpp)_3$	TEOS	0–100% O_2	Fluorescence	[3]
Oxygen	Platinum octaethylporphyrin	TEOS	0.5–100% O_2	Fluorescence	[4]
Oxygen	$Ru(dpp)_3$	MTEOS, ETEOS		Fluorescence	[5]
Oxygen	$Ir(piq)_2(acac)$	TFP-TriMOS, n-propyl-TriMOS	0–100% O_2, 0.009% O_2	Fluorescence	[6]
Oxygen (in vivo)	$Ru(dpp)_3$ + Oregon green 488-dextran	TEOS	–	Fluorescence	[7]
Oxygen (in vivo)	$Ru(dpp)_3$	MTEOS	–	Fluorescence	[8]
Carbon dioxide	$Ru(dpp)_3$ + HPTS	MTEOS	0–100% CO_2, 0.08% CO_2	Fluorescence	[10]
Carbon dioxide	HPTS + Titanium dioxide nanoparticles	MTEOS	0.03–30% CO_2, 0.008% CO_2	Fluorescence	[11]
Carbon dioxide	Titanium dioxide nanoparticles	$TiCl_4$, isopropanol	–	Resistive gas sensor	[13]
Carbon dioxide	$La_{1-x}Sr_xFeO_3$	Lanthanum nitrate, strontium nitrate, ferric nitrate	–	Resistive gas sensor	[14]
Ammonia	Zinc oxide nanofibres	Zinc acetate dihydrate	–	Resistive gas sensor	[16]
Ammonia	Molybdenum trioxide	Molybdenum isopropoxide	–	Resistive gas sensor	[17]
Ammonia	Bromocresol green	Ph-TriEOS, MTEOS, DiMe-DiM-OS, APTES	0–80 ppm NH_3	Reflectance	[18]
Hydrogen	Titanium dioxide	Titanium tetra-isopropoxide	–	Resistive gas sensor	[19]
Hydrogen	Zinc oxide	Zinc acetate, zinc nitrate	–	Resistive gas sensor	[20]

(continued)

Table 5.1 (continued)

Analyte	Reagent	Sol-gel precursor/additives	Linear range/detection limit	Principle of detection	Refs.
Nitric oxide	Manganese myoglobin	TMOS	–	Optical Absorption	[22]
Nitric oxide	Carbon nanotubes	TEOS, TWEEN	–	Electrochemical	[23]
Nitric oxide	Tungsten oxide	Tungsten isopropoxide	–	Resistive gas sensor	[24]
Nitrogen dioxide	Tungsten oxide	Tungsten hexachloride	5–100 ppm NO_2	Resistive gas sensor	[25]
Nitrogen dioxide	Polyethylimine carbon quantum dots	TEOS	2–10 ppm NO_2, 250 ppb NO_2	Fluorescence	[26]
Formaldehye	Acetylacetone	TEOS	0.03 ppm Formaldehyde	Optical absorption	[28]
Benzene, toluene, xylene	–	TMOS, APTES	10–100 ppb	Optical absorption	[29]
Hydrogen sulphide	Tin oxide	Tin oxalate	–	Surface acoustic wave	[30]
Humidity	Methylene blue	TEOS	0.062% RH	Optical absorbance	[32]
Humidity	Alumina	Aluminium butoxide	2.5–25 ppm	Capacitive	[33]
Humidity	Mesoporous silica	TEOS, $EO_{20}PO_{70}EO_{20}$	–	Resonance	[34]
Humidity	Zinc oxide	Zinc acetate	–	QCM	[35]

Ru(bpy)$_3$ tris(bipyridyl) ruthenium(II), *Ru(dpp)$_3$* Ru(II) tris (4,7-diphenyl-1,10 phenanthroline), *TFP-TriMOS* 3,3,3-trifluoropropyltrimethoxysilane, *N propyl-TriMOS* propyltrimethoxysilane, *Ir (piq) acac* Bis(1-phenylisoquinoline)(acetylacetonate)iridium(III), *TEOS* tetraethoxyorthosilicate, *MTEOS* Methyltriethoxysilane, *ETEOS = HPTS* 1-hydroxypyrene-3,6,8-trisulfonate, *Ph-TriEOS* phenyl-triethoxysilane, *DiMe–DiMOS* dimethyl-dimethoxysilane, *APTES* aminopropyl-triethoxysilane, *TMOS* tetramethylorthosilicate, *TWEEN* polyoxyethylene (20) sorbitan monooleate, $EO_{20}PO_{70}EO_{20}$ poly (ethylene oxide)-poly (propylene oxide)-poly (ethylene oxide) block copolymer, *QCM* Quartz Crystal microbalance

Table 5.2 Ion sensors prepared using sol-gel methods

Analyte	Reagent	Sol-gel precursor/additives	Linear range/detection limit	Principle of detection	Refs.
Al^{3+}	DNMA, Gold nanoparticles	MPTS	5×10^{-10} to 5×10^{-2} M, 0.054 ppb	Potentiometric	[36]
Hg^{2+}	Urease	TMOS	0.005–0.5 mM	Conductometric	[39]
Cu^{2+}	Urease	TMOS	0.02–1.0 mM	Conductometric	[39]
Cd^{2+}	Urease	TMOS	1.0–10 mM	Conductometric	[39]
Pb^{2+}	Urease	TMOS	0.9 to 10 mM	Conductometric	[39]
Cd^{2+}	SH-SAMMS	–	2.5–100 ppb	Anodic stripping voltammetry	[40]
Pb^{2+}	SH-SAMMS	–	0–1.5 ppb	Anodic stripping voltammetry	[40]
Cu^{2+}	PAR	TEOS, APTES	5–80 μg L^{-1}	Optical absorption	[41]
Cu^{2+}	Amine grafted silica monoliths	TMOS	–	Colorimetric	[43]
Cr^{6+}	2-[2-(trimethoxysilyl)ethyl]-pyridine	TMOS	10–300 ppb	Optical absorption	[44]
Hg^{2+}	Azobenzene receptor	TEOS	1–10 μM	Optical absorption	[45]
NO_2^-	Rhodamine 6G	TEOS	40–120 ppb, 7 ppb	Optical absorption	[46]
NO_2^-	Silicomolybdate–methylsilicate–graphite composite	MTMOS	5×10^{-6} to 1.5×10^{-2} M, 1×10^{-6} M	Amperometric	[47]
F^-	Terbium complex	TEOS	1 μM	Fluorescence	[48]
I^-	Ag(I)-GSH supramolecular hydrogels	–	4.23×10^{-4} M	Optical	[49]
F^-	FITC	APTMS	0–150 mM	Fluorescence	[50]

DNMA 2,20-dihydroxy-1-naphthylidine-10-naphthyl methyl amine, *MPTS* 3-(mercaptopropyl)trimethoxysilane, *SH-SAMMS* thiolated self-assembled monolayers on mesoporous silica, *PAR* 4-(2-pyridylazo)resorcinol, *MTMOS* methyltrimethoxysilane, *FITC* fluorescein isothiocyanate, *APTMS* 3-Aminopropyltrimethoxysilane synthesis are all contributory factors in the rising interest in sol-gel processing techniques. These sensors can be used for in vivo measurements in medical diagnosis, for environmental monitoring and for analysis in the food industry.

However, work still needs to be done to improve sensor performance. Issues such as microstructural stability, leaching, reversibility, response time, repeatability, sensitivity and selectivity all need to be optimised. It is hoped that resolution of these issues will lead to commercially viable, mass produced sol-gel-based sensors.

References

1. Monton, M.R.N., Forsberg, E.M., Brennan, J.D.: Tailoring sol-gel derived silica materials for optical biosensing. Chem. Mater. **24**, 796–811 (2012)
2. Vinogradov, V., Avnir, D.: Exceptional thermal stability of therapeutical enzymes entrapped in alumina sol-gel matrices. J. Mater. Chem. B **2**(19), 2868–2873 (2014)
3. MacCraith, B.D., McDonagh, C., O'Keeffe, G., Keyes, E.T., Vos, J.G., O'Kelly, B., McGilp, J.F.: Fibre optic oxygen sensor based on fluorescence quenching of evanescent-wave excited ruthenium complexes in sol-gel derived porous coating. Analyst **118**, 385–388 (1993)
4. Lee, S.-K., Okura, I.: Porphyrin-doped sol-gel glass as a probe for oxygen sensing. Anal. Chim. Acta **342**, 181–188 (1997)
5. McDonagh, C., MacCraith, B.D., McEvoy, A.K.: Tailoring of sol-gel films for optical sensing of oxygen in gas and aqueous phase. Anal. Chem. **70**, 45–50 (1998)
6. Xiong, Y., Ye, Z., Xu, J., Zhu, Y., Chen, C., Guan, Y.: An integrated micro-volume fiber-optic sensor for oxygen determination in exhaled breath based on iridium (III) complexes immobilized in fluorinated xerogels. Analyst **138**, 1819–1827 (2013)
7. Xu, H., Aylott, J.W., Kopelman, R., Miller, T.J., Philbert, M.A.: A real-time ratiometric method for the determination of molecular oxygen inside living cells using sol-gel based spherical optical nanosensors with applications to rat C6 glioma. Anal. Chem. **73**, 4124–4133 (2001)
8. Korzeniowska, B., Raspe, M., Wencel, D., Woolley, R., Jalink, K., McDonagh, C.: Development of organically modified silica nanoparticles for monitoring the intracellular level of oxygen using a frequency domain FLIM platform. RSC Adv. **5**, 36938 (2015)
9. Koo, Y.E.L., Cao, Y.F., Kopelman, R., Koo, S.M., Brausel, M., Philbert, M.A.: Real-time measurments of dissolved oxygen inside live cells by organically modified silicate fluorescent nanosensors. Anal. Chem. **76**, 2498–2505 (2004)
10. Von Bültzingslöwen, C., McEvoy, A.K., McDonagh, C., MacCraith, B.D., Klimant, I., Krausec, C., Wolfbeis, O.S.: Sol-gel based optical carbon dioxide sensor employing dual luminophore referencing for application in food packaging technology. Analyst **127**, 1478–1483 (2002)
11. Dansby Sparks, R., Jin, J., Mechery, S.J., Sampathkumaran, U., Owen, T.W., Yu, B.D., Goswami, K., Hong, K., Grant, J., Xue, Z.L.: Fluorescent-dye-doped sol-gel sensor for highly sensitive carbon dioxide gas detection below atmospheric concentrations. Anal. Chem. **82**, 593–600 (2010)
12. Seiyama, T., Kato, A., Fujiishi, K., Nagatani, M.: A new detector for gaseous components using semiconductive thin films. Anal. Chem. **34**, 1502–1503 (1962)
13. Bahar, M., Gholami, M., Azim Araghi, M.E.: Sol-gel synthesized titania nanoparticles deposited on porous polycrystalline silicon: improved carbon dioxide sensor properties. Mater. Sci. Semicond. Process. **26**, 491–500 (2014)
14. Fan, K., Qin, H., Wang, L., Ju, L., Hu, J.: CO_2 gas sensors based on La_{1-x} Sr_xFeO_3 nanocrystalline powders. Sens. Actuators B **177**, 265–269 (2013)

15. National Research Council (US) Committee on Acute Exposure Guideline Levels. Acute Exposure Guideline Levels for Selected Airborne Chemicals, Vol. 6. National Academies Press (US), Washington (DC) (2008) (Vol. 2, Ammonia Acute Exposure Guideline)
16. Senthil, T., Anandhan, S.: Structure–property relationship of sol-gel electrospun ZnO nanofibers developed for ammonia gas sensing. J. Colloid Interface Sci. **432**, 285–296 (2014)
17. Prasad, A.K., Kubinski, D.J., Gouma, P.I.: Comparison of sol-gel and ion beam deposited MoO_3 thin film gas sensors for selective ammonia detection. Sens. Actuators B **93**, 25–30 (2003)
18. Markovics, A., Kovacs, B.: Fabrication of optical chemical ammonia sensors using anodized alumina supports and sol gel method. Talanta **109**, 101–106 (2013)
19. Hazra, A., Das, S., Kanungo, J., Sarkar, C.K., Basu, S.: Studies on a resistive gas sensor based on sol-gel grown nanocrystalline p-TiO_2 thin film for fast hydrogen detection. Sens. Actuators B **183**, 87–95 (2013)
20. Hou, Y., Soleimanpour, A.M., Jayatissa, A.H.: Low resistive aluminum doped nanocrystalline zinc oxide for reducing gas sensor application via sol-gel process. Sensors and Actuators B **177**, 761–769 (2013)
21. Culotta, E., Koshland, D.E.: NO news is good news. Science **258**, 1862 (1992)
22. Lan, E.H., Dave, B.C., Fukuto, J.M., Dunn, B., Zink, J.I., Valentine, J.S.: Synthesis of sol-gel encapsulated heme proteins with chemical sensing Properties. J. Mater. Chem. **9**, 45–53 (1999)
23. Yu, M.R., Suyambrakasam, G., Wua, R.J., Chavali, M.: Preparation of organic–inorganic (SWCNT/TWEEN–TEOS) nano hybrids and their NO gas sensing properties. Sens. Actuators B **161**, 938–947 (2012)
24. Gouma, P.I., Kalyanasundaram, K.: A selective nanosensing probe for nitric oxide. Appl. Phys. Lett. **93**, 244102 (2008)
25. Mane, A.T., Kulkarni, S.B., Navale, S.T., Ghanwat, A.A., Shinde, N.M., Kim, J.H., Patil, V. B.: NO_2 sensing properties of nanostructured tungsten oxide thin films. Ceram. Int. **40**, 16495 (2014)
26. Wang, R., Li, G., Dong, Y., Chi, Y., Chen, G.: Carbon quantum dot-functionalized aerogels for NO_2 gas sensing. Anal. Chem. **85**, 8065–8069 (2013)
27. Lim, S.H., Feng, L., Kemling, J.W., Musto, C., Suslick, K.S.: An optoelectronic nose for detection of toxic gases. Nat. Chem. **1**, 562–567 (2009)
28. Bunkoeda, O., Davis, F., Kanatharanaa, P., Thavarungkula, P., Higson, S.P.J.: Sol-gel based sensor for selective formaldehyde determination. Anal. Chim. Acta **659**, 251–257 (2010)
29. Hue, J., Dupoy, M., Bordy, T., Rousier, R., Vignoud, S., Schaerer, B., Tran Thi, T.H., Rivron, C., Mugherli, L., Karpeda, P.: Benzene and xylene detection by absorbance in the range of 10–100 ppb application: quality of indoor air. Sens. Actuators B **189**, 194–198 (2013)
30. Luo, W., Deng, J., Fu, Q., Zhou, D., Hu, Y., Gong, S., Zheng, Z.: Nanocrystalline SnO_2 film prepared by the aqueous solgel method and its application as sensing films of the resistance and SAW H_2S sensor. Sens. Actuators B **217**, 119–128 (2015)
31. Tan, K.M., Tay, C.M., Tjin, S.C., Chan, C.C., Rahardjo, H.: High relative humidity measurement using gelatin coated long-period grating sensors. Sens. Actuators B **110**, 335–341 (2005)
32. Zhao, Z., Duan, Y.: A low cost fiber-optic humidity sensor based on silica sol–gel film. Sens. Actuators B **160**, 1340–1345 (2011)
33. Islam, T., Kumar, L., Khan, S.A.: A novel sol–gel thin film porous alumina based capacitive sensor for measuring trace moisture in the range of 2.5–25 ppm. Sens. Actuators B **173**, 377–384 (2012)
34. Lee, H.J., Park, K.K., Kupnik, M., Melosh, N.A., Khuri-Yakub, B.T.: Mesoporous thin-film on highly-sensitive resonant chemical sensor for relative humidity and CO_2 detection (2012)
35. Ates, T., Tatar, C., Yakuphanoglu, F.: Preparation of semiconductor ZnO powders by sol–gel method: humidity sensors. Sens. Actuators A **190**, 153–160 (2013)
36. Mashhadizadeh, M.H., Khani, H.: Sol–Gel-Au nano-particle modified carbon paste electrode for potentiometric determination of sub ppb level of Al(III). Anal. Methods **2**, 24–31 (2010)

37. Arvand, M., Asadollahzadeh, S.A.: Ion-selective electrode for aluminum determination in pharmaceutical substances, tea leaves and water samples. Talanta **75**, 1046–1054 (2008)
38. Feng, L., Zhang, Y., Wen, L., Chen, L., Shen, Z., Guan, Y.: Discrimination of trace heavy-metal ions by filtration on sol–gel membrane arrays. Chem. Eur. J. **17**, 1101–1104 (2011)
39. Lee, S.M., Lee, W.Y.: Determination of heavy metal ions using conductometric biosensor based on Sol-gel-immobilized urease. Bull. Korean Chem. Soc. **23**, 1169 (2002)
40. Yantasee, W., Charnhattakorn, B., Fryxell, G.E., Lin, Y., Timchalk, C., Addleman, R.S.: Detection of Cd, Pb, and Cu in non-pretreated natural waters and urine with thiol functionalized mesoporous silica and nafion composite electrodes. Anal. Chim. Acta. **620**, 55–63 (2008)
41. Jeronimo, P.C.A., Araujo, A.A., Conceicao, M., Montenegro, B.S.M., Pasquini, C., Raimundo Jr., I.M.: Direct determination of copper in urine using a sol–gel optical sensor coupled to a multicommutated flow system. Anal. Bioanal. Chem. **380**, 108–114 (2004)
42. Alcock, N.W.: Trace elements. In: Kaplan, L.A., Pesce, A.J. (eds.) Clinical Chemistry-Theory, Analysis and Correlation, 3rd edn. Mosby, St.Louis (1996)
43. Carrington, N.A., Xue, Z.L.: Inorganic sensing using organofunctional sol-gel materials. Acc. Chem. Res. **40**, 343–350 (2007)
44. Carrington, N.A., Thomas, G.H., Rodman, D.L., Beach, D.B., Xue, Z.: Optical determination of Cr(VI)using regenerable, functionalized sol-gel monoliths. Anal. Chim. Acta **581**, 232–240 (2007)
45. Kim, Eunjeong, Seo, Sungmin, Seo, Moo Lyong, Jung, Jong Hwa: Functionalized monolayers on mesoporous silica and on titania nanoparticles for mercuric sensing. Analyst **135**, 149–156 (2010)
46. Dhanya, S., Joy, J., Rao, T.P.: Fabrication and characterization of rhodamine 6G entrapped sol–gel film test strip for virtually specific and sensitive sensing of nitrite. Sens. Actuaturs B **173**, 510–516 (2012)
47. Wang, P., Wang, X., Bi, L., Zhu, G.: Renewable-surface amperometric nitrite sensor based on sol–gel-derived silicomolybdate–methylsilicate–graphite composite material. Analyst **125**, 1291–1294 (2000)
48. Wang, Q., Tan, C., Chen, H., Tamiaki, H.: A new fluoride luminescence quencher based on a nanostructured covalently bonded terbium hybrid material. J. Phys. Chem. C (2010)
49. Shen, J.S., Li, D.H., Cai, Q.G., Jiang, Y.B.: Highly selective iodide-responsive gel–sol state transition in supramolecular hydrogels. J. Mater. Chem. **19**, 6219–6224 (2009)
50. Appiah Ntiamoah, R., Jadhav, A.H., Puguan, J.M.C., Momade, F.W.Y., Kim, H.: A silica nanoparticle supported fluorescence "turnon" fluoride ion sensing system with tunable structure and sensitivity. RSC Adv. **5**, 30526 (2015)
51. Wang, E., Chow, K.F., Kwan, V., Chin, T., Wong, C.: Fast and long term optical sensors for pH based on sol–gels. Anal. Chim. Acta **495**, 45–50 (2003)
52. Nivens, D.A., Zhang, Y., Angel, S.M.: A fiber-optic pH sensor prepared using a base-catalyzed organo-silica sol-gel. Anal. Chim. Acta **376**, 235–245 (1998)
53. Jeon, D., Yoo, W.J., Seo, J.K., Shin, S.H., Han, K.T., Kim, S.G., Park, J.Y., Lee, B.: Fiber-optic pH sensor based on sol-gel film immobilized with neutral red. Opt. Rev. **20**, 209–213 (2013)
54. Lobnik, A., Majcen, N., Niederreiter, K., Uray, G.: Optical pH sensor based on the absorption of antenna generated europium luminescence by bromothymolblue in a sol-gel membrane. Sens. Actuators B **74**, 200–206 (2001)
55. Huang, W.D., Cao, H., Deb, S., Chiao, M., Chiao, J.C.: A flexible pH sensor based on the iridium oxide sensing film. Sens. Actuators A **169**, 1–11 (2011)
56. Silva, G.M., Lemos, S.G., Picrifka, L.A., Marreto, P.D., Rosario, A.V., Pereira, E.C.: Development of low-cost metal oxide pH electrodes based on the polymeric precursor method. Anal. Chim. Acta **616**, 36–41 (2008)

57. Gill, I., Ballesteros, A.: Encapsulation of biologicals within silicate, siloxane and hybrid sol-gel polymers: an efficient and generic approach. J. Am. Chem. Soc. **120**, 8587–8598 (1998)
58. Park, C.B., Clark, D.S.: Sol-gel encapsulated enzyme arrays for high-throughput screening of biocatalytic activity. Biotech. Bioeng. **78**, 229–235 (2002)
59. Ansari, A.A., Solanki, P.R., Malhotra, B.D.: Sol gel derived nanostructured cerium oxide film for glucose sensor. Appl. Phys. Lett. **92**, 263901 (2008)
60. Ansari, A.A., Pandey, P., Malhotra, B.D.: Sol-gel derived nanoporous cerium oxide-titanium oxide platform for glucose sensor. Adv. Sci. Eng. Med. **5**, 1–7 (2013)
61. Xu, C., Ren, J., Feng, L., Xu, X.Q.: H_2O_2 triggered sol–gel transition used for visual detection of glucose. Chem. Commun. **48**, 3739–3741 (2012)
62. Yao, H., Shum, A.J., Cowan, M., Lähdesmäki, I., Parviz, B.A.: A contact lens with embedded sensor for monitoring tear glucose level. Biosens. Bioelectron. **2011**(26), 3290–3296 (2011)
63. Baca, J.T., Taormina, C.R., Feingold, E., Finegold, D.N., Grabowski, J.J., Asher, S.A.: Mass spectral determination of fasting tear glucose concentrations in nondiabetic volunteers. Clin. Chem. **53**, 1370–1372 (2007)
64. Khoo, S.B., Chen, F.: Studies of sol-gel ceramic film incorporating methylene blue on glassy carbon: an electrocatalytic system for the simultaneous determination of ascorbic and uric acids. Anal. Chem. **74**, 5734–5741 (2002)
65. Hossain, S.M.Z., Luckham, R.E., Smith, A.M., Lebert, J.M., Davies, L.M., Pelton, R.H., Filipe, C.D.M., Brennan, J.D.: Development of a bioactive paper sensor for detection of neurotoxins using piezoelectric inkjet printing of sol-gel-derived bioinks. Anal. Chem. **81**, 5474–5483 (2009)
66. Yu, D., Volponi, J., Chhabra, S., Brinker, C.J., Mulchandani, A., Singh, A.K.: Aqueous sol–gel encapsulation of genetically engineered *Moraxella* spp. cells for the detection of organophosphates. Biosens. Bioelectron. **20**, 1433–1437 (2005)
67. Ponamoreva, O.N., Kamanina, O.A., Alferov, V.A., Machulin, A.V., Rogova, T.V., Arlyapov, Alferov S.V., Suzina, N.E., Ivanova, E.P.: Yeast-based self-organized hybrid bio-silica sol-gels for the design of biosensors. Biosens. Bioelectron. **67**, 321–326 (2015)
68. Carrasquilla, C., Lau, P.S., Li, Y., Brennan, J.D.: Stabilizing structure-switching signaling rna aptamers by entrapment in sol–gel derived materials for solid-phase assays. J. Am. Chem. Soc. **134**, 10998–11005 (2012)
69. Solanki, P.R., Patel, M.K., Kaushik, A., Pandey, M.K., Kotnala, R.K., Malhotra, B.D.: Sol-gel derived nanostructured metal oxide platform for bacterial detection. Electroanalysis **23**, 2699–2708 (2011)
70. Gubala, A.J.: Multiplex real-time PCR detection of *Vibrio cholera*. J. Microbiol. Meth. **65**, 278 (2006)
71. Zeisel, S.H., da Costa, K.A.: Choline: an essential nutrient for public health. Nutr. Rev. **67**, 615–623 (2009)
72. Shaw, G., Carmichael, S., Yang, W., Selvin, S., Schaffer, D.: Periconceptional dietary intake of choline and betaine and neural tube defects in offspring. Am. J. Epidemiol. **160**, 102–109 (2004)
73. Mazurenko, I., Tananaiko, O., Biloivan, O., Zhybak, M., Pelyak, I., Zaitsev, V., Etienne, M., Walcarius, A.: Amperometric biosensor for choline based on gold screen-printed electrode modified with electrochemically-deposited silica biocomposite. Electroanalysis (2015)
74. George, S., Brat, P., Alter, P., Amiot, M.J.: Rapid determination of polyphenols and vitamin C in plant-derived products. J. Agric. Food Chem. **53**, 1370–1373 (2005)
75. Casero, E., Petit-Domínguez, M.D., Vázquez, L., Ramírez-Asperilla, I., Parra-Alfambra, A. M., Pariente, F., Lorenzo, E.: Laccase biosensors based on different enzyme immobilization strategies for phenolic compounds determination. Talanta **115**, 401–408 (2013)
76. Pena Vázquez, E., Maneiro, E., Pérez-Conde, C., Moreno-Bondi, Costas E.: Microalgae fiber optic biosensors for herbicide monitoring using sol–gel technology. Biosens. Bioelectron. **24**, 3538–3543 (2009)
77. European Union, Brussels: Water framework directive (2000/60/EC) (2000)

78. Zhu, L., Gamez, G., Chen, H., Chingin, K., Zenobi, R.: Rapid detection of melamine in untreated milk and wheat gluten by ultrasound-assisted extractive electrospray ionization mass spectrometry (EESI-MS). Chem. Commun. **5**, 556–558 (2009)
79. Coghlan, A.: Reuters. Melamine 'widespread' in China's food chain. New Scientist (2008)
80. Brown, C.A., Jeong, K.S., Poppenga, R.H., Puschner, B., Miller, D.M., Ellis, A.E., Kang, K. I., Sum, S., Cistola, A.M., Brown, S.A.: Outbreaks of renal failure associated with melamine and cyanuric acid in dogs and cats in 2004 and 2007. J. Vet. Diagn. Invest. **19**, 525–531 (2007)
81. Xu, G., Zhang, H., Zhong, M., Zhang, T., Lu, X., Kan, X.: Imprinted sol–gel electrochemical sensor for melamine direct recognition and detection. J. Electroanal. Chem. **713**, 112–118 (2014)
82. Azahar Ali, M., Srivastava, S., Solanki, P.R., Agrawal, V.V., John, R., Malhotra, B.D.: Nanostructured anatase-titanium dioxide based platform for application to microfluidics cholesterol biosensor. Appl. Phys. Lett. **101**, 084105 (2012)
83. Ahn, K.Y., Kwon, K., Huh, J., Kim, G.T., Lee, E.B., Park, D., Lee, J.: A sensitive diagnostic assay of rheumatoid arthritis using three-dimensional ZnO nanorod structure. Biosens. Bioelectron. **28**, 378–385 (2011)
84. Chooduma, A., Kanatharana, P., Wongniramaikul, W., NicDaeid, N.: A sol–gel colorimetric sensor for methamphetamine detection. Sens. Actuators B **215**, 553–560 (2015)
85. Liu, L., Gill, S.K., Gao, Y., Hope-Weeks, L.J., Cheng, K.H.: Exploration of the use of novel SiO_2 nanocomposites doped with fluorescent Eu^{3+}/sensitizer complex for latent fingerprint detection. Forensic Sci. Int. **176**(2008), 163–172 (2008)
86. Beyazkilic, P., Yildirim, A., Bayindir, M.: Formation of pyrene excimers in mesoporous ormosil thin films for visual detection of nitro-explosives. ACS Appl. Mater. Interfaces **6**, 4997–5004 (2014)

Chapter 6
Sol-Gel Processed Cathode Materials for Lithium-Ion Batteries

Vinodkumar Etacheri

6.1 Introduction

Nowadays, electricity production from renewable sources and reduction of dependence of transportation on fossil fuels are the two main challenges faced by modern society. Since energy generation from sustainable energy sources fluctuates considerably, development of suitable energy storage devices is necessary. Use of electric vehicles to reduce carbon footprint also requires efficient energy storage devices namely batteries. Despite the advances in the energy harvesting technologies such as photovoltaic cells, wind turbines, and photothermal receivers, advances in the area of energy storage devices are still lagging behind [1–3]. Consequently, development of environmental friendly batteries with high power/energy density and long cycle life is a major challenge in modern electrochemistry research. Commercialization of rechargeable Li-ion batteries happened in early 1990s, and only this system has the practical energy/power density and cycle life required for modern electric vehicles [1]. Current generation rechargeable Li-ion batteries contain $LiCoO_2$ cathode, graphite anode, and an organic electrolyte (Fig. 6.1) [1]. The main cell-reaction involves reversible Li-ion intercalation/deintercalation between the layered electrodes during charge–discharge process. Rechargeable Li-ion batteries have also attracted noteworthy interest in the recent years due to their excellent electrochemical properties compared to other energy storage devices [1, 4, 5]. They are successfully implemented in portable electronics, medical implants, electric vehicles, etc. [1, 6–13].

Though rechargeable Li-ion batteries are superior to other secondary batteries (Ni-MH, Lead-acid, etc.), their energy and power densities are not enough for next-generation energy demanding applications. One of the most efficient methods to improve the performance of rechargeable Li-ion batteries is by increasing the

V. Etacheri (✉)
IMDEA Materials Institute, C/Eric Kandel 2, 28906 Getafe, Madrid, Spain
e-mail: vinodkumar.etacheri@imdea.org

S.C. Pillai and S. Hehir (eds.), *Sol-Gel Materials for Energy, Environment and Electronic Applications*, Advances in Sol-Gel Derived Materials and Technologies,
DOI 10.1007/978-3-319-50144-4_6

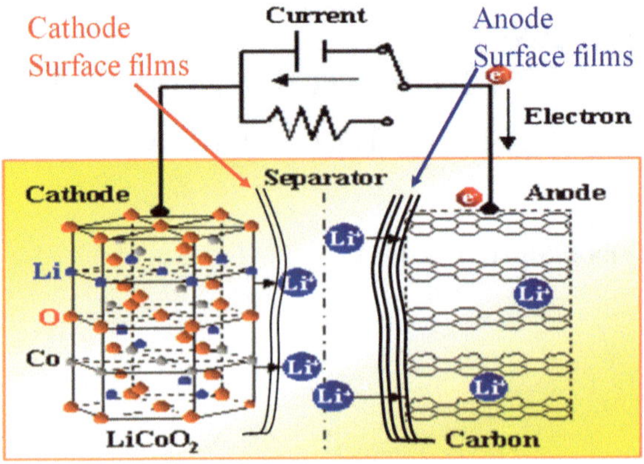

Fig. 6.1 Schematic of most commonly used rechargeable Li-ion battery containing graphite anode and $LiCoO_2$ cathode. Reproduced with permission from Ref. [1]. Copyright 2011 Royal Society of Chemistry

specific capacity of the cathode materials. The electrochemical performance of cathodes (specific capacity, rate performance, and cycling stability) highly depends on the synthetic method employed, which can be used to control the material properties (particle size, phase purity, crystallinity, and cation disorder). Usually, these electrochemically active materials are synthesized through high-temperature heat treatment that leads to excessive particle growth and poor electrochemical performance. Sol-gel process is demonstrated as a potential route for the fabrication of high-performance phase-pure, surface modified, and hybrid cathode materials. Various efforts and advancements in this area of research for the improvement of Li-ion battery electrochemistry are herein summarized.

6.2 Sol-Gel Process

Sol-gel process has its origin from the early works on colloidal chemistry by Thomas Graham in 1861. Sols and gels were two forms of matter that have been known to exist naturally, for a long time. And for long time, sols and gels were only of pure scientific interest. Sol-gel process is an effective method for the synthesis of high surface materials. Sols and gels have extremely high surface areas, which made them increasingly interesting in the field of functional materials for catalysis, energy storage, and conversion [14]. The different steps involved in the sol-gel process are presented in Fig. 6.2. Depending on the precursor type, two types of the sol-gel process can be distinguished; first one is alcohol-based process and the precursor is a metal alkoxide; and next one is aqueous-based process and the precursor is an inorganic metal salt [15]. Sol-gel method has several advantages compared to other synthetic techniques. For instance, particle size, homogeneity, purity, stoichiometry, and dopant concentration can be easily controlled. Additionally, modification of the sol-gel process with suitable organic compounds allows the *in situ* formation of conductive carbon during heat treatment, which is

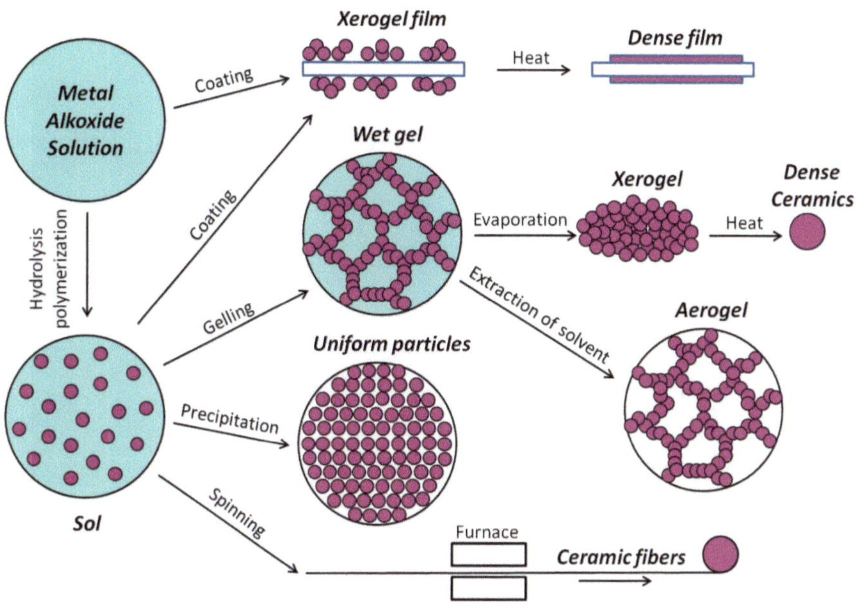

Fig. 6.2 Different steps involved in a sol-gel process

beneficial for obtaining desired electronic properties [16, 17]. Moreover, processing temperature and time required for material synthesis can be significantly lowered using the sol-gel method.

Recently, the sol-gel method has been applied in the synthesis of very large surface area mesoporous metal oxides, phosphates, sulfates, etc. [18–20]. Even small amount of dopants can be inserted into preferred positions through the sol-gel method due to molecular-level mixing of dispersed ions [16, 17]. Usually organic compounds such as ethylene glycol, ascorbic acid, and citric acid are used as chelating agents for the in situ generation of carbon [17, 21]. Ionic and neutral surfactants are successfully employed in this synthetic method as templates to prepare mesoporous materials [22–27]. Non-surfactant organic compounds such as diolates [18, 28, 29] and glycerine [30, 31] can also be used as pore formers. Sol-gel methods coupled with hydrothermal routes also offer a convenient method for the synthesis of mesoporous materials [31, 32].

(i) **Alcohol-based process**

Alkoxides are used as precursors in the alcohol-based sol-gel process. There exist metal–oxygen bonds in these alkoxides, and due to the large difference in electronegativity between metal and oxygen, the bonds become highly polar and extremely reactive. A simultaneous hydrolysis and condensation reaction proceeds, which ultimately leads to the formation of a gel. Various steps involved in this alcohol-based sol-gel process are discussed in detail in the following sections.

Scheme 6.1 Hydrolysis reaction of metal alkoxide

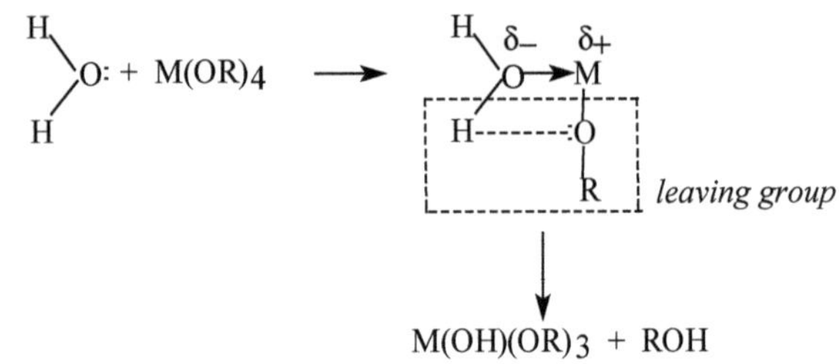

$$M_1\text{-}OX{+}M_2\text{-}OY \longrightarrow M_1\text{-}OX\text{-}M_2{+}OY$$

$$M_1\text{-}OX{+}M_2\text{-}OY \longrightarrow M_1\text{-}OX\text{-}M_2\text{-}OY$$

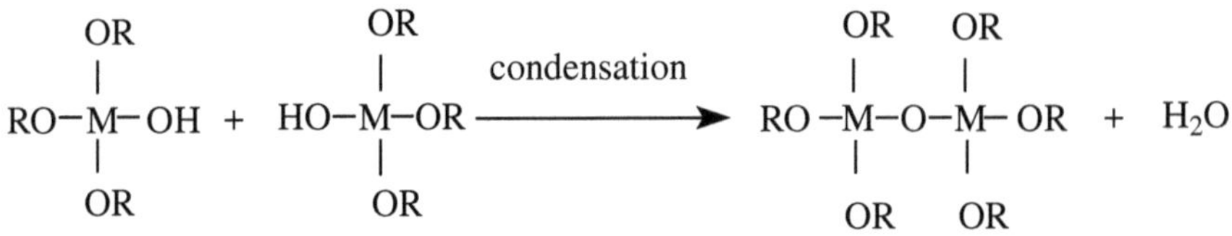

Scheme 6.2 Condensation reaction of metal alkoxide

(a) *Hydrolysis*: Hydrolysis involves the replacement reactive OR group by an OH group, that is the deprotonation of a solvated metal cation; the reaction can be represented as in Scheme 6.1.
(b) *Condensation*: Two partially hydrolyzed molecules can link together in a condensation reaction. Condensation involves removal of a molecule of water or alcohol and forms a metal–oxygen–metal bond (Scheme 6.2).

Condensation proceeds by either of two nucleophilic mechanisms depending on the co-ordination of the metal. When the preferred co-ordination is satisfied the condensation occurs by nucleophilic substitution. When the preferred co-ordination is not satisfied condensation can occur by nucleophilic addition [33]. The gel formed as a result of hydrolysis and condensation can be converted into oxide nanomaterials through proper heat treatment.

(ii) **Aqueous-based process**

In aqueous sol-gel method, sol is usually obtained by the dispersion of freshly prepared hydrated oxides, and the process consists of the following steps [34].

(a) *Precipitation*: This involves the hydrolysis of inorganic metal salts, which hydrolyze rapidly to get a gelatinous precipitate by the addition of a base followed by washing to remove excess electrolyte.

(b) *Peptization*: This is the process of redispersing the precipitate into a colloid, which can be coagulated. This refers to a direct disintegration of substance into particles of colloid size by an added agent known as peptizing agent around pH 3 [34].
(c) *Aging*: Aging is a process in which the obtained sol or gel is allowed to remain in a definite atmosphere so that all the reactions and processes attain completion. An as-prepared gel is aged in the mother liquor for the hydrolysis condensation reactions to be complete. Similar processes are used (aging in different solvents and conditions) to strengthen the gel. Aging leads to an equilibrium size distribution by the depletion of smaller particles and subsequent growth of larger ones. Aging also produces coarser particles with better physical properties. The gel obtained after aging can be converted into oxide nanomaterials through proper heat treatment [33, 34].

6.3 Sol-Gel Synthesis of Cathode Materials

6.3.1 Conventional Cobaltite Cathodes

Since the commercialization of rechargeable Li-ion batteries, $LiCoO_2$ with a theoretical capacity of 140 mAh g^{-1} is used as the cathode material. $LiNiO_2$, $LiMnO_2$, and $LiMn_2O_4$ have also attracted serious attention due to their Li insertion/extraction properties [35–37]. Their satisfactory performance resulted from the layered structure, which enables rapid Li-ion diffusion and a high working voltage (3.9 V vs. Li/Li^+) [38, 39]. However, their useful capacity is always limited to 50% of the theoretical capacity due to reversible extraction of only 0.5 Li atoms per unit structure [40]. This cathode can be dissolved in the electrolyte as a result of HF (formed by the reaction between $LiPF_6$ and water) attack and results in capacity fade [41, 42]. Other drawbacks of this cathode material are toxicity, expensiveness, low electronic conductivity, and limited availability of cobalt [43–46]. Various strategies are employed to overcome these limitations of conventional layered cathode materials. Surface modification is one of them, in which individual particles of $LiCoO_2$ are coated with a thin layer of metal oxide such as ZrO_2, Al_2O_3, and TiO_2 [47–50]. These coatings maintain Li-ion diffusion pathways and improve electrode stability by preventing reaction between the electrolyte decomposition product and Co^{4+} ions of the active material.

Complex sol-gel methods are reported for the synthesis of Zr- and Mg-doped $LiCoO_2$ cobaltites, which resulted in the formation of composite electrode containing ZrO_2, MgO, and Co_3O_4 inactive metal oxides [50]. Improved cycling stability, bulk, and interfacial transport properties of this composite electrode can be explained by the simultaneous effect of Zr^{4+} doping and morphology modification (by inhibiting grain growth) resulting from inactive ZrO_2/MgO dispersion. Surface of inactive oxides can also perform as interconnects for Li-ion diffusion, thereby enhancing the charge–discharge performance [51, 52]. An alternative method is to dope $LiCoO_2$ with aliovalent cations like Mg^{2+} that can change the valance state of individual ions and improve the electronic conductivity. It has been also proved that the dopant remains unreactive throughout the electrochemical charge–discharge process [50, 53–56]. Doping with other cations such as Mo^{6+} Zr^{4+} Ti^{4+} and V^{5+} resulted in performance enhancements, which strongly depended on the synthetic method used [57–59]. Mixed cobaltites (Ni–Mn, Ni–Co, and Ni–Co–Mn type) have been also doped with Zr^{4+} to obtain improved performance [60, 61]. In a sol-gel process to dope aliovalent ions, Zr and Mg precursors are mixed with Co and Li salts and a citric acid complexing agent followed by gelation and heat treatment. Obtained cathode nanopowders consist of undoped/Zr-doped cobaltites with oxides of Zr, Mg, or Co. Zr and Mg doping also resulted in the reduction of average particle size of these cathode materials (Fig. 6.3a–c). Electrochemistry of these composites highly depends on their morphology and composition. For instance, ultrafine $LiCo_{0.94}Zr_{0.06}O_2/MgO/ZrO_2$ composites exhibited best electrochemical performance only at higher cut-off potentials and current rates rather than low rates and potentials (Fig. 6.3d–e). Electrochemical and spectroscopic studies proved that Zr-doping caused reduced electrode polarization electronic conductivities in

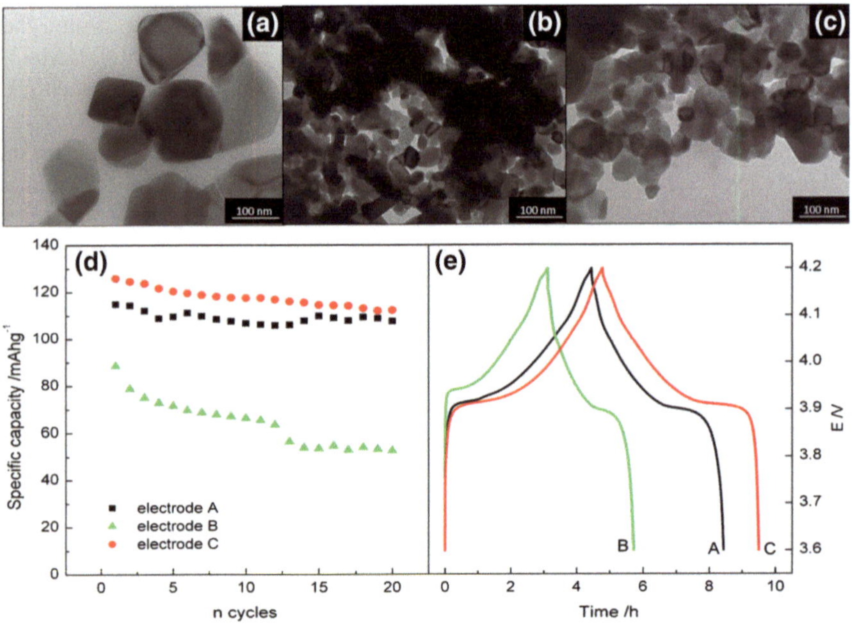

Fig. 6.3 TEM images of **a** $LaiCoO_2$, **b** $LiCo_{0.92}Zr_{0.08}O_2/ZrO_2$, and **c** $LiCo_{0.94}Zr_{0.06}O_2/MgO/ZrO_2$. **d** Galvanostatic cycling performance and **e** corresponding third-cycle voltage profiles of (A) $LiCoO_2$, (B) $LiCo_{0.92}Zr_{0.08}O_2/ZrO_2$, and (C) $LiCo_{0.94}Zr_{0.06}O_2/MgO/ZrO_2$ at a charge/discharge rate of C/5. Reproduced with permission from Ref. [50]. Copyright 2012 Elsevier Science

comparison to commercial cobaltite with larger grain size. Though several synthetic methods and modifications are reported for cobaltite cathode material, performance enhancement is nominal.

6.3.2 Polyanion-Type Cathodes

Due to the low electrochemical performance of modified layered cobaltite cathodes, extensive research has been devoted to the development of alternative electrode materials [62, 63]. Cathode materials based on phosphate polyanion building blocks have been recently suggested as potential cathode materials that can replace $LiCoO_2$ for Li-ion batteries (Fig. 6.4). Among these materials $LiFePO_4$ is the first one to be identified and systematically investigated [64]. In 1997 Goodenough and coworkers proposed olivine structured lithium iron phosphate ($LiFePO_4$) as a potential cathode material [65]. Advantages of this new generation of cathodes include high specific

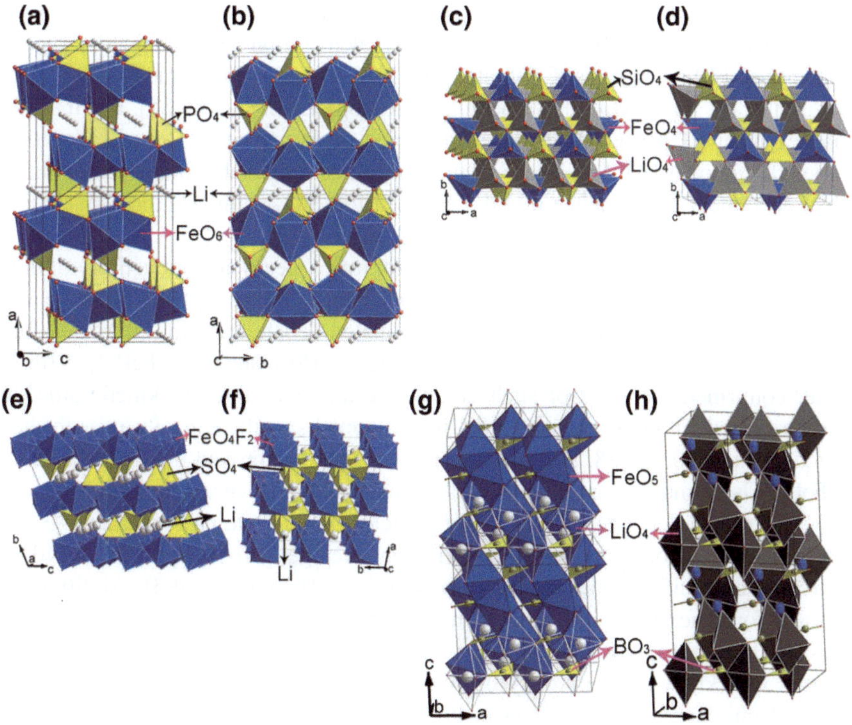

Fig. 6.4 Crystal structure of polyanion-type cathode materials **a**, **b** $LiFePO_4$ viewed along *b* and *c*-axis, respectively, **c** Li_2FeSiO_4 (space group P*mn*2_1) viewed along the *c*-axis, **d** Li_2FeSiO_4 (space group P*mnb*) viewed along the *c*-axis., **e**, **f** $LiFeSO_4F$ viewed along the *a*- and *c*-axis, respectively, **g**, **h** $LiFeBO_3$ viewed along the *b*-axis with and without FeO_5 bipyramid respectively. Reproduced with permission from Ref. [67]. Copyright 2011 Royal Society of Chemistry

capacity (170 mAh g^{-1}), improved coulombic efficiency, flat charge–discharge potential, low coat, and environmental compatibility [65, 66]. Strong covalent bonding between oxygen and phosphorous atoms in phosphate results in the superior performance of olivine-type cathodes [67]. Oxygen removal from this cathode structure is limited in this case due to the existence of highly stable phosphorous–metal bonds [68]. Working voltage of this electrode is enhanced by the inductive effect of phosphate functional groups [69]. In addition, volume changes only by 6.81% upon Li-ion extraction and insertion, which ensures superior cathode stability during charge–discharge of batteries.

However, the main drawback of these olivine families of electrodes is their low electrical conductivity (10^{-9}–10^{-10} Scm^{-1}) and poor Li-ion diffusion, which hinder the electrochemical performance of $LiFePO_4$ in secondary Li-ion batteries [66, 70]. Numerous strategies, such as particle size reduction, carbon coating, metal dispersion, cation/anion doping, and optimization of synthetic methods, have proven to be efficient in overcoming the above-mentioned drawbacks [70–79]. For example, Yamada et al. proved that particle size control through annealing temperature control is crucial for obtaining high-performance $LiFePO_4$ cathode [80]. Instead, Chung et al. suggested that electronic conductivity of $LiFePO_4$ could be significantly improved by doping with supervalent cations, and attained a specific capacity of 140 mAh g^{-1} at C/10 rate [70, 81]. A number of ions such as Ti^{4+}, Cr^{3+}, V^{3+}, Ni^{2+}, Al^{3+}, and Mn^{2+} have been used to dope $LiFePO_4$ [82–85]. For instance, Gao et al. and Chen et al. reported improved performance of Fe- and Co-doped $LiFePO_4$ cathodes [83, 86]. These results confirmed remarkable improvement in the electrochemical performance of 1% Fe-doped $LiFePO_4$, with specific capacities of 114.8 and 104.2 mAh g^{-1} at 10 and 20 °C, respectively.

Nevertheless, Nazar et al. attributed the improved electrical conductivity of these doped materials to the formation of iron phosphide during high-temperature synthesis [87]. These controversial results stimulated debate regarding the need for a conductive agent to improve the electrochemical performance of $LiFePO_4$. Armand et al. confirmed the need of such conductive additives through kinetic study of carbon-coated $LiFePO_4$ particles, and they attained specific capacity near to theoretical limit at 80 °C [88, 89]. Though previous studies proved that Li diffusion in $LiFePO_4$ is primarily one-dimensional along the *b*-axis, recent modeling studies proved a strong dependence of smaller particle size and rate performance [90, 91]. For instance, carbon-free $LiFePO_4$ with an average particle size of $\sim$140 nm delivered a specific capacity of 140 mAh g^{-1} at 10C rate [92, 93]. Additionally, high-performance nanosized (20–40 nm) carbon-free $LiFePO_4$ has been also synthesized through polyol process and various sol-gel methods [79, 94–97].

It is well known that porosity of electrode materials plays an important role in their electrochemical performance [94, 98–101]. Several methods have been developed for the pore size control of cathode materials [102–106]. Pores in the active electrode material act as channels for Li-ion supply and material kinetic is decided by the distance between the pores [94]. As mentioned earlier, previous reports proposed that nanostructure, carbon coating, and supervalent ion doping are required for attaining full capacity of $LiMPO_4$ cathodes at room temperature [70,

77, 107–109]. Yamada et al. suggested that internal friction during charge–discharge process and slow kinetics limit the electrochemical performance of $LiMPO_4$-type cathodes [110]. On the contrary, Dominko et al. developed a citric acid-modified sol-gel method for the porosity-controlled synthesis of porous micron-sized $LiMPO_4$/C hybrids (M=Fe and/or Mn), which perform similar to the nanoparticle electrode [94]. They explained the porous microstructure of micron-sized particles as an inverse picture of nanoparticulate arrangement (Fig. 6.5a). In this study, citrate anions are implemented as building blocks for transparent gels containing lithium, transition metal, and phosphate. Further heat treatment of xerogels with molecular-level homogeneity resulted in the formation of porous $LiMPO_4$/C composite cathode (Fig. 6.5b–c). In the citric acid-modified sol-gel method, *in-situ* decomposition of citrate ions cause gas generation and a consequent pore formation. Pure carbon formed during citrate degradation is uniformly coated on individual cathode particle. Additionally, Fe^{III} impurities are efficiently reduced to Fe^{II} by citrate decomposition products [111]. During electrochemical charge–discharge process, electrolyte filled in the pores assisted fast ionic conduction, and carbon coating resulted in increased electronic conduction.

Several macromolecular carbon precursors such as polyvinyl alcohol (PVA), stearic acid, and polyvinylpyrrolidone (PVP) have been recently employed in modified sol-gel process [112–114]. All these additives act as both source of carbon

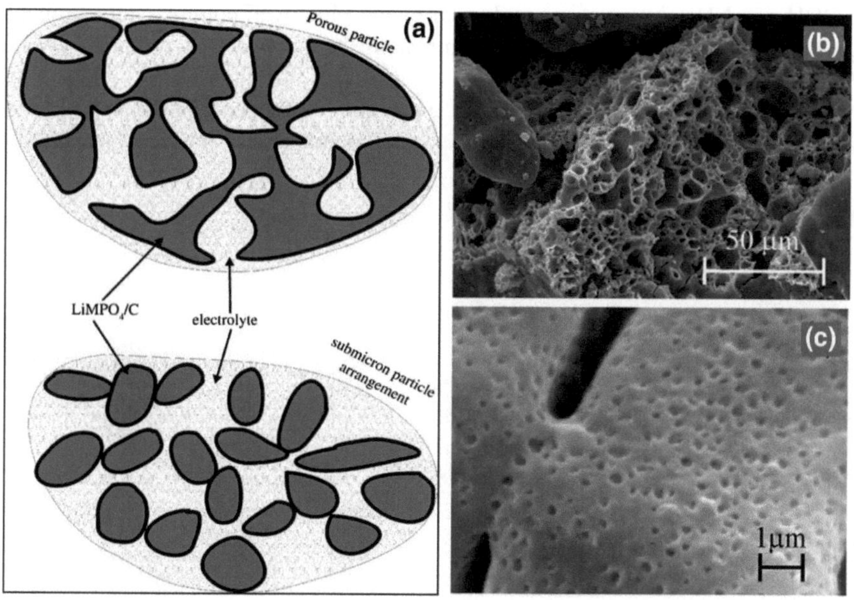

Fig. 6.5 **a** Schematic of porous submicron particle arrangement and **b**, **c** SEM images of $LiFePO_4$/C composites prepared through citric acid-modified sol-gel method. Reproduced with permission from Ref. [94]. Copyright 2006 Elsevier Science

coating the cathode particles and excellent dispersing agents reducing the particle size/agglomeration at the same time [66, 115, 116]. For instance, gelatin-modified sol-gel method (Fig. 6.6) was found to be highly effective for preparing carbon-coated $LiFePO_4$ nanoparticles [66]. Electrochemical performances of these cathode nanoparticles with a thin layer of carbon coating (2–5 nm thickness) and a tap density of 1.42 g cm^{-3} are excellent, exhibiting a high discharge capacity of 128 mAh g^{-1} at 5C and capacity retention of 95.2% at 0.2C after 100 cycles. This improved electrochemical performance also point toward the necessity to prepare nanosized $LiFePO_4/C$ particles for enhanced electrochemical performance.

Unlike citric acid-based sol-gel method, gelatin-modified sol-gel method is simple and does not require temperature control between 60 and 90 °C, use of additives, and pH adjustments due to reversible thermo-gelation behavior of gelatin [74, 94, 117, 118]. Conversion of gelatin into a gel on cooling is easy to control and beneficial for the synthesis of nanoscale particles. In this synthetic method, electrostatic interaction between gelatin and precursors resulted in their uniform dispersion. On cooling this solution below the gel point, hydrogen bond cross-links are formed between the three adjacent left-handed spiral segments of gelatin, forming a three-dimensional net-like polypeptide molecule [119, 120]. Water molecules are then trapped between the gaps of the 3-D superhelix network and hydrogen bonded to CO or NH groups. Uniform dispersion of precursors in gelatin was further evidenced by the narrow size distribution (100–300 nm) of $LiFePO_4$ particles. Role of gelatin as a good carbon source was also confirmed by a residual carbon content of 8.6 wt%. Gelatin also promotes $LiFePO_4$ crystallization due to the presence of COOH and NH_2 groups on its molecular chain. Moreover, commercially available gelatin is an inexpensive and environmentally benign material produced by the hydrolysis of collagen [66].

Surfactant-assisted sol-gel methods were also developed for the synthesis of olivine cathode with ultrafine particle size. For example, Choi et al. reported a

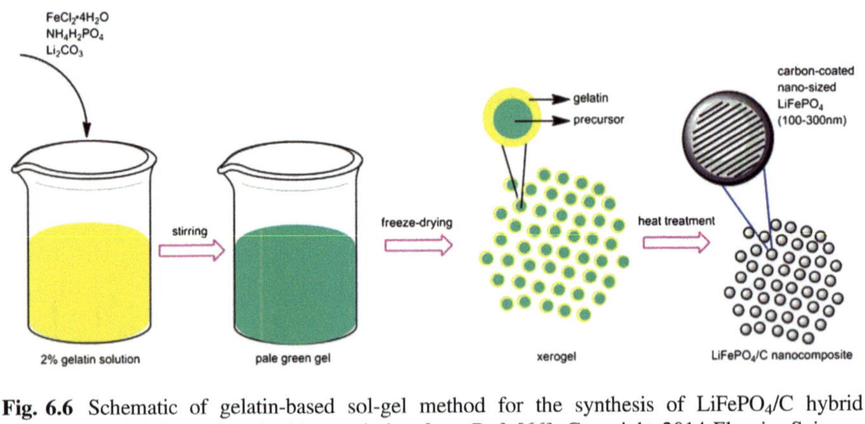

Fig. 6.6 Schematic of gelatin-based sol-gel method for the synthesis of $LiFePO_4/C$ hybrid cathode material. Reproduced with permission from Ref. [66]. Copyright 2014 Elsevier Science

Lauric acid-modified ethanol-based sol-gel method for the fabrication of $LiFePO_4$ with an average particle size of <65 nm and surface area ranging from 8 to 36 m^2 g^{-1} (depending on the amount of surfactant) [79]. These cathodes exhibited specific capacity of 157 and 123 mAh g^{-1} at charge–discharge rates of 1C and 10C, respectively, with less than 0.08% capacity fade per cycle. Surfactant-assisted sol-gel method has been also reported for submicron-sized $LiFePO_4$/C and $LiFe_{0.99}Ru_{0.01}PO_4$/C cathodes (Fig. 6.7) [83]. In this method, Lauric acid surfactant effectively prevents particle growth/agglomeration of $LiFePO_4$ particles.Li-ion diffusion kinetics is improved as a result of Ru doping, and $LiFe_{0.99}Ru_{0.01}PO_4$/C exhibits outstanding high-rate performances (107 mAh g^{-1} at 10C) and specific capacity (162 mAh g^{-1} at 0.1C) compared to pure $LiFePO_4$/C. Thus it is clear that surfactant-assisted sol-gel method offers a robust, simple, and inexpensive method for the synthesis of $LiFePO_4$.

Another member of the olivine family, $LiMnPO_4$ also received great deal of attention and investigated as a promising cathode material. However, charge transport rates are poor, which restricted their commercialization in secondary Li-ion batteries. Several successful attempts were made by previous researchers to improve the electrochemical performance of $LiMnPO_4$ either by carbon coating or by doping with transition metal ions. In the case of $LiMn_{0.9}Fe_{0.1}PO_4$, Ye et al. reported an improved discharge capacity of 149.2mAh g^{-1} at 0.1C rate. Carbon coating of $LiMn_{0.71}Fe_{0.29}PO_4$ enhanced both electronic conductivity and electrochemical performance. Wang et al. employed a sol-gel method to grow $LiMn_{0.75}Fe_{0.25}PO_4$ on graphene for improved electrochemical performance. *N*,*N*-dimethyl formamide (DMF) acts as an effective dispersing agent in modified sol-gel method, and the developed $LiMn_{0.5}Fe_{0.5}PO_4$/C materials demonstrated excellent electrochemical properties. Among the various Fe-doped compositions prepared,

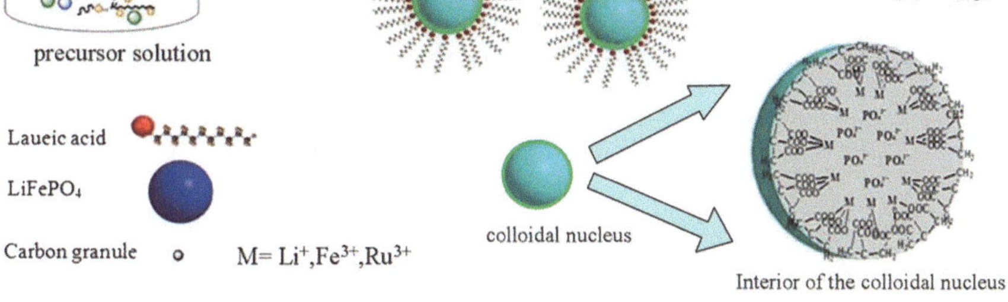

Fig. 6.7 Schematic of the synthesis of Ru-doped $LiFePO_4$/C hybrid cathode. Reproduced with permission from Ref. [83]. Copyright 2014 Wiley VCH

best specific capacity (132 mAh g^{-1} at 1C) and cycling stabilities (96.3% after 50 cycles) are identified for the composition $LiMn_{0.5}Fe_{0.5}PO_4/C$.

Exploiting the full potential of olivine-type polyanion cathodes are often hindered by the reduced high voltage stability of organic solvents. Hence, alternative Li-ion batteries based on LiOH electrolyte and spinel cathodes have received great attention [121–125]. Main advantages of this system are high charge–discharge rates and lower voltage drops resulting from improved (up to twofold higher) ionic conductivity of aqueous electrolytes than those of organic electrolytes. However, limited voltage (~2 V) is the main disadvantage of the aqueous Li-ion battery system. In order to compensate this low-voltage issue, charge–discharge rates and storage capacity of this system must be considerably higher to those of conventional nonaqueous Li-ion batteries. Olivines synthesized by high-temperature solid-state reaction are usually agglomerates of irregularly distributed particles due to prolonged high-temperature treatment [16]. Additionally, final product often contains silica and alumina impurities from crucibles, which further reduces the low electronic and ionic conductivity of spinel cathodes. Kandhasamy et al. proposed PVP-modified sol-gel method (Fig. 6.8) to overcome the intrinsic drawbacks of olivine cathodes [17]. When the homo-polymer chelating agent PVP is added to metal colloidal solution, imide groups strongly interact with each metal colloid unit, stabilizing the particle growth and increasing nucleus formation [126, 127]. An organic interface is formed between the cathode particles and electrolyte on heat treatment of this complex, increasing the electrolyte penetration (ionic mobility)

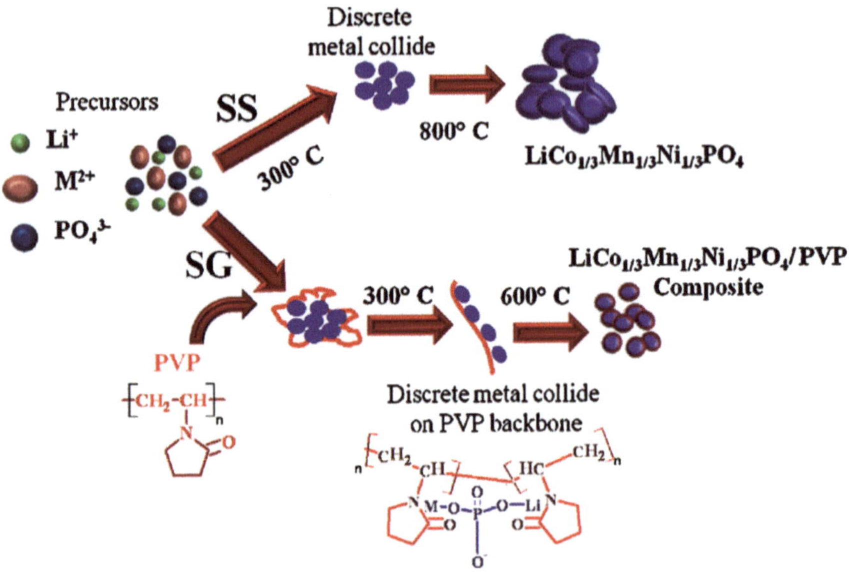

Fig. 6.8 Schematic of the synthetic mechanism of $LiCo_{1/3}Mn_{1/3}Ni_{1/3}PO_4$ through PVP-assisted sol-gel and solid-state method. Reproduced with permission from Ref. [17]. Copyright 2012 Elsevier Science

and electronic conductivity [128, 129].$LiCo_{1/3}Mn_{1/3}Ni_{1/3}PO_4$ cathode prepared by PVP-modified sol-gel method consists of ultrafine carbon-coated primary particles, which outperformed identical olivine cathode prepared through solid-state method in an aqueous Li-ion battery.

Goodenough et al. for the first time reported a new class of NASICON (sodium super ionic conductor)-based cathode material in 1997 [130]. $Li_2FeTi(PO_4)_3$, $Li_2NaV_2(PO_4)_3$, $LiFeNb(PO_4)_3$, $TiNb(PO_4)_3$, and $LiNa_2FeV(PO_4)_3$ are main members of this phosphate-based cathode. However, monoclinic $Li_3V_2(PO4)_3$, a derivative of $Li_2NaV_2(PO_4)_3$ reported by Nazar et al., is regarded as the most promising cathode material for commercial batteries [131–133]. This material was extensively studied due to its high specific capacity (197 mAh g^{-1}) due to extraction of all three Li-ions from the structure, unique 3D structure allowing fast Li-ion migration, high power density, long cycle life, and ease of synthesis [131–133]. Qiao et al. and Tang et al. reported Li-ion coefficient of 10^{-8}–10^{-9} cm^2 s^{-1} for $Li_3V_2(PO_4)_3$, which is notably higher than that of $LiFePO_4$ [134].

Nevertheless, in the voltage range of 3.0–4.8 V, extraction of third Li-ion is kinetically difficult due to the low electronic conductivity of $Li_3V_2(PO_4)_3$. Pristine $Li_3V_2(PO_4)_3$ only showed room temperature electronic conductivity of 10^{-7} Scm^{-1}, which is much lower than that of $LiCoO_2$ [133, 135]. Therefore, $Li_3V_2(PO_4)_3$ is often cycled in the voltage range of 3.0–4.3 V delivering a theoretical capacity of 133 mAh g^{-1}, which limits its practical applications. Several approaches including carbon coating, particle size reduction, and supervalent ion doping have been demonstrated to improve the electronic conductivity of $Li_3V_2(PO_4)_3$. Among these methods, carbon coating is reported as the most effective that is capable of increasing the electronic conductivity of $Li_3V_2(PO_4)_3$ to 10^{-1} S cm^{-1} [136]. Carbon coating also causes energy density and tap density increase of these phosphate cathodes. More than 10% of carbon contents were reported in previous reports of $Li_3V_2(PO_4)_3/C$ hybrids [137, 138]. For optimum electrochemical performance (specific capacity and energy density), it is necessary to prepare cathodes with low carbon content. Though other methods such as solid-state and microwave synthesis are reported for $Li_3V_2(PO_4)_3/C$ cathode material, particle size control, mass production, and fine tuning of carbon content are main challenges [139–142].

Sol-gel synthetic method is very promising for $Li_3V_2(PO4)_3$ due to fine particle size and uniform carbon coating. Usually, this route for the synthesis of high-rate $Li_3V_2(PO_4)_3/C$ cathode material employs special additives such as glycine, oleic acid, 1,2-propylene glycol, and carbon black. Yan et al. reported a novel sol-gel synthetic method without using complex carbon sources. In this self-catalyzed method, addition of distilled water into a mixture of V_2O_5, Li_2CO_3, and $NH_4H_2PO_4$ cause the partial hydrolysis (Scheme 6.3), and the resulting H^+ decomposes Li_2CO_3 to liberate CO_2 and H_2O, which further hydrolyze $NH_4H_2PO_4$. In this case H_2O acts as both reaction initiator and by-product. The obtained solution was then mixed with citric acid and heat treated to form $Li_3V_2(PO_4)_3/C$ composite. This cathode exhibited excellent specific capacities and cycling stabilities even at high rate of 10C. Yuan et al. developed a spontaneous chemical reaction-based fast sol-gel process to fabricate high-performance carbon-coated $Li_3V_2(PO_4)_3$ cathode

Scheme 6.3 Self-catalyzed synthesis of $Li_3V_2(PO_4)_3/C$ composite. Reproduced with permission from Ref. [295]. Copyright 2012 Elsevier Science

$$NH_4H_2PO_4 + H_2O \longrightarrow NH_4^+ + H^+ + HPO_4^{2-}$$

$$Li_2CO_3 + 2H^+ \longrightarrow 2Li^+ + CO_2 + H_2O$$

$$V_2O_5 + NH_4^+ + 5/8\, C_6H_8O_6 + 3Li+ + 3HPO_4^{2-} \longrightarrow$$

$$Li_3V_2(PO_4)_3 + 7/4\, CO + NH_3 + 2CO_2 + 3H_2O$$

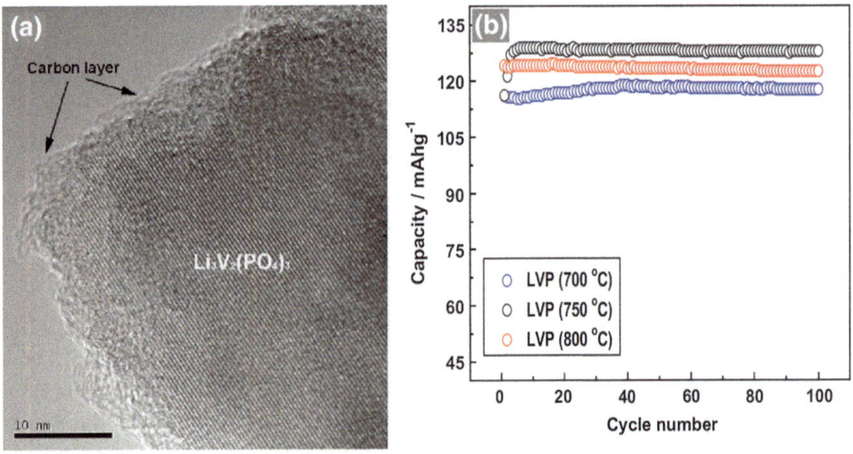

Fig. 6.9 **a** HRTEM image of $Li_3V_2(PO_4)_3/C$ sample synthesized at 750 °C, and **b** galvanostatic cycling performance at 5C rate of $Li_3V_2(PO_4)_3/C$ samples prepared at various temperatures. Reproduced with permission from Ref. [143]. Copyright 2012 Elsevier Science

(Fig. 6.9a) [143]. This method simplified the complex steps involved in conventional sol-gel process, and has a potential for the commercial production of $Li_3V_2(PO_4)_3/C$ with low carbon content. The synthetic process involved dissolution of stoichiometric amounts of Li_2CO_3, V_2O_5, and $NH_4H_2PO_4$ in deionized water followed by mixing with citric acid (carbon source). Dry gel obtained by heating the precursor solution was then heat treated at different temperatures to form $Li_3V_2(PO_4)_3/C$ hybrid cathode. Excellent electrochemical performances are identified for this carbon-coated cathode (Fig. 6.9b), which is ascribed to its enhanced electronic conductivity resulting from uniform nanosized carbon coating.

Another novel approach to enhance the electrochemical performance of olivine cathodes is to form a hybrid electrode with $Li_3V_2(PO_4)_3$ fast ion conductor [144–149]. This results in the improved ionic conductivity of the composite $xLiFePO_4 \cdot yLi_3V_2(PO_4)_3$ due to significantly faster Li-ion migration (10^{-10}–10^{-9} cm^2 s^{-1}) in monoclinic $Li_3V_2(PO_4)_3$ due to its 3D framework than those in $LiFePO_4$ (10^{-16}–10^{-14} cm^2 s^{-1}) [150–153]. During heat treatment for the composite preparation, Fe or/and Li-ions of $LiFePO_4$ may be substituted by V, and partial V or/and Li-ions in $Li_3V_2(PO_4)_3$ may be substituted by Fe [154–156]. This mutual doping caused the improved electronic conductivity of composite than either $Li_3V_2(PO_4)_3$ or $LiFePO_4$. Furthermore, power density of the composite is also better than the individual components due to high voltage redox plateaus of $Li_3V_2(PO_4)_3$ (4.55, 4.08, 3.68, and 3.62) compared to $LiFePO_4$ (3.4 V).

Chemical reduction, lithiation, solid-state, spray drying, and rheological phase reaction are usually employed for the synthesis of $xLiFePO_4 \cdot yLi_3V_2(PO_4)_3$ composites [144, 146–149, 157]. These studies clearly demonstrated that $Li_3V_2(PO_4)_3$ additive can marginally improve the electrochemical performance of submicron-sized electrode. For instance, Wang et al. reported first discharge capacities of 151 and 131 mAh g^{-1} at 1 C rate for $8LiFePO_4 \cdot Li_3V_2(PO_4)_3$ and $LiFePO_4$, respectively [149]. To further improve the electrochemical performance of $xLiFePO_4 \cdot yLi_3V_2(PO_4)_3$ it is necessary to reduce the particle size, which will effectively reduce the electron/Li-ion transport distances [97, 158–160]. Zhong et al. for the first time reported a sol-gel method to synthesize nanosized $9LiFePO_4 \cdot Li_3V_2(PO_4)_3/C$ hybrid with excellent high-rate capacity and cycling stability [145]. In this citric acid-modified sol-gel method, V_2O_5, $NH_4H_2PO_4$, Li_2CO_3, and ferric citrate act as raw materials. XRD investigation proved that most of the vanadium and iron precursors resulted in the formation of corresponding phases, and only small fractions of V and Fe ions enter into the lattice of $LiFePO_4$ and $Li_3V_2(PO_4)_3$, respectively. Li-ion diffusion coefficient (10^{-10} cm^2 s^{-1}) and electronic conductivity (6.615×10^{-3} Scm^{-1}) of this hybrid cathode are one- and threefold, respectively, higher than those of $LiFePO_4/C$. High-rate performance and cycling stabilities are exceptional with an initial discharge capacity of 131.3 mAh g^{-1} at 10C rate and capacity retention of 95.1% after 200 cycles.

The effects of fluorine doping on the electrochemical performance of phosphate-based cathodes are investigated by Arroyo y de Dompablo et al. [161]. Increased Li insertion voltage by fluorine introduction into the phosphate structure is confirmed by the computational results. Barker et al. for the first time explored the electrochemical performance of lithium vanadium fluorophosphates (LiV-PO_4F) as cathode material [162]. Increased thermal stability and power density of this material's main benefits allowed its use as a Li-ion battery cathode [163, 164]. Additionally, discharge plateau of $LiVPO_4F$ appeared at 4.2 V, which is higher than that of the $LiFePO_4$ (3.5 V), $Li_3V_2(PO_4)_3$ (3.9 V), and $LiCoO_2$ (3.9 V) [164]. Despite of these advantages, its electric conductivity is considerably lower than that of $LiCoO_2$ and $LiMn_2O_4$, and mitigation of this unfavorable factor remains as the main challenge in the research area of $LiVPO_4F$ cathode [165]. Similar to the case of many other cathode materials, an efficient way to improve the performance is carbon coating and particle size reduction [165–167]. Synthesis of $LiVPO_4F/C$ hybrid through high-temperature solid-state method is usually tedious and cause $Li_3V_2(PO_4)_3$ impurity formation [168]. Xiong et al. developed a mechanical-blended assisted sol-gel method for the rapid low-temperature synthesis of ultrafine $LiVPO_4F/C$ hybrid [164]. This composite electrode exhibited excellent electrochemical performance even under elevated temperatures, which can be credited to its small particle size and uniform carbon coating.

Furthermore to maintain high performance, cost of production and safety of Li-ion batteries should be also considered seriously [169]. Barker et al. suggested cheaper sodium-based cathodes for Li-ion hybrid batteries, and this triggered extensive research in the area of sodium-based cathode materials [170, 171]. Sodium vanadium fluorophosphate ($NaVPO_4F$) is a promising candidate of the new generation cathode materials. This material has a tetragonal crystal structure, and is consistent with aluminum fluorophosphates cathode suggested by Meins et al. [172, 173]. $NaVPO_4F$ demonstrated excellent performance in Li-ion and Na-ion batteries [174, 175]. Electrochemical Li-/Na-ion storage mechanism of this material is based on the V^{3+}/V^{4+} redox couple. Another monoclinic phase of $NaVPO_4F$ cathode was reported by Zhuo et al. [176]. In addition, $NaVPO_4F$, $Na_{1.5}VOPO_4F_{0.5}$, and $Na_3V_2(PO_4)_2F_3$ also find considerable interest due to higher working potential and lower cost [68, 169–171, 174, 177]. Previous researchers also employed carbon doping to improve the conductivity and electrochemical performance of $NaVPO_4F$ cathodes [178, 179]. Usually, solid-state and hydrothermal approaches have been employed for the synthesis of sodium vanadium fluorophosphates. However, both of them are time consuming, and complex procedure is required for the synthesis of VPO_4, which is a required reaction intermediate phase [68, 170, 171, 174–177, 179]. In addition, hydrothermal method involves the use of high pressure, and low crystallinity of the cathode material often causes its unfavorable electrochemical degradation [174]. There have been limited efforts to develop sol-gel methods for the synthesis of fluorophosphates, especially for $NaVPO_4F$ [169, 180]. Zhao et al. developed an ethanol-based sol-gel method for the synthesis of $NaVPO_4F$, which exhibited remarkable performance, cycling ability, and high-rate performance in a hybrid Li-ion battery [178].

In addition to phosphate-based cathodes, polyanionic orthosilicate Li_2MSiO_4 (M=Fe, Mn, etc.) cathodes have also proved as a potential cathode material due to excellent electrochemical/structural stability (resulting from strong Si–O bonds), safety, and inexpensiveness [181–187]. A major factor hindering the commercialization of Li_2FeSiO_4 is the intrinsic low electronic conductivity and slow lithium diffusion [185]. Luckily, both these limitations can be successfully mitigated by coating cathode particles with carbon and reducing particle size, respectively [188, 189]. Other approaches for overcoming limitations of Li_2FeSiO_4 include fabrication of porous structures and doping with foreign cations [190–192]. Yan et al. developed a modified sol-gel method to fabricate nanostructured Li_2FeSiO_4/C hybrid cathode ($\sim$50 nm spherical particles uniformly dispersed in a continuous carbon matrix) in which polyethylene glycol (PEG) is used as surfactant and L-ascorbic acid as the carbon additive [193]. Use of PEG (a non ionic surfactant) effectively reduces the particle size growth/agglomeration by adsorbing on the colloidal particles, and plays a key role in deciding the particle morphology [194]. This method is superior to other sol-gel methods that involve ball mill mixing of carbon precursors with pre-synthesized Li_2FeSiO_4 [195, 196]. As a result, Li_2FeSiO_4/C samples prepared by PEG-modified sol-gel method exhibited excellent electrochemical performance (138 mAh g^{-1} at 1C rate, and 95% capacity retention after 40 cycles) compared to samples prepared without PEG. Enhanced performances are attributed to smaller particle size facilitating shorter Li-ion diffusion, and continuous carbon film-assisting electronic conduction and reduced side reaction with the electrolyte.

6.3.3 *Layered Metal Oxides*

As mentioned earlier, increasing the specific capacity of cathodes is the most effective method to increase the energy density of Li-ion batteries. In recent years, layered structured material $LiNi_{1-x-y}Co_xMn_yO_2$ (NCM) has received great consideration as Li-ion battery cathode due to the effective integration of the advantages of $LiNiO_2$ (high capacity), $LiCoO_2$ (good cycle performance), and $LiMnO_2$ (low price) [197–200]. $LiNi_{1/3}Co_{1/3}Mn_{1/3}O_2$ is one of the most widely studied layered materials that are capable of delivering a specific capacity of 160 mAh g^{-1} upon charging to 4.3 V [199–206]. This cathode material is currently being used in a variety of commercial Li-ion batteries to partially replace $LiCoO_2$. However, this layered cathode does not offer safety and cost benefits of $LiMn_2O_4$ and $LiFePO_4$ cathodes that are critical for electric vehicle batteries. Considering the fact that specific capacity of NCM strongly dependent on the Ni-content (main active redox species), increasing Ni-content is an efficient approach to improve their charge–discharge capacity [207–209]. This resulted in the development of low-cost and high capacity (200 mAh g^{-1}) Ni-rich NCM materials, $Li[Ni_{0.8}Co_{0.1}Mn_{0.1}]O_2$, that are suitable for hybrid and electric vehicle batteries [207, 210, 211]. In this material about 4% Ni^{2+} occupies Li sites, and its long-term cycling performance is impressive [207, 212, 213]. At slow charge–discharge rates (0.1–0.2C), it retained 94% after 50 cycles, and 83.5% for 70 cycles at a higher C-rate of 0.5C.

Since the rate performance and cycling stability of NCM electrodes are highly dependent on the particle size/morphology, phase purity, degree of crystallinity, and cation disorder, synthesis methods play a crucial role in their electrochemical performance. Synthesis of low Ni-content NCM materials ($LiNi_{1/3}Co_{1/3}Mn_{1/3}O_2$) is easy, which can be achieved through sol-gel, solution–combustion, co-precipitation, and spray pyrolysis methods [201–206]. Nevertheless, it is difficult to synthesize high-quality Ni-rich NCM materials due to incomplete oxidation of Ni^{2+} to Ni^{3+}. Lithium deficiency and high degree of cation disorder resulting from the incomplete oxidation deteriorate their structural stability and electrochemical performance [214–216]. Additionally, Ni-rich NCM materials are vulnerable to oxidation (LiOH and Li_2CO_3 formation) on exposure to air for a long time and during calcination, which can be accelerated at high temperature [210, 217, 218]. Pure oxygen atmosphere rather than air is required to prepare phase-pure Ni-rich NCM materials [211, 216]. It is challenging and costly to use pure oxygen atmosphere, and synthesis of high-quality $Li[Ni_{0.8}Co_{0.1}Mn_{0.1}]O_2$ under air atmosphere is critical for the commercialization of this material. Usually Ni-rich NCM materials are synthesized through co-precipitation method, and sol-gel methods are seldom reported.

Although sol-gel method is not suitable for the commercial scale synthesis, it offers benefits of good stoichiometry small particle size and high degree of homogeneity compared to other methods. Lu et al. reported a facile citric acid-modified sol-gel method for the synthesis of $Li[Ni_{0.8}Co_{0.1}Mn_{0.1}]O_2$ [200]. In this method, stoichiometric amounts of $LiNO_3$, $Ni(NO_3)_2{\cdot}6H_2O$, $Co(NO_3)_2{\cdot}6H_2O$ and $Mn(NO_3)_2{\cdot}4H_2O$ are dissolved in deionized water followed by mixing with citric acid solution, pH adjustment with ammonia, drying, and calcination of the obtained gel in air. The effect of calcination temperature and synthetic method on the cathode crystal structure and its effect on the electrochemical performance are also systematically investigated. Rietveld XRD refinements of sol-gel sample proved lower site exchange between Li^+ and Ni^{2+} (2.44%) in comparison to the co-precipitation sample (4.02%). Low-degree of particle agglomeration is also observed for this sample compared to $Li[Ni_{0.8}Co_{0.1}Mn_{0.1}]O_2$ prepared through co-precipitation route (Fig. 6.10). When investigated the electrochemical performance of this sol-gel-derived cathode, it outperformed similar composition obtained through co-precipitation method. Improved electrochemical performance of this

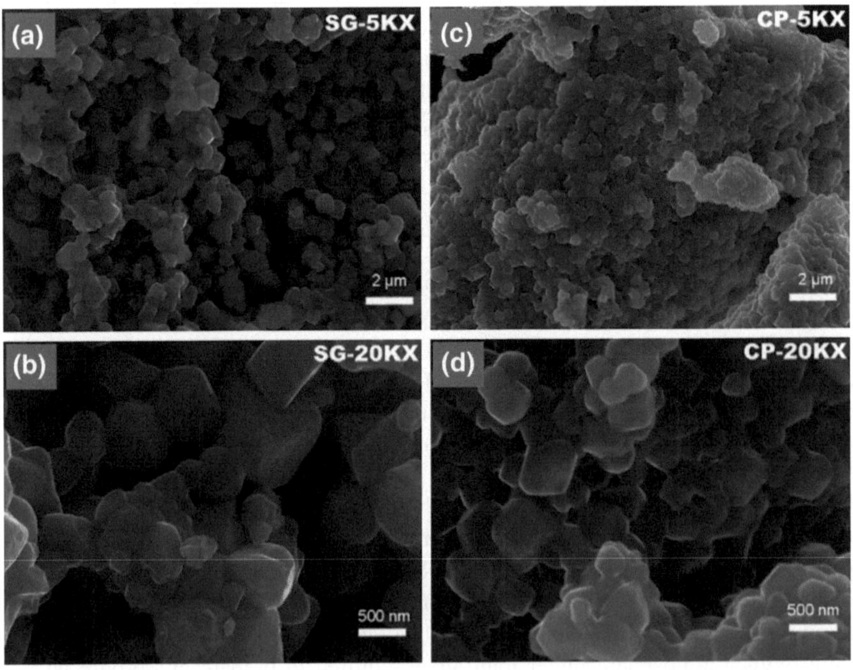

Fig. 6.10 FESEM images of $Li[Ni_{0.8}Co_{0.1}Mn_{0.1}]O_2$ powders synthesized via **a**, **b** sol-gel and **c**, **d** co-precipitation route. Reproduced with permission from Ref. [200]. Copyright 2013 Elsevier Science

sol-gel-derived cathode material is attributed to the low Li/Ni disorder, and increased electrode electrolyte contact resulting from high surface area.

Li-rich-layered materials recently made breakthroughs in the Li-ion battery research due to their extremely high capacity of 250 mAh g^{-1} [219–221]. Nevertheless, practical applications of these cathodes are debatable because of low coulombic efficiency, cycling/rate performance [222–226]. Doping with foreign ions, synthesis of nanostructures and surface modifications have been implemented by several researchers to achieve the commercialization of Li-rich-layered cathode materials [221, 224–231]. Among these approaches, surface coating with metal oxides, fluorides, and phosphates is regarded as the most effective method [224, 226, 232–237]. Owing to their excellent ionic conductivity, Li-ion conductors such as $LiTaO_3$, $LiAlO_2$, Li_3VO_4, Li_2O-B_2O_3, $LiAlSiO_4$, and $LiAlSi_2O_6$ are also promising candidates for surface coating [238–242]. Often complex synthetic methods are used for coating the cathode surface, which reduces the possibility of scale-up and commercialization of these capable cathodes. Sun et al. for the first time employed a sol-gel method to coat Li-rich-layered cathode Li$(Li_{0.17}Ni_{0.2}Co_{0.05}Mn_{0.58})O_2$ (LNCMO) with amorphous $LiAlSiO_4$ (Fig. 6.11) [238]. Cathode composition containing an optimum amount of Li-ion conductor (2 wt%) exhibited improved specific capacity (273 mAh g^{-1} at 0.1C), cycling stability (84.59% retention after 200 cycles), rate performance (173.3 mAh g^{-1} at 5C rate), and initial coulombic efficiency (84%) compared to the pristine material. Superior thermal stability is also identified for the LAS-coated material. Enhanced electrochemical performance and thermal stability are endorsed to the thin uniform amorphous coating that inhibits crystal structure change without inhibiting Li-ion diffusion.

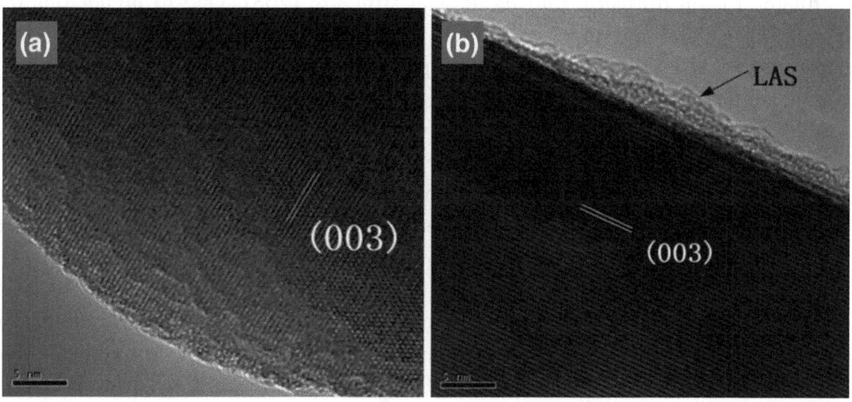

Fig. 6.11 HRTEM images of **a** $Li(Li_{0.17}Ni_{0.2}Co_{0.05}Mn_{0.58})O_2$, and **b** $LiAlSiO_4$-coated Li$(Li_{0.17}Ni_{0.2}Co_{0.05}Mn_{0.58})O_2$ cathode. Reproduced with permission from Ref. [238]. Copyright 2015 Elsevier Science

6.3.4 Spinel Metal Oxides

Pure spinel $LiMn_2O_4$ is a capable cathode material compared to other layered oxide cathode materials such as $LiNiO_2$ and $LiCoO_2$, due to ease of preparation, low cost, and environmental friendliness [243–245]. It also has high energy density, reasonable specific capacity, high thermal stability, and low self-discharge rate [7, 246, 247]. Nanostructured $LiMn_2O_4$ spinel recently demonstrated excellent rate performance and long-term cycling stability at room temperature [248, 249]. Nevertheless, specific capacity of $LiMn_2O_4$ steadily decreases upon extended cycling at elevated temperature [250, 251]. Several factors such as Jahn–Teller distortion, slow manganese dissolution into the organic electrolyte, lattice instability, and a two-phase unstable reaction increased particle size of active material [244, 252–254]. A number of attempts to improve the capacity retention by overcoming Jahn–Teller distortion by doping with divalent, trivalent, and tetravalent ions such as Al, Cr, Cu, Fe, Ni, Ti, Zn, Co, and Ga have been reported [255–257]. Lee and Song et al. reported excellent capacity retention of $LiMn_2O_4$ through partial doping of cations, and Se doping was highly effective in reducing the capacity fade in the 3V region [254, 258, 259]. Doping part of the manganese with additional Li has been also explored [260]. Thackeray et al. pointed out stabilization of the spinel phase resulting from the partial substitution of Mn by other cations [257, 261]. Self-propagating combustion synthesis is the most widely used method for preparing spinel cathodes, which often produce highly agglomerated particles [262]. Implementation of low-temperature synthesis methods such as chemical precipitation, hydrothermal, and sol-gel methods is also necessary to control the particle size of cathode materials to improve long-term cycling stability [263–266].

Thirunakaran et al. for the first time reported a glutamic acid-modified sol-gel method (Fig. 6.12) for a tri-cation-doped stable spinel cathode $LiMg_xSn_yAl_zMn_{2-x-y-z}O_4$ (x, y, z = Mg, Sn, Al) [267]. Large ionic radius (~7.2 nm) of Mg^{2+} ions facilitates Li-ion diffusion into the crystal structure [268]. Due to higher binding energy, Sn^{2+} stabilizes spinel structure by forming shorter and stronger Sn–O bonds compared to Mn–O bonds [269]. During electrochemical cycling, Al^{3+} facilitates charge compensation of Mn^{3+} and mitigates the Jahn–Teller distortion [270]. It should be also noted that similar radii of Sn^{2+} and Al^{3+} ions (0.69 and 5.9 nm) are beneficial for stabilizing the spinel structure. Triple doping of $LiMn_2O_4$ also balances the change of the Mn^{3+}/Mn^{4+} couple during lithiation–delithiation cycle, and thereby enhancing the electrochemical performance [267]. Phthalic acid is also used as a chelating agent for the sol-gel synthesis of pure and Cr-doped $LiMn_2O_4$ nanoparticles [271]. Synthetic method involved mixing of phthalic acid with aqueous solutions of Li, Mn, and Cr precursors followed by heating and calcination of the obtained gel in air. This process reduced heating time of synthesis and facilitated the low-temperature (250–650 °C) formation of $LiCr_{0.10}Mn_{1.90}O_4$ particles. Chelating effect of phthalic acid effectively reduced the particle growth and resulted in nanoparticles with uniform morphology and size distribution (average grain size below 100 nm). Single-phase and highly crystalline $LiCr_{0.10}Mn_{1.90}O_4$

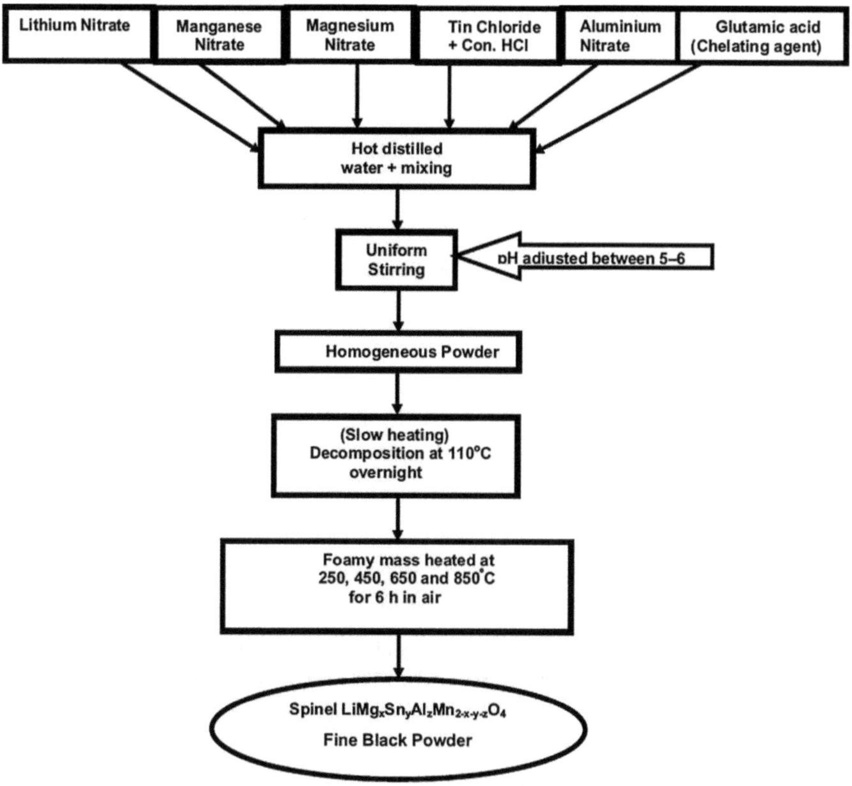

Fig. 6.12 Flow chart of the synthesis of $LiMg_xSn_yAl_zMn_{2-x-y-z}O_4$ through glutamic acid-modified sol-gel method. Reproduced with permission from Ref. [267]. Copyright 2011 Elsevier Science

prepared through this modified sol-gel method excelled in cycling stability and columbic efficiency in comparison to undoped $LiMn_2O_4$.

In another modified sol-gel process (Fig. 6.13), citric acid is used as the complexing agent and simple salts of lithium, iron, and manganese as precursors resulting in $LiMn_2O_4$, $LiFe_{0.5}Mn_{1.5}O_4$, $Li_{1.4}Mn_{1.7}O_4$, and $Li_{1.27}Mn_{1.73}O_4$ nanopowders [272]. In this method, stoichiometric amounts of manganese, lithium, and iron precursors in deionized water are mixed with citric acid (complexing agent) and ethylene glycol (polymerizing agent) followed by stirring at a constant temperature. Calcination of the gel obtained after drying resulted in the formation of pure and cation-doped spinel nanopowders with excellent electrochemical performance.As mentioned earlier, spinel $LiMn_2O_4$ showed very poor high-temperature electrochemical performance. Previous studies proved that partial substitution of Ni for Mn of $LiMn_2O_4$ is effective to stabilize the high-temperature capacities and voltage plateau [273]. However, rate performance and cycling stabilities remained as low as the Ni-free spinel. Wang et al. reported a sol-gel method to prepare

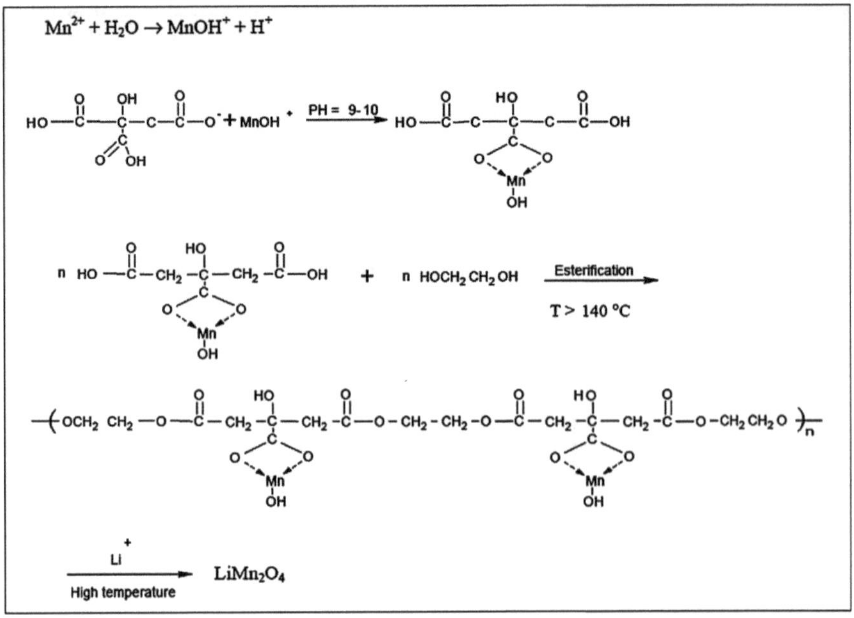

Fig. 6.13 Possible reactions during the sol-gel synthesis of $LiMn_2O_4$ spinel, which used citric acid as complexing agent and ethylene glycol as polymerizing agent. Reproduced with permission from Ref. [272]. Copyright 2011 Elsevier Science

Co-doped $LiMn_2O_4$ with enhanced cycling stability and rate performance [274]. In this synthetic method, stoichiometric amounts of acetate precursor solutions are mixed with the chelating agent citric acid followed by ammonia addition and heating to 80 °C to form a clear liquid. The resulting solution is then dried under vacuum to form a gel, and then calcined at 400 °C. A series of high surface area spinel $LiMn_{2-x}Co_xO_4$ ($x = 0.1$, 0.3, 0.5) with smaller lattice parameters are prepared by this method. When proper dopant contents are selected Co-doped $LiMn_2O_4$ satisfies high capacity, rate performance, and cycling stability requirements at 200 and 300 °C. Specific capacities of 219 mAh g^{-1} at 200 °C and 307 mAh g^{-1} at 300 °C are obtained for $LiMn_{1.7}Co_{0.3}O_4$ and $LiMn_{1.5}Co_{0.5}O_4$, respectively, at a high current density of 30 mA cm^{-2}. This enhanced high-temperature electrochemical performances are ascribed to the improved electrical conductivity resulted from Co-doping.

Modified sol-gel methods, especially Pechini process for the low-temperature synthesis of cathode materials, often involve several tedious steps of mixing chelating and polymerizing agents (citric acid, ethylene glycol, etc.) with reactants to obtain a polymeric precursor [275]. Removal of ethylene glycol under reduced pressure is a critical step that induces polymerization and gelation, which limits the application of this method for large-scale material synthesis. Subramania et al. invented a novel sol-gel thermolysis method suitable for the scalable synthesis of cubic spinel $LiSr_xMn_{2-x}O_4$ (x = 0.10, 0.15, 0.20, 0.25) nanopowders (Fig. 6.14) [261]. This processing combines chemical gelation and thermal process, in which organic polymer prevents particle growth due to its long chain structure and also acts as an excellent fuel. Nitrate and acetate precursors of Li, Mn, and Sr are used in conjunction with polyvinyl alcohol (PVA) and urea to form a sol, which on slow heating at 100 °C form a gel. Pyrolysis of this precursor sample at 340 °C resulted in the formation of phase-pure and submicron $LiSr_xMn_{2-x}O_4$ powder. This flexible method allows precise control of morphology and particle size of nanocrystalline materials. Spinel cathodes synthesized through sol-gel thermolysis demonstrated excellent electrochemical performances.

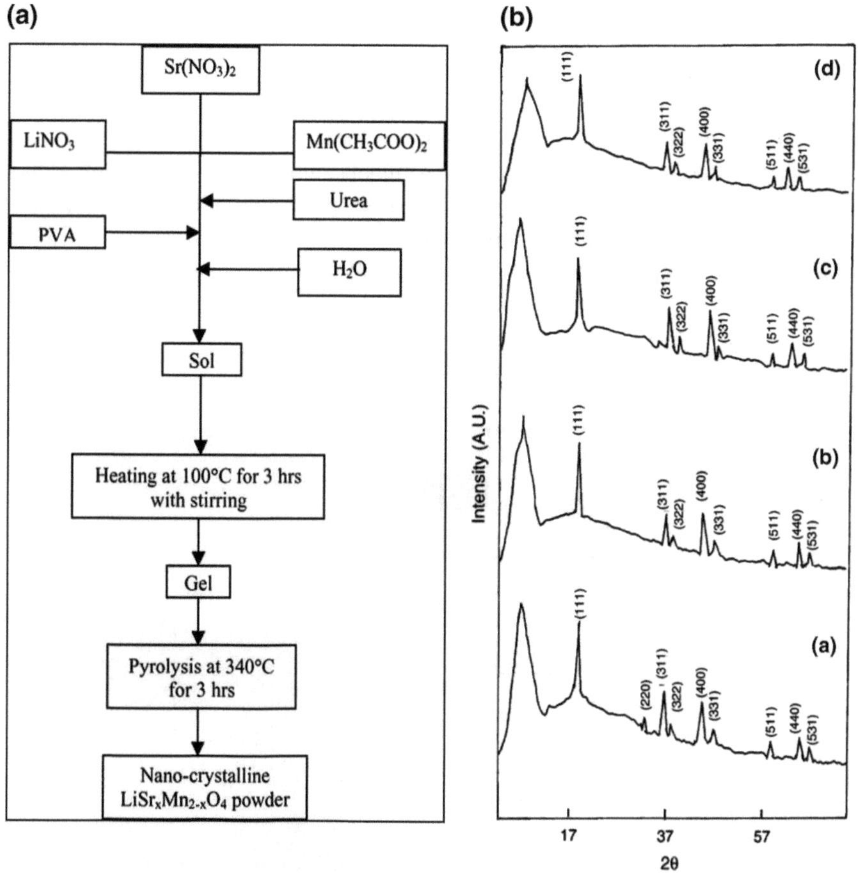

Fig. 6.14 **a** Flow chart of the sol-gel thermolysis synthesis of $LiSr_xMn_{2-x}O_4$ and **b** X-ray diffraction patterns for cubic spinel $LiSr_xMn_{2-x}O_4$ (x = 0.10, 0.15, 0.20, 0.25) obtained by sol-gel thermolysis. Reproduced with permission from Ref. [261]. Copyright 2006 Elsevier Science

Another method to improve the intrinsic low electrical conductivity of $LiMn_2O_4$ spinel is fabrication of a hybrid electrode with carbonaceous current collectors. Such additives are reported to have a positive impact on the electrochemical performance of electrode materials including $LiFePO_4$, $LiCoO_2$, Co_3O_4, TiO_2, and SnO_2 [276–280]. Direct sol-gel syntheses of hybrid cathode materials for Li-ion batteries are seldom reported. Liu et al. developed a sol-gel method to synthesize $LiMn_2O_4$-multiwalled carbon nanotube (MWCNT) nanocomposite followed by low-temperature calcination (Fig. 6.15a) [281]. In this study, morphology of MWCNT is preserved during the sol-gel processing and they are functionalized with $LiMn_2O_4$ nanoparticles (Fig. 6.15b, c). Due to their high electrical conductivity, large surface area, and 1D structure, MWCNTs are considered an excellent carbonaceous additive for electrode materials. In addition to the intrinsic properties of MWCNT and nanocrystals, unique interfacial properties also arise from their electrical and thermal interactions. Remarkable electrochemical performances are identified for the sol-gel-derived $LiMn_2O_4$-MWCNT hybrid electrodes, which retained 99% of specific capacity after 20 charge–discharge cycles. Electrochemical impedance measurements proved that reduced charge transfer resistance due to excellent electrical conductivity of MWCNTs is responsible for the increased high-rate performance and cycling stability of hybrid electrodes.

Though $LiNi_{0.5}Mn_{1.5}O_4$ is an appealing spinel cathode due to 4.7 V high voltage plateau (higher energy density), Ni^{4+} undergoes oxidation and side reaction with electrolyte at high temperature compared to conventional $LiMn_2O_4$ spinel system [282–287]. Recently, researchers focused on stabilizing $LiNi_{0.5}Mn_{1.5}O_4$ at high

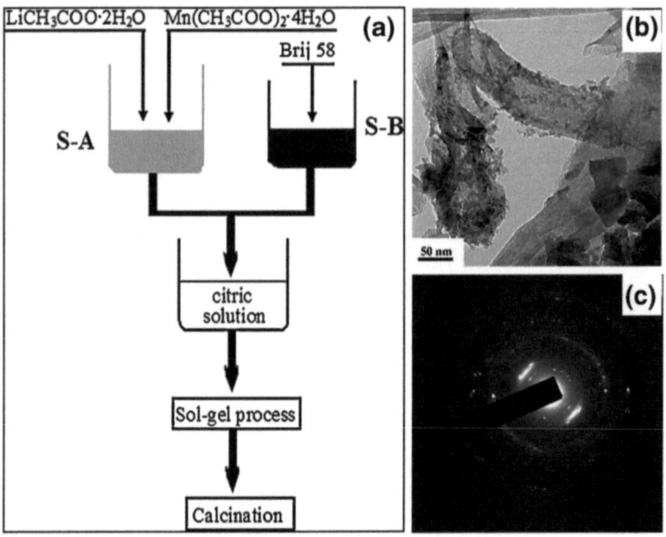

Fig. 6.15 **a** Schematic of the sol-gel synthesis, **b** HRTEM image and **c** SAED pattern of MWCNT-$LiMn_2O_4$ composite. Reproduced with permission from Ref. [281]. Copyright 2010 Elsevier Science

temperature by surface coating with several oxides and phosphates (ZnO, ZrO_2, ZrP_2O_7, ZrP_2O_7, Al_2O_3, $AlPO_4$, SiO_2, Bi_2O_3, and BiOF) [288–292]. These results proved the effectiveness of inorganic surface coatings to improve the electrochemical performance at elevated temperature by protecting the active cathode material from HF attack, reducing interfacial resistance and pH control between the electrolyte and cathode. Lee et al. developed a citric acid-modified sol-gel method to coat $LiNi_{0.5}Mn_{1.5}O_4$ cathode with $ZnAl_2O_4$, (Fig. 6.16a, b) and the resulting hybrid cathode exhibited 87% capacity retention after 100 cycles at 1C rate at 60 °C (Fig. 6.16c, d). While the pristine cathode experienced significant degradation in capacity on prolonged cycling (57% retention) [288], $ZnAl_2O_4$ was selected as the preferred coating material due to its high temperature/chemical stability, hydrophobic nature, and low surface acidity [293, 294]. These findings recommend a route for the further development of high-temperature stable spinel cathode materials.

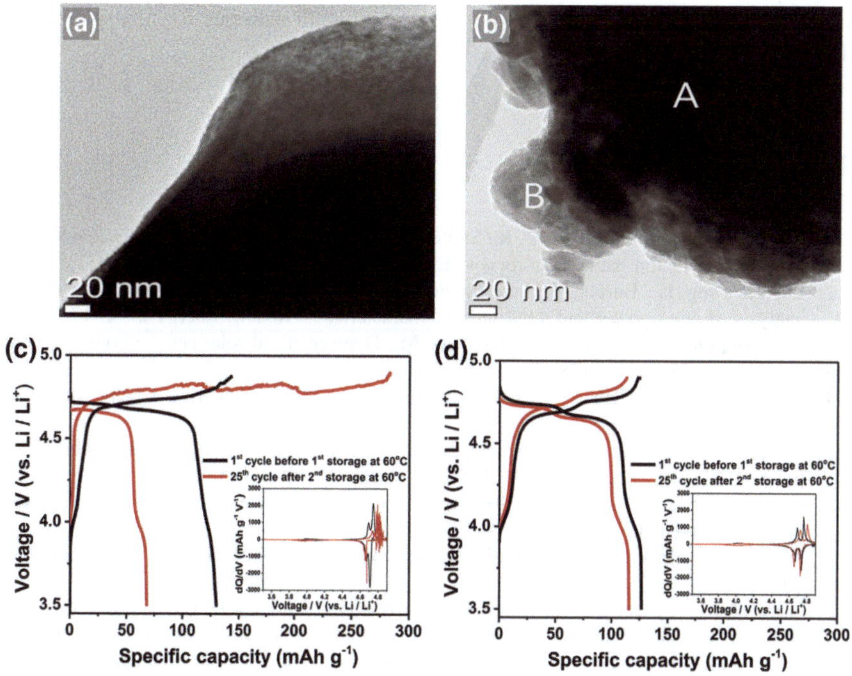

Fig. 6.16 HRTEM images of **a** pristine $LiNi_{0.5}Mn_{1.5}O_4$, and **b** $ZnAl_2O_4$-coated $LiNi_{0.5}Mn_{1.5}O_4$ synthesized by a sol-gel method. Charge–discharge voltage profiles of **c** pristine $LiNi_{0.5}Mn_{1.5}O_4$ and **d** $ZnAl_2O_4$-coated $LiNi_{0.5}Mn_{1.5}O_4$. Reproduced with permission from Ref. [288]. Copyright 2014 Elsevier Science

6.4 Conclusions

Without a doubt, sol-gel method is a highly promising route for the synthesis of Li-ion battery cathode materials with desired physical properties and electrochemical performances. Improved electrochemical performances of sol-gel-derived Li-ion battery cathode materials are clearly demonstrated in recent years. In addition to the well-known benefits such as homogeneity, small particle size, high surface area, and low processing temperature, sol-gel methods also offer better control over porosity and morphology through selection of suitable chelating and polymerizing agents. Furthermore, modified sol-gel method effectively overcomes the major drawbacks of Li-ion cathode materials, namely low electronic conductivity and transition metal ion leaching by doping, composite formation with carbon, metal nanostructures, fast Li-ion conductors, uniform coating of the active material with carbon and metal oxides, etc. Recent advances in the area of modified sol-gel process permit scalable materials synthesis, which offers a route for the commercialization of several Li-ion battery cathode materials. In conclusion, sol-gel process allows the efficient synthesis of wide range of cathode materials that facilitate the development of high-performance Li-ion batteries for next-generation electrochemical energy storage.

References

1. Etacheri, V., Marom, R., Elazari, R., Salitra, G., Aurbach, D.: Challenges in the development of advanced Li-Ion batteries: a review. Energy Environ. Sci. **4**, 3243–3262 (2011)
2. Lu, L., Yang, H., Burnett, J.: Investigation on wind potential on Hong Kong islands—an analysis of wind power and wind turbine characteristics. Renew. Energy **27**, 1–12 (2002)
3. Nazeeruddin, M.K., Baranoff, E., Graetzel, M.: Dye-sensitized solar cells: a brief overview. Sol. Energy **85**, 1172–1178 (2011)
4. Goodenough, J.B., Kim, Y.: Challenges for rechargeable Li batteries. Chem. Mater. **22**, 587–603 (2010)
5. Marom, R., Amalraj, S.F., Leifer, N., Jacob, D., Aurbach, D.: A review of advanced and practical lithium battery materials. J. Mater. Chem. **21**, 9938–9954 (2011)
6. Goodenough, J.B., Kim, Y.: Challenges for rechargeable batteries. J. Power Sources **196**, 6688–6694 (2011)
7. Whittingham, M.S.: Lithium batteries and cathode materials. Chem. Rev. **104**, 4271–4301 (2004)
8. Armand, M., Tarascon, J.M.: Building better batteries. Nature **451**, 652–657 (2008)
9. Etacheri, V., Seisenbaeva, G.A., Caruthers, J., Daniel, G., Nedelec, J.M., Kessler, V.G., Pol, V.G.: Ordered network of interconnected SnO_2 nanoparticles for excellent lithium-ion storage. Adv. Energy Mater. **5**, 1401289–1401297 (2015)
10. Kang, B., Ceder, G.: Battery materials for ultrafast charging and discharging. Nature **458**, 190–193 (2009)
11. Bruce, P.G., Scrosati, B., Tarascon, J.M.: Nanomaterials for rechargeable lithium batteries. Angew. Chem. Int. Ed. **47**, 2930–2946 (2008)
12. Manthiram, A.: Materials challenges and opportunities of lithium ion batteries. J. Phys. Chem. Lett. **2**, 176–184 (2011)

13. Etacheri, V., Yourey, J.E., Bartlett, B.M.: Chemically bonded TiO_2-bronze nanosheet/reduced graphene oxide hybrid for high-power lithium ion batteries. ACS Nano **8**, 1491–1499 (2014)
14. Pierre, A.C.: Introduction to sol gel processing. Kluwer Academic Press, Boston (1990)
15. Yoldas, A.E.: Alumina sol preparation from alkoxides. Am. Ceramic. Soc. Bull. **54**, 289–290 (1975)
16. Xian-Jun, Z., Hong-Hao, C., Hui, Z., Han-Xing, L., Dai-Ling, Y., Yun-Hong, Z.: Synthesis and characteristics of $LiNi_{0.85}Co_{0.15}O_2$ cathode materials by particulate sol-gel method for lithium ion batteries. Chin. J. Chem. **23**, 491–495 (2005)
17. Kandhasamy, S., Pandey, A., Minakshi, M.: Polyvinylpyrrolidone assisted sol-gel route $LiCo_{1/3}Mn_{1/3}Ni_{1/3}PO_4$ composite cathode for aqueous rechargeable battery. Electrochim. Acta **60**, 170–176 (2012)
18. Zhang, H., Reller, A.: Nanocrystalline iron-doped mesoporous titania and its phase transition. J. Mater. Chem. **11**, 2537–2541 (2001)
19. Stone, V., Davis, R.: Synthesis characterization and photocatalytic activity of titania and niobia mesoporous molecular sieves. Chem. Mater. **10**, 1468–1474 (1998)
20. Antonelli, D., Ying, J.: Synthesis of a stable hexagonally packed mesoporous niobium oxide molecular sieve through a novel ligand assisted templating mechanism. Angew. Chem. Int. Ed. **34**, 2014–2017 (1995)
21. Liu, H., Wu, Y.P., Rahm, E., Holze, R., Wu, H.Q.: Cathode materials for lithium ion batteries prepared by sol-gel methods. J. Solid State Electrochem. **8**, 450–466 (2004)
22. Sato, S., Oimatsu, S., Takahashi, R., Sodesawa, T., Nozaki, F.: Pore size regulation of TiO_2 by use of a complex of titanium tetraisopropoxide and stearic acid. Chem. Commun. **22**,19–2220 (1997)
23. Putnam, R., Nakagawa, N., McGrath, K., Yao, N., Aksay, I., Gruner, S.: Titanium dioxide-surfactant mesophases and Ti-TMS1. Chem. Mater. **9**, 2690–2693 (1997)
24. Thieme, M., Schuth, F.: Preparation of a mesoporous high surface area titanium oxo phosphate via a non-ionic surfactant route. Microporous Mesoporous Mater. **27**, 193–200 (1999)
25. Cabrera, S., Haskouri, J.E., Beltrán-Porter, A., Beltrán-Porter, D., Marcos, M., Amorós, P.: Enhanced surface area in thermally stable pure mesoporous TiO_2. Solid State Sci. **2**, 513–518 (2000)
26. Singhal, B., Porwal, A., Sharma, A., Ameta, R., Ameta, S.: Photocatalytic degradation of cetylpyridinium chloride over titanium dioxide powder. J. Photochem. Photobiol. A **108**, 85–88 (1997)
27. Cassiers, K., Linssen, T., Mathieu, M., Bai, Y., Zhu, H., Cool, P.: Surfactant-directed synthesis of mesoporous titania with nanocrystalline anatase walls and remarkable thermal stability. J. Phys. Chem. B **108**, 3713–3721 (2004)
28. Thoms, H., Epple, M., Fröba, M., Wong, J., Reller, A.: Metal diolates: useful precursors for tailor-made oxides prepared at low temperatures. J. Mater. Chem. **8**, 1447–1451 (1998)
29. Saadoun, L., Ayllón, J., Jiménez-Becerril, J., Peral, J., Domenech, X., Rodríguez-Clemente, R.: 1, 2 Diolates of titanium as stable precursors for the preparation of photoactive high surface area titania. Appl. Catal. B **21**, 269–277 (1999)
30. Zheng, J., Qiu, K., Wei, Y.: Sol-gel synthesis of mesoporous titania using nonsurfactant organic compounds as templates. Mol. Cryst. Liq. Cryst. **354**, 183–194 (2000)
31. Wang, C., Li, Q., Wang, R.: Synthesis and characterization of mesoporous iron-doped TiO_2. J. Mater. Chem. **39**, 1899–1901 (2004)
32. Wang, C., Xi, H., Wang, R.: Synthesis of mesoporous Ce-doped TiO_2 with high thermal stability. Chem. Lett. **33**, 20–21 (2004)
33. Kingery, H., Bown, K., Uhimann, D.R.: Introduction to Ceramics. Wiley, Newyork (1996)
34. Segal, A.L.: Sol-gel processing: routes to oxide ceramics using colloidal dispersions of hydrous oxides and alkoxide intermediates. J. Non. Cryst. Solids **63**, 183–191 (1984)
35. Nohma, T., Kurokawa, H., Uehara, M., Takahashi, M., Nishio, K., Saito, T.: Electrochemical characteristics of $LiNiO_2$ and $LiCoO_2$ as a positive material for lithium secondary batteries. J. Power Sources **54**, 522–524 (1995)

36. Capitaine, F., Gravereau, P., Delmas, C.: A new variety of $LiMnO_2$ with a layered structure. Solid State Ionics **89**, 197–202 (1996)
37. Thackeray, M.M.: Spinel electrodes for lithium batteries. J. Am. Ceram. Soc. **82**, 3347–3354 (1999)
38. Mitzushima, K., Jones, P.C., Wiseman, P.J., Goodenough, J.B.: Li_xCoO_2 ($0 < x < -1$): a new cathode material for batteries of high energy density. Mater. Res. Bull. **15**, 783–799 (1980)
39. Ozawa, K.: Lithium-ion rechargeable batteries with $LiCoO_2$ and carbon electrodes: the $LiCoO_2$/C system. Solid State Ionics **69**, 212–221 (1994)
40. Alcantara, R., Lavela, P., Tirado, J.L., Stoyanova, R., Zhecheva, E.: Structure and electrochemical properties of boron-doped $LiCoO_2$. J. Solid State Chem. **134**, 265–273 (1997)
41. Kim, Y.J., Cho, J., Kim, T.J., Parka, B.: Suppression of cobalt dissolution from the $LiCoO_2$ cathodes with various metal-oxide coatings. J. Electrochem. Soc. **150**, A1723–A1725 (2003)
42. Amatucci, G.G., Tarascon, J.M., Klein, L.C.: Cobalt dissolution in $LiCoO_2$-based non-aqueous rechargeable batteries. Solid State Ionics **83**, 167–173 (1996)
43. Dokko, K., Mohamedi, M., Fujita, Y., Itoh, T., Nishizawa, M., Umeda, M., Uchida, I.: Kinetic characterization of single particles of $LiCoO_2$ by AC impedance and potential step methods. J. Electrochem. Soc. **148**, A422–A426 (2001)
44. Ménetriér, M., Saudoune, I., Levasseur, S., Delmas, C.: The insulator-metal transition upon lithium deintercalation from $LiCoO_2$: electronic properties and 7Li NMR study. J. Mater. Chem. **9**, 1135–1140 (1999)
45. Molenda, J., Stoklosa, A., Bak, T.: Modification in the electronic structure of cobalt bronze Li_xCoO_2 and the resulting electrochemical properties. Solid State Ionics **53**, 53–58 (1989)
46. Shibuya, M., Nishina, T., Matsue, T., Uchida, I.: In situ conductivity measurements of $LiCoO_2$ film during lithium insertion/extraction by using interdigitated microarray electrodes. J. Electrochem. Soc. **143**, 3157–3160 (1996)
47. Chen, Z., Dahn, J.R.: Effect of a ZrO_2 coating on the structure and electrochemistry of Li_xCoO_2 when cycled to 4.5 V. Electrochem. Solid-State Lett. **5**, A213–A216 (2002)
48. Cho, J., Kim, Y.J., Kim, T.J., Park, B.: Novel $LiCoO_2$ cathode material with Al_2O_3 coating for a Li ion cell. Chem. Mater. **12**, 3788–3791 (2000)
49. Cho, J., Kim, Y.J., Kim, T.J., Park, B.: $LiCoO_2$ cathode material that does not show a phase transition from hexagonal to monoclinic phase. J. Electrochem. Soc. **148**, A1110–A1115 (2001)
50. Nobili, F., Croce, F., Tossici, R., Meschini, I., Reale, P., Marassi, R.: Sol-gel synthesis and electrochemical characterization of Mg-/Zr-doped $LiCoO_2$ cathodes for Li-ion batteries. J. Power Sources **197**, 276–284 (2012)
51. Jamnik, J., Daminko, R., Erjavec, B., Remskar, M., Pintar, A., Gaberscek, M.: Stabilizers of particle size and morphology: a road towards high-rate performance insertion materials. Adv. Mater. **21**, 2715–2719 (2009)
52. Hu, Y.S., Guo, Y.G., Dominko, R., Gaberscek, M., Jamnik, J., Maier, J.: Improved electrode performance of porous $LiFePO_4$ using RuO_2 as an oxidic nanoscale interconnect. Adv. Mater. **19**, 1963–1966 (2007)
53. Mladenov, M., Stoyanova, R., Zhecheva, E., Vassilev, S.: Effect of Mg doping and MgO-surface modification on the cycling stability of $LiCoO_2$ electrodes. Electrochem. Commun. **3**, 410–416 (2001)
54. Thirunakaran, R., Kalaiselvi, N., Periasamy, P., Renganathan, N.G.: Mg substituted $LiCoO_2$ for reversible lithium intercalation. Ionics **9**, 388–394 (2003)
55. Levasseur, S., Ménétrier, M., Delmas, C.: On the $Li_xCo_{1-y}Mg_yO_2$ system upon deintercalation: electrochemical, electronic properties and 7Li MAS NMR studies. J. Power Sources **112**, 419–427 (2002)
56. Julien, C., Camacho-Lopez, M.A., Mohan, T., Chitra, S., Kalyani, P., Gopukumar, S.: Combustion synthesis and characterization of substituted lithium cobalt oxides in lithium batteries. Solid State Ionics **135**, 241–248 (2000)
57. Ganesan, M.: Studies on the effect of titanium addition on $LiCoO_2$. Ionics **15**, 609–614 (2009)

58. Kim, H.S., Ko, T.K., Na, B.K., Cho, W.I., Chao, B.W.: Electrochemical properties of $LiM_xCo_{1-x}O_2$ [M=Mg, Zr] prepared by sol-gel process. J. Power Sources **138**, 232–239 (2004)
59. Needham, S.A., Wang, G.X., Liu, H.K., Drodz, V.A., Liu, R.S.: Synthesis and electrochemical performance of doped $LiCoO_2$ materials. J. Power Sources **174**, 828–831 (2007)
60. Bang, H.J., Park, B.C., Sun, Y.K.: Synthesis and electrochemical properties of Li $[Ni_{0.45}Co_{0.1}Mn_{0.45-x}Zr_x]O_2$ (x = 0, 0.02) via co-precipitation method. J. Power Sources **174**, 565–568 (2007)
61. Oh, S.H., Lee, S.M., Cho, W.I., Cho, B.W.: Electrochemical characterization of zirconium-doped $LiNi_{0.8}Co_{0.2}O_2$ cathode materials and investigations on deterioration mechanism. Electrochim. Acta **51**, 3637–3644 (2006)
62. Cho, J., Kim, Y., Kim, B., Lee, J., Park, B.: A breakthrough in the safety of lithium secondary batteries by coating the cathode material with $AlPO_4$ nanoparticles. Angew. Chem. Int. Ed. **42**, 1618–1621 (2003)
63. Lu, Y., Mansour, A.N., Yabuuchi, N., Shao-Horn, Y.: Probing the origin of enhanced stability of $AlPO_4$ nanoparticle coated $LiCoO_2$ during cycling to high voltages: combined XRD and XPS studies. Chem. Mater. **21**, 4408–4424 (2009)
64. Barker, J., Saidi, M.Y., Gover, R.K.B., Burns, P., Bryan, A.: The effect of Al substitution on the lithium insertion properties of lithium vanadium fluorophosphate, $LiVPO_4F$. J. Power Sources **174**, 927–931 (2007)
65. Padhi, A.K., Nanjundaswamy, K.S., Goodenough, J.B.: Phospho-olivines as positive-electrode materials for rechargeable lithium batteries. J. Electrochem. Soc. **144**, 1188–1194 (1997)
66. Gao, M., Liu, N., Li, Z., Wang, W., Li, C., Zhang, H., Chen, Y., Yu, Z., Huang, Y.: A gelatin-based sol-gel procedure to synthesize the $LiFePO_4$/C nanocomposite for lithium ion batteries. Solid State Ionics **258**, 8–12 (2014)
67. Gong, Z., Yang, Y.: Recent advances in the research of polyanion-type cathode materials for Li-ion batteries. Energy Environ. Sci. **4**, 3223–3242 (2011)
68. Gover, R.K.B., Bryan, A., Burns, P., Barker, J.: The electrochemical insertion properties of sodium vanadium fluorophosphate, $Na_3V_2(PO_4)_2F_3$. Solid State Ionics **177**, 1495–1500 (2006)
69. Li, Y., Zhou, Z., Ren, M., Gao, X., Yan, J.: Electrochemical performance of nanocrystalline $Li_3V_2(PO_4)_3$/carbon composite material synthesized by a novel sol-gel method. Electrochim. Acta **51**, 6498–6502 (2006)
70. Chung, S.Y., Bloking, J.T., Chiang, Y.M.: Electronically conductive phospho-olivines as lithium storage electrodes. Nat. Mater. **2**, 123–128 (2002)
71. Wang, Y., Wang, Y., Hosono, E., Wang, K., Zhou, H.: The design of a $LiFePO_4$/carbon nanocomposite with a core-shell structure and its synthesis by an in situ polymerization restriction method. Angew. Chem. Int. Ed. **47**, 7461–7465 (2008)
72. Lai, C., Xu, Q., Ge, H., Zhou, G., Xie, J.: Improved electrochemical performance of $LiFePO_4$/C for lithium-ion batteries with two kinds of carbon sources. Solid State Ionics **179**, 1736–1739 (2008)
73. Wang, J., Sun, X.: Understanding and recent development of carbon coating on $LiFePO_4$ cathode materials for lithium-ion batteries. Energy Environ. Sci. **5**, 5163–5185 (2012)
74. Beninati, S., Damen, L., Mastragostino, M.: Fast sol-gel synthesis of $LiFePO_4$/C for high power lithium-ion batteries for hybrid electric vehicle application. J. Power Sources **194**, 1094–1098 (2009)
75. Shim, J., Striebel, K.A.: Cycling performance of low-cost lithium ion batteries with natural graphite and $LiFePO_4$. J. Power Sources **119–121**, 955–958 (2003)
76. Striebel, K.A., Guerfi, A., Shim, J., Armand, M., Gauthier, M., Zaghib, K.: $LiFePO_4$/gel/natural graphite cells for the BATT program. J. Power Sources **119–121**, 951–954 (2003)
77. Huang, H., Yin, S.C., Nazar, L.F.: Approaching theoretical capacity of $LiFePO_4$ at room temperature at high rates. Electrochem. Solid-State Lett. **4**, A170–A172 (2001)

78. Yang, S., Song, Y., Zavalij, P.Y., Whittingham, M.S.: Reactivity, stability and electrochemical behavior of lithium iron phosphates. Electrochem. Commun. **4**, 239–244 (2002)
79. Choi, D., Kumta, P.N.: Surfactant based sol-gel approach to nanostructured $LiFePO_4$ for high rate Li-ion batteries. J. Power Sources **163**, 1064–1069 (2007)
80. Yamada, A., Chung, S.C., Hinokuma, K.: Optimized $LiFePO_4$ for lithium battery cathodes. J. Electrochem. Soc. **148**, A224–A229 (2001)
81. Chung, S.Y., Chiang, Y.M.: Microscale measurements of the electrical conductivity of doped $LiFePO_4$. Electrochem. Solid-State Lett. **6**, A278–A281 (2003)
82. Xie, H., Zhou, Z.T.: Physical and electrochemical properties of mix-doped lithium iron phosphate as cathode material for lithium ion battery. Electrochim. Acta **51**, 2063–2068 (2006)
83. Gao, Y., Li, L., Peng, H., Wei, Z.: Surfactant-assisted sol-gel synthesis of nanostructured ruthenium-doped lithium iron phosphate as a cathode for lithium-ion batteries. ChemElectroChem **1**, 2146–2152 (2014)
84. Vujkovic, M., Jugovic, D., Mitric, M., Stojkovic, I., Cvjetićanin, N., Mentus, S.: The $LiFe_{(1-x)}V_xPO_4$/C composite synthesized by gel-combustion method, with improved rate capability and cycle life in aerated aqueous solutions. Electrochim. Acta **109**, 835–842 (2013)
85. Shin, H.C., Park, S.B., Jang, H., Chung, K.Y., Cho, W.I., Kim, C.S., Cho, B.W.: Rate performance and structural change of Cr-doped $LiFePO_4$/C during cycling. Electrochim. Acta **53**, 7946–7950 (2008)
86. Gao, H.Y., Jiao, L.F., Yang, J.Q., Qi, Z., Wang, Y.J., Yuan, H.T.: High rate capability of Co-doped $LiFePO_4$/C. Electrochim. Acta **97**, 143–149 (2013)
87. Herle, P.S., Ellis, B., Coombs, N., Nazar, L.F.: Nano-network electronic conduction in iron and nickel olivine phosphates. Nat. Mater. **3**, 147–152 (2004)
88. Ravet, N., Besner, S., Simoneau, M., Vallée, A., Armand, M.: Hydro-Québec, Canadian Patent 2, 270, 771
89. Ravet, N., Goodenough, J.B., Besner, S., Simoneau, M., Hovington, P., Armand, M.: The 196th Electrochemical Society and the Electrochemical Society of Japan Meeting Abstracts (Abstract 127), pp. 99–102 (1999)
90. Chen, G., Song, X., Richardson, T.J.: Electron microscopy study of the $LiFePO_4$ to $FePO_4$ phase transition. Electrochem. Solid-State Lett. **9**, A295–A298 (2006)
91. Srinivasan, V., Newman, J.: Discharge model for the lithium iron-phosphate electrode. J. Electrochem. Soc. **151**, A1517–A1529 (2004)
92. Delacourt, C., Poizot, P., Levasseur, S., Masquelier, C.: Size effects on carbon-free $LiFePO_4$ powders. Electrochem. Solid-State Lett. **9**, A352–A355 (2006)
93. Delacourt, C., Poizot, P., Morcrette, M., Tarascon, J.M., Masquelier, C.: One-step low-temperature route for the preparation of electrochemically active $LiMnPO_4$ powders. Chem. Mater. **16**, 93–99 (2004)
94. Dominko, R., Bele, M., Gaberscek, M., Remskar, M., Hanzel, D., Goupil, J.M., Pejovnik, S., Jamnik, J.: Porous olivine composites synthesized by sol-gel technique. J. Power Sources **153**, 274–280 (2006)
95. Yang, J.S., Xu, J.J.: Synthesis and characterization of carbon-coated lithium transition metal phosphates $LiMPO_4$ (M = Fe, Mn Co, Ni) prepared via a nonaqueous sol-gel route. J. Electrochem. Soc. **153**, A716–A723 (2006)
96. Sanchez, M.A.E., Brito, G.E.S., Fantini, M.C.A., Goya, G.F., Matos, J.R.: Synthesis and characterization of $LiFePO_4$ prepared by sol-gel technique. Solid State Ionics **177**, 497–500 (2006)
97. Kim, D.H., Kim, J.: Synthesis of $LiFePO_4$ nanoparticles in polyol medium and their electrochemical properties. Electrochem. Solid-State Lett. **9**, A439–A442 (2006)
98. Etacheri, V., Wang, C., O'Connell, M.J., Chan, C.K., Pol, V.G.: Porous carbon sphere anodes for enhanced lithium-ion storage. J. Mater. Chem. A **3**, 9861–9868 (2015)
99. Etacheri, V., Kuo, Y., Ven, A.V., Bartlett, B.M.: Mesoporous TiO_2–B microflowers composed of (110) facet-exposed nanosheets for fast reversible lithium-ion storage. J. Mater. Chem. A **1**, 12028–12032 (2013)

100. Liu, H., Bi, Z., Sun, X., Unocic, R., Paranthaman, P., Dai, S., Brown, G.: Mesoporous TiO_2–B microspheres with superior rate performance for lithium ion batteries. Adv. Mater. **23**, 3450–3454 (2011)
101. Dylla, A.G., Lee, J.A., Stevenson, K.J.: Influence of mesoporosity on lithium-ion storage capacity and rate performance of nanostructured TiO_2(B). Langmuir **28**, 2897–2903 (2012)
102. Qiu, H., Zhu, K., Li, H., Li, T., Zhang, T., Yue, H., Wei, Y., Du, F., Wang, C., Chen, G., Zhang, D.: Mesoporous Li_2FeSiO_4@ordered mesoporous carbon composites cathode material for lithium-ion batteries. Carbon **87**, 365–373 (2015)
103. Kim, S.J., Lee, Y.W., Hwang, B.M., Kim, S.B., Kim, W.S., Cao, G., Park, K.W.: Mesoporous composite cathode materials prepared from inverse micelle structures for high performance lithium ion batteries. RSC Adv. **4**, 11598–11604 (2014)
104. Jung, H., Shin, J., Chae, C., Lee, J.K., Kim, J.: FeF_3/ordered mesoporous carbon (OMC) nanocomposites for lithium ion batteries with enhanced electrochemical performance. J. Phys. Chem. C **117**, 14939–14946 (2013)
105. Wang, G., Liu, H., Liu, J., Qiao, S., Lu, G.M., Munroe, P., Ahn, H.: Mesoporous $LiFePO_4$/C nanocomposite cathode materials for high power lithium ion batteries with superior performance. Adv. Mater. **22**, 4944–4948 (2010)
106. Gummow, R.J., He, Y.: Mesoporous manganese-deficient lithium manganese silicate cathodes for lithium-ion batteries. RSC Adv. **4**, 11580–11584 (2014)
107. Franger, S., Cras, F.L., Burbon, C., Rouault, H.: $LiFePO_4$ synthesis routes for enhanced electrochemical performance. Electrochem. Solid-State Lett. **5**, A231–A233 (2002)
108. Singhal, A., Skandan, G., Amatucci, G., Badway, F., Ye, N., Manthiram, A., Ye, H., Xu, J. J.: Nanostructured electrodes for next generation rechargeable electrochemical devices. J. Power Sources **129**, 38–44 (2004)
109. Panero, S., Scrosati, B., Wachtler, M., Croce, F.: Nanotechnology for the progress of lithium batteries R&D. J. Power Sources **129**, 90–95 (2004)
110. Yamada, A., Chung, S.C.: Crystal chemistry of the Olivine-type Li ($Mn_yFe_{1-y})PO_4$ and $(Mn_yFe_{1-y})PO_4$ as possible 4 V cathode materials for lithium batteries. J. Electrochem. Soc. **148**, A960–A967 (2001)
111. Dominko, R., Bele, M., Gaberscek, M., Remskar, M., Hanzel, D., Pejovnik, S., Jamnik, J.: Impact of the carbon coating thickness on the electrochemical performance of $LiFePO_4$/C composites. J. Electrochem. Soc. **152**, A607–A610 (2005)
112. Huang, G., Li, W., Sun, H., Wang, J., Zhang, J., Jiang, H., Zhai, F.: Polyvinylpyrrolidone (PVP) assisted synthesized nano-$LiFePO_4$/C composite with enhanced low temperature performance. Electrochim. Acta **97**, 92–98 (2013)
113. Cao, Y., Duan, J., Hu, G., Jiang, F., Peng, Z., Du, K., Guo, H.: Synthesis and electrochemical performance of nanostructured $LiMnPO_4$/C composites as lithium-ion battery cathode by a precipitation technique. Electrochim. Acta **98**, 183–189 (2013)
114. Huang, Y., Ren, H., Peng, Z., Zhou, Y.: Synthesis of $LiFePO_4$/carbon composite from nano-$FePO_4$ by a novel stearic acid assisted rheological phase method. Electrochim. Acta **55**, 311–315 (2009)
115. Hamamoto, K., Fukushima, M., Mamiya, M., Yoshizawa, Y., Akimoto, J., Suzuki, T., Fujishiro, Y.: Morphology control and electrochemical properties of $LiFePO_4$/C composite cathode for lithium ion batteries. Solid State Ionics **225**, 560–563 (2012)
116. Liu, R., Zhang, H., Huang, Y., Wang, W., Li, Z., Yu, Z., Wang, A., Yuan, K.: In situ gelatin carbonation to prepare a binder-free $LiFePO_4$ cathode for high-power lithium ion batteries. Electrochim. Acta **78**, 563–568 (2012)
117. Jeong, B., Kim, S.W., Bae, Y.H.: Thermosensitive sol-gel reversible hydrogels. Adv. Drug Delivery Rev. **64**, 154–162 (2012)
118. Kim, J.K., Choi, J.W., Chauhan, G.S., Ahn, J.H., Hwang, G.C., Choi, J.B., Ahn, H.J.: Enhancement of electrochemical performance of lithium iron phosphate by controlled sol-gel synthesis. Electrochim. Acta **53**, 8258–8264 (2008)
119. Djabourov, M., Leblond, J., Papon, P.: Gelation of aqueous gelatin solutions. I. Structural investigation. J. Phys. France **49**, 319–332 (1988)

120. Joly-Duhamel, C., Hellio, D., Ajdari, A., Djabourov, M.: All gelatin networks: 2. The master curve for elasticity. Langmuir **18**, 7158–7166 (2002)
121. Minakshi, M., Singh, P., Sharma, N., Blackford, M., Ionescu, M.: Lithium extraction–insertion from/into $LiCoPO_4$ in aqueous batteries. Ind. Eng. Chem. Res. **50**, 1899–1905 (2011)
122. Minakshi, M., Singh, P., Appadoo, D., Martin, D.E.: Synthesis and characterization of olivine $LiNiPO_4$ for aqueous rechargeable battery. Electrochim. Acta **56**, 4356–4360 (2011)
123. Minakshi, M., Singh, P., Thurgate, S., Prince, K.: Electrochemical behavior of Olivine-type $LiMnPO_4$ in aqueous solutions. Electrochem. Solid State Lett. **9**, A471–A474 (2006)
124. Minakshi, M., Singh, P., Thurgate, S., Prince, K.: Redox behavior and surface characterization of $LiFePO_4$ in lithium hydroxide electrolyte. J. Power Sources **158**, 646–649 (2006)
125. Minakshi, M., Sharma, N., Ralph, D., Appadoo, D., Nallathamby, K.: Synthesis and characterization of $Li(Co_{0.5}Ni_{0.5})PO_4$ cathode for Li-Ion aqueous battery applications. Electrochem. Solid State Lett. **14**, A86–A89 (2011)
126. Lee, J.M., Jun, Y.D., Kim, D.W., Lee, Y.H., Oh, S.G.: Effects of PVP on the formation of silver–polystyrene heterogeneous nanocomposite particles in novel preparation route involving polyol process: molecular weight and concentration of PVP. Mater. Chem. Phys. **114**, 549–555 (2009)
127. Bonet, F., Delmas, V., Grugeon, S., Urbina, R.H., Silvert, P.Y., Elhsissen, K.T.: Synthesis of monodisperse Au, Pt, Pd, Ru and Ir nanoparticles in ethylene glycol. Nanostruct. Mater. **11**, 1277–1284 (1999)
128. Du, Y.K., Yang, P., Mou, Z.G., Hua, N.P., Jiang, L.: Thermal decomposition behaviors of PVP coated on platinum nanoparticles. J. Appl. Polym. Sci. **99**, 23–26 (2006)
129. Lim, S.H., Cho, J.: PVP-Assisted ZrO_2 coating on $LiMn_2O_4$ spinel cathode nanoparticles prepared by MnO_2 nanowire templates. Electrochem. Commun. **10**, 1478–1481 (2008)
130. Padhi, A.K., Nanjundaswamy, K.S., Masquelier, C., Goodenough, J.B.: Mapping of transition metal redox energies in phosphates with NASICON structure by lithium intercalation. J. Electrochem. Soc. **144**, 2581–2586 (1997)
131. Huang, H., Yin, S.C., Kerr, T., Taylor, N., Nazar, L.F.: Nanostructure composites: a high capacity, fast rate $Li_3V_2(PO_4)_3$/carbon cathode for rechargeable lithium batteries. Adv. Mater. **14**, 1525–1528 (2002)
132. Yin, S.C., Grondey, H., Strobel, P., Huang, H., Nazar, L.F.: Charge ordering in lithium vanadium phosphates: electrode materials for lithium-ion batteries. J. Am. Chem. Soc. **125**, 326–327 (2003)
133. Yin, S.C., Strobel, P.S., Grondey, H., Nazar, L.F.: $Li_{2.5}V_2(PO_4)_3$: a room-temperature analogue to the fast-ion conducting high-temperature phase of $Li_3V_2(PO_4)_3$. Chem. Mater. **16**, 1456–1465 (2004)
134. Rui, X.H., Jin, Y., Feng, X.Y., Zhang, L.C., Chen, C.H.: A comparative study on the low temperature performance of $LiFePO_4$/C and $Li_3V_2(PO_4)_3$/C cathodes for lithium ion batteries. J. Power Sources **196**, 2109–2114 (2011)
135. Tukamoto, H., West, A.R.: Electronic conductivity of $LiCoO_2$ and its enhancement by magnesium doping. J. Electrochem. Soc. **144**, 3164–3168 (1997)
136. Wang, J.W., Liu, J., Yang, G.L., Zhang, X.F., Yan, X.D., Pan, X.M., Wang, R.S.: Electrochemical performance of $Li_3V_2(PO_4)_3$/C cathode material using a novel carbon source. Electrochim. Acta **54**, 6451–6454 (2009)
137. Rui, X.H., Li, C., Liu, J., Cheng, T., Chen, C.H.: Composites with high-rate capability prepared by a maltose-based sol-gel route. Electrochim. Acta **55**, 6761–6767 (2010)
138. Pan, A.Q., Liu, J., Zhang, J.G., Xu, W., Cao, G.Z., Nie, Z.M., Arey, B.W., Liang, S.Q.: Nanostructured $Li_3V_2(PO_4)_3$/carbon composite for high-rate lithium-ion batteries. Electrochem. Commun. **12**, 1674–1677 (2010)
139. Fu, P., Zhao, Y.M., Dong, Y.Z., An, X.N., Shen, G.P.: Synthesis of $Li_3V_2(PO_4)_3$ with high performance by optimized solid-state synthesis routine. J. Power Sources **162**, 651–657 (2006)
140. Zhu, X.J., Liu, Y.X., Geng, L.M., Chen, L.B.: Synthesis and performance of lithium vanadium phosphate as cathode materials for lithium ion batteries by a sol-gel method. J. Power Sources **184**, 578–582 (2008)

141. Yang, G., Liu, H.D., Ji, H.M., Chen, Z.Z., Jiang, X.F.: Temperature-controlled microwave solid-state synthesis of $Li_3V_2(PO_4)_3/C$ as cathode materials for lithium batteries. J. Power Sources **195**, 5374–5378 (2010)
142. Qiao, Y.Q., Tu, J.P., Xiang, J.Y., Wang, X.L., Mai, Y.J., Zhang, D., Liu, W.L.: Effects of synthetic route on structure and electrochemical performance of $Li_3V_2(PO_4)_3/C$ cathode materials. Electrochim. Acta **56**, 4139–4145 (2011)
143. Yuan, W., Yan, J., Tang, Z., Sha, O., Wang, J., Mao, W., Ma, L.: Synthesis of $Li_3V_2(PO_4)_3$ cathode material via a fast sol-gel method based on spontaneous chemical reactions. J. Power Sources **201**, 301–306 (2012)
144. Zheng, J.C., Li, X.H., Wang, Z.X., Li, J.H., Li, L.J., Wu, L., Guo, H.J.: Characteristics of $xLiFePO_4{\cdot}y\ Li_3V_2(PO_4)_3$ electrodes for lithium batteries. Ionics **15**, 753–759 (2009)
145. Zhong, S., Wu, L., Liu, J.: Sol-gel synthesis and electrochemical properties of $9LiFePO_4{\cdot}Li_3V_2(PO_4)_3/C$ composite cathode material for lithium ion batteries. Electrochim. Acta **74**, 8–15 (2012)
146. Zhang, L.L., Liang, G., Ignatov, A., Croft, M.C., Xiong, X.Q., Hung, I.M., Huang, Y.H., Hu, X.L., Zhang, W.X., Peng, Y.L.: Effect of vanadium incorporation on electrochemical performance of $LiFePO_4$ for lithium-ion batteries. J. Phys. Chem. C **115**, 13520–13527 (2011)
147. Xiang, J.Y., Tu, J.P., Zhang, L., Wang, X.L., Zhou, Y., Qiao, Y.Q., Lu, Y.: Improved electrochemical performances of $9LiFePO_4{\cdot}Li_3V_2(PO_4)/C$ composite prepared by a simple solid-state method. J. Power Sources **195**, 8331–8335 (2010)
148. Yang, M.R., Ke, W.H., Wu, S.H.: Improving electrochemical properties of lithium iron phosphate by addition of vanadium. J. Power Sources **165**, 646–650 (2007)
149. Wang, L., Li, Z., Xu, H., Zhang, K.: Studies of $Li_3V_2(PO_4)_3$ additives for the $LiFePO_4$ based Li ion batteries. J. Phy. Chem. C **112**, 308–312 (2008)
150. Rui, X.H., Ding, N., Liu, J., Li, C., Chen, C.H.: Analysis of the chemical diffusion coefficient of lithium ions in $Li_3V_2(PO_4)_3$ cathode material. Electrochim. Acta **55**, 2384–2390 (2010)
151. Pan, A., Liu, J., Zhang, J.G., Xu, W., Cao, G., Nie, Z., Arey, B.W., Liang, S.: Nano-structured $Li_3V_2(PO_4)_3$/carbon composite for high-rate lithium-ion batteries. Electrochem. Commun. **12**, 1674–1677 (2010)
152. Yang, G., Liu, H., Ji, H., Chen, Z., Jiang, X.: Microwave solid-state synthesis and electrochemical properties of carbon-free $Li_3V_2(PO_4)_3$ as cathode materials for lithium batteries. Electrochim. Acta **55**, 2951–2957 (2010)
153. Qiao, Y.Q., Wang, X.L., Zhou, Y., Xiang, J.Y., Zhang, D., Shi, S.J., Tu, J.P.: Electrochemical performance of carbon-coated $Li_3V_2(PO_4)_3$ cathode materials derived from polystyrene-based carbon-thermal reduction synthesis. Electrochem. Commun. **56**, 510–516 (2010)
154. Ma, J., Li, B., Du, H., Xu, C., Kang, F.: The effect of vanadium on physicochemical and electrochemical performances of $LiFePO_4$ cathode for lithium battery. J. Electrochem. Soc. **158**, A26–A32 (2011)
155. Sun, C.S., Zhou, Z., Xu, Z.G., Wang, D.G., Wei, J.P., Bian, X.K., Yan, J.: Improved high-rate charge/discharge performances of $LiFePO_4/C$ via V-doping. J. Power Sources **193**, 841–845 (2009)
156. Ren, M., Zhou, Z., Li, Y., Gao, X.P., Yan, J.: Preparation and electrochemical studies of Fe-doped $Li_3V_2(PO_4)_3$ cathode materials for lithium ion batteries. J. Power Sources **162**, 1357–1362 (2006)
157. Zheng, J.C., Li, X.H., Wang, Z.X., Niu, S.S., Liu, D.R., Wu, L., Li, L.J., Li, J.H., Guo, H.J.: Novel synthesis of $LiFePO_4$–$Li_3V_2(PO_4)_3$ composite cathode material by aqueous precipitation and lithiation. J. Power Sources **195**, 2935–2938 (2010)
158. Zhang, W.J.: Structure and performance of $LiFePO_4$ cathode materials: a review. J. Power Sources **196**, 2962–2970 (2011)
159. Fey, G.T., Chen, Y.G., Kao, H.M.: Electrochemical properties of $LiFePO_4$ prepared via ball-milling. J. Power Sources **189**, 169–178 (2009)
160. Gaberscek, M., Dominko, R., Jamnik, J.: Is small particle size more important than carbon coating? An example study on $LiFePO_4$ cathodes. Electrochem. Commun. **9**, 2778–2783 (2007)

161. Dompablo, M.E.A.Y.D., Amador, U., Tarascon, J.M.: A computational investigation on fluorinated-polyanionic compounds as positive electrode for lithium batteries. J. Power Sources **174**, 1251–1257 (2007)
162. Barker, J., Saidi, M.Y., Swoyer, J.L.: Electrochemical insertion properties of the novel lithium vanadium fluorophosphate, $LiVPO_4F$. J. Electrochem. Soc. **150**, A1394–A1398 (2003)
163. Liu, J.Q., Zhong, S.K., Wu, L., Wan, K., Lu, F.: Electrochemical performance of $LiVPO_4F/C$ synthesized by different methods. Trans. Nonferrous Met. Soc. China **22**, 157–161 (2012)
164. Xiong, Z., Zhang, G., Xiong, J., Yang, X., Zhang, Y.: Modified sol-gel synthesis of nanosized $LiVPO_4F/C$ cathode material with mechanical blending assist. Mater. Lett. **111**, 214–216 (2013)
165. Qiao, X., Yang, J., Wang, Y., Chen, Q., Zhang, T., Liu, L., Wang, X.: Electrochemical performance of $LiVPO_4F/C$ composite cathode prepared through amorphous vanadium phosphorus oxide intermediate. J. Solid State Electrochem. **16**, 1211–1217 (2012)
166. Gong, H., Yu, Y., Li, T., Mei, T., Xing, Z., Zhu, Y., Qian, Y., Shen, X.: Solvothermal synthesis of $LiFePO_4/C$ nanopolyhedrons and microellipsoids and their performance in lithium-ion batteries. Mater. Lett. **66**, 374–376 (2012)
167. Qiao, Y.Q., Wang, X.L., Xiang, J.Y., Zhang, D., Liu, W.L., Tu, J.P.: Electrochemical performance of $Li_3V_2(PO_4)_3/C$ cathode materials using stearic acid as a carbon source. Electrochim. Acta **56**, 2269–2275 (2011)
168. Zhou, F., Zhao, X., Dahn, J.R.: Reactivity of charged $LiVPO_4F$ with 1M $LiPF_6$ EC:DEC electrolyte at high temperature as studied by accelerating rate calorimetry. Electrochem. Commun. **11**, 589–591 (2009)
169. Jiang, T., Chen, G., Li, A., Wang, C., Wei, Y.: Sol-gel preparation and electrochemical properties of $Na_3V_2(PO_4)_2F_3/C$ composite cathode material for lithium ion batteries. J. Alloys Compd. **478**, 604–607 (2009)
170. Barker, J., Gover, R.K.B., Burns, P., Bryan, A.J.: A lithium-ion cell based on a sodium insertion material. Electrochem. Solid-State Lett. **9**, A190–A192 (2006)
171. Barker, J., Gover, R.K.B., Burns, P., Bryan, A.J.: $Li_{4/3}Ti_{5/3}O_4$ || Na_3V_2 (PO_4) $2F_3$: an example of a hybrid-ion cell using a non-graphitic anode. J. Electrochem. Soc. **154**, A882–A887 (2007)
172. Meins, J.M.L., Bohnke, O., Courbion, G.: Ionic conductivity of crystalline and amorphous $Na_3Al_2(PO_4)_2F_3$. Solid State Ionics **111**, 67–75 (1998)
173. Meins, J.M.L., Crosnier-Lopez, M.P., Hemon-Ribaud, A., Courbion, G.: Phase transitions in the $Na_3M_2(PO_4)_2F_3$ family (M = Al^{3+}, V^{3+}, Cr^{3+}, Fe^{3+}, Ga^{3+}): synthesis, thermal, structural, and magnetic studies. J. Solid State Chem. **148**, 260–277 (1999)
174. Barker, J., Saidi, M.Y., Swoyer, J.L.: A comparative investigation of the Li insertion properties of the novel fluorophosphate phases, $NaVPO_4F$ and $LiVPO_4F$. J. Electrochem. Soc. **151**, A1670–A1677 (2004)
175. Barker, J., Saidi, M.Y., Swoyer, J.L.: A sodium-ion cell based on the fluorophosphate compound $NaVPO_4F$ electrochem. Solid State Lett. **6**, A1–A4 (2003)
176. Zhuo, H., Wang, X., Tang, A., Liu, Z., Gamboa, S., Sebastian, P.J.: The preparation of $NaV_{1-x}Cr_xPO_4F$ cathode materials for sodium-ion battery. J. Power Sources **160**, 698–703 (2006)
177. Sauvage, F., Quarez, E., Tarascon, J.M., Baudrin, E.: Crystal structure and electrochemical properties vs. Na^+ of the sodium fluorophosphate $Na_{1.5}VOPO_4F_{0.5}$. Solid State Sci. **8**, 1215–1221 (2006)
178. Zhao, J., He, J., Ding, X., Zhou, J., Ma, Y.O., Wu, S., Huang, R.: A novel sol-gel synthesis route to $NaVPO_4F$ as cathode material for hybrid lithium ion batteries. J. Power Sources **195**, 6854–6859 (2010)
179. Zhang, C.X., He, J., Zhao, G., Zhao, J.: Electrochemical characteristics of C-doped $NaVPO_4F$ cathode material for sodium-ion battery. Chin. J. Inorg. Chem. **23**, 649–654 (2007)
180. Li, Y., Zhou, Z., Gao, X.P., Yan, J.: A novel sol-gel method to synthesize nanocrystalline $LiVPO_4F$ and its electrochemical Li intercalation performances. J. Power Sources **160**, 633–637 (2006)

181. Dominko, R., Conte, D.E., Hanzel, D., Gaberscek, M., Jamnik, J.: Impact of synthesis conditions on the structure and performance of Li_2FeSiO_4. J. Power Sources **178**, 842–847 (2008)
182. Dominko, R.: Li_2MSiO_4 (M=Fe and/or Mn) cathode materials. J. Power Sources **184**, 462–468 (2008)
183. Lv, D., Wen, W., Huang, X., Bai, J., Mi, J., Wu, S., Yang, Y.: A novel Li_2FeSiO_4/C composite: synthesis, characterization and high storage capacity. J. Mater. Chem. **21**, 9506–9512 (2011)
184. Muraliganth, T., Stroukoff, K.R., Manthiram, A.: Microwave-solvothermal synthesis of nanostructured Li_2MSiO_4/C (M=Mn and Fe) cathodes for lithium-ion batteries. Chem. Mater. **22**, 5754–5761 (2010)
185. Huang, X., Li, X., Wang, H., Pan, Z., Qu, M., Yu, Z.: Synthesis and electrochemical performance of Li_2FeSiO_4/C as cathode material for lithium batteries. Solid State Ionics **181**, 1451–1455 (2010)
186. Deng, C., Zhang, S., Fu, B.L., Yang, S.Y., Ma, L.: Characterization of Li_2MnSiO_4 and Li_2FeSiO_4 cathode materials synthesized via a citric acid assisted sol-gel method. Mater. Chem. Phys. **14**, 14–17 (2010)
187. Ensling, D., Stjerndahl, M., Nyten, A., Gustafsson, T., Thomas, J.O.: A comparative XPS surface study of Li_2FeSiO_4/C cycled with LiTFSI- and $LiPF_6$-based electrolytes. J. Mater. Chem. **19**, 82–88 (2009)
188. Zhang, S., Deng, C., Yang, S.Y.: Preparation of nano-Li_2FeSiO_4 as cathode material for lithium-ion batteries. Electrochem. Solid-State Lett. **12**, A136–A139 (2009)
189. Kam, K.C., Gustafsson, T., Thomas, J.O.: Synthesis and electrochemical properties of nanostructured Li_2FeSiO_4/C cathode material for Li-ion batteries. Solid State Ionics **192**, 356–359 (2010)
190. Fan, X.Y., Li, Y., Wang, J.J., Gou, L., Zhao, P., Li, D.L., Huang, L., Sun, S.G.: Synthesis and electrochemical performance of porous Li_2FeSiO_4/C cathode material for long-life lithium-ion batteries. J. Alloys Compd. **493**, 77–80 (2010)
191. Deng, C., Zhang, S., Yang, S.Y., Fu, B.L., Ma, L.: Synthesis and characterization of $Li_2Fe_{0.97}M_{0.03}SiO_4$ (M=Zn^{2+}, Cu^{2+}, Ni^{2+}) cathode materials for lithium ion batteries. J. Power Sources **196**, 386–392 (2011)
192. Rangappa, D., Murukanahally, K.D., Tomai, T., Unemoto, A., Honma, I.: Ultrathin nanosheets of Li_2MSiO_4 (M=Fe, Mn) as high-capacity Li-ion battery electrode. Nano Lett. **12**, 1146–1151 (2012)
193. Yan, Z., Cai, S., Zhou, X., Zhao, Y., Miao, L.: Sol-gel synthesis of nanostructured Li_2FeSiO_4/C as cathode material for lithium ion battery. J. Electrochem. Soc. **159**, A894–A898 (2012)
194. Liu, Y., Cao, C., Li, J.: Enhanced electrochemical performance of carbon nanospheres–$LiFePO_4$ composite by PEG based sol-gel synthesis. Electrochim. Acta **55**, 3921–3926 (2010)
195. Gong, Z.L., Li, Y.X., He, G.N., Li, J., Yang, Y.: Nanostructured Li_2FeSiO_4 electrode material synthesized through hydrothermal-assisted sol-gel process. Electrochem. Solid-State Lett. **11**, A60–A63 (2008)
196. Guo, H.J., Cao, X., Li, X.Q., Li, L.M., Li, X.H., Wang, Z.X., Peng, W.J., Li, Q.H.: Optimum synthesis of $Li_2Fe_{1-x}Mn_xSiO_4$/C cathode for lithium ion batteries. Electrochim. Acta **55**, 8036–8042 (2010)
197. Lu, Z., MacNeil, D.D., Dahn, J.R.: Layered Li $[Ni_xCo_{1-2x}Mn_x]$ O_2 cathode materials for lithium-ion batteries. Electrochem. Solid State Lett. **4**, A200–A203 (2001)
198. Ohzuku, T., Makimura, Y.: Layered lithium insertion material of $LiCo_{1/3}Ni_{1/3}Mn_{1/3}O_2$ for lithium-ion batteries. Chem. Lett. **30**, 642–643 (2001)
199. Makimura, Y., Ohzuku, T.: Lithium insertion material of $LiNi_{1/2}Mn_{1/2}O_2$ for advanced lithium-ion batteries. J. Power Sources **119–121**, 156–160 (2003)

200. Lu, H., Zhou, H., Svensson, A.M., Fossdal, A., Sheridan, E., Lu, S., Vullum-Bruer, F.: High capacity $Li[Ni_{0.8}Co_{0.1}Mn_{0.1}]O_2$ synthesized by sol-gel and co-precipitation methods as cathode materials for lithium-ion batteries. Solid State Ionics **249–250**, 105–111 (2013)
201. Park, S.H., Shin, H.S., Myung, S.T., Yoon, C.S., Amine, K., Sun, Y.K.: Synthesis of nanostructured $Li[Ni_{1/3}Co_{1/3}Mn_{1/3}]O_2$ via a modified carbonate process. Chem. Mater. **17**, 6–8 (2005)
202. Zhang, X., Jiang, W.J., Mauger, A., Gendrond, F., Juliend, C.M.: Minimization of the cation mixing in $Li_{1+x}(NMC)_{1-x}O_2$ as cathode material. J. Power Sources **195**, 1292–1301 (2010)
203. Idemoto, Y., Matsui, T.: Thermodynamic stability, crystal structure, and cathodic performance of $Li_x(Mn_{1/3}Co_{1/3}Ni_{1/3})O_2$ depend on the synthetic process and Li content. Solid State Ionics **179**, 625–635 (2008)
204. Cho, S.W., Ryu, K.S.: Sulfur anion doping and surface modification with $LiNiPO_4$ of a $LiNi_{0.5}Mn_{0.3}Co_{0.2}O_2$ cathode. Mater. Chem. Phys. **135**, 533–540 (2012)
205. Parka, S.H., Yoon, C.S., Kang, S.G., Kim, H.S., Moon, S.I., Sun, Y.K.: Synthesis and structural characterization of layered $Li[Ni_{1/3}Co_{1/3}Mn_{1/3}]O_2$ cathode materials by ultrasonic spray pyrolysis method. Electrochim. Acta **49**, 557–563 (2004)
206. Oh, S.W., Park, S.H., Park, C.W., Sun, Y.K.: Structural and electrochemical properties of layered $Li[Ni_{0.5}Mn_{0.5}]_{1-x}Co_xO_2$ positive materials synthesized by ultrasonic spray pyrolysis method. Solid State Ionics **171**, 167–172 (2004)
207. Lee, K.S., Myung, S.T., Amine, K., Yashiro, H., Sun, Y.K.: Structural and electrochemical properties of layered $Li[Ni_{1-2x}Co_xMn_x]O_2$ ($x = 0.1–0.3$) positive electrode materials for Li-ion batteries. J. Electrochem Soc. **154**, A971–A977 (2007)
208. Saavedra-Arias, J.J., Karan, N.K., Pradhan, D.K., Kumar, A., Nieto, S., Thomas, R., Katiyar, R.S.: Synthesis and electrochemical properties of $Li(Ni_{0.8}Co_{0.1}Mn_{0.1})O_2$ cathode material: ex situ structural analysis by Raman scattering and X-ray diffraction at various stages of charge–discharge process. J. Power Sources **183**, 761–765 (2008)
209. Chen, C.H., Wang, C.J., Hwang, B.J.: Electrochemical performance of layered Li $[Ni_xCo_{1-2x}Mn_x]O_2$ cathode materials synthesized by a sol-gel method. J. Power Sources **146**, 626–629 (2005)
210. Eom, J., Kim, M.G., Cho, J.: Storage characteristics of $LiNi_{0.8}Co_{0.1+x}Mn_{0.1-x}O_2$ ($x = 0$, 0.03, and 0.06) cathode materials for lithium batteries. J. Electrochem Soc. **155**, A239–A245 (2008)
211. Woo, S.U., Park, B.C., Yoon, C.S., Myung, S.T., Prakash, J., Sun, Y.K.: Improvement of electrochemical performances of $Li[Ni_{0.8}Co_{0.1}Mn_{0.1}]O_2$ cathode materials by fluorine substitution. J. Electrochem Soc. **154**, A649–A655 (2007)
212. Kim, M.H., Shin, H.-S., Shin, D., Sun, Y.K.: Synthesis and electrochemical properties of $Li[Ni_{0.8}Co_{0.1}Mn_{0.1}]O_2$ and $Li[Ni_{0.8}Co_{0.2}]O_2$ via co-precipitation. J. Power Sources **159**, 1328–1333 (2006)
213. Woo, S.W., Myung, S.T., Bang, H., Kim, D.W., Sun, Y.K.: Improvement of electrochemical and thermal properties of $Li[Ni_{0.8}Co_{0.1}Mn_{0.1}]O_2$ positive electrode materials by multiple metal (Al, Mg) substitution. Electrochim. Acta **54**, 3851–3856 (2009)
214. Cho, J., Jung, H., Park, Y., Kim, G., Lim, H.S.: Electrochemical properties and thermal stability of $LiaNi_{1-x}CO_xO_2$ cathode materials. J. Electrochem. Soc. **147**, 15–20 (2000)
215. Bianchi, V., Bach, S., Belhomme, C., Farcy, J., Pereira-Ramos, J.P., Caurant, D., Baffier, N., Willmann, P.: Electrochemical investigation of the Li insertion–extraction reaction as a function of lithium deficiency in $Li_{1-x}Ni_{1+x}O_2$. Electrochim. Acta **46**, 999–1011 (2001)
216. Hu, G., Liu, W., Peng, Z., Du, K., Cao, Y.: Synthesis and electrochemical properties of $LiNi_{0.8}Co_{0.15}Al_{0.05}O_2$ prepared from the precursor $Ni_{0.8}Co_{0.15}Al_{0.05}OOH$. J. Power Sources **198**, 258–263 (2012)
217. Liu, H.S., Zhang, Z.R., Gong, Z.L., Yang, Y.: Origin of deterioration for $LiNiO_2$ cathode material during storage in air. Electrochem. Solid State Lett. **7**, A190–A193 (2004)
218. Shizuka, K., Kiyohara, C., Shima, K., Takeda, Y.: Effect of CO_2 on layered $Li_{1+z}Ni_{1-x-y}Co_xM_yO_2$ (M=Al, Mn) cathode materials for lithium ion batteries. J. Power Sources **166**, 233–238 (2007)

219. Armstrong, A.R., Holzapfel, M., Novák, P., Johnson, C.S., Kang, S.H., Thackeray, M.M., Bruce, P.G.: Demonstrating oxygen loss and associated structural reorganization in the lithium battery cathode $Li[Ni_{0.2}Li_{0.2}Mn_{0.6}]O_2$. J. Am. Chem. Soc. **128**, 8694–8698 (2006)
220. Johnson, C.S., Li, N., Lefief, C., Vaughey, J.T., Thackeray, M.M.: Synthesis, characterization and electrochemistry of lithium battery electrodes: $xLi_2MnO_3{\cdot}(1-x)$ $LiMn_{0.333}Ni_{0.333}Co_{0.333}O_2$ $(0 \leq x \leq 0.7)$. Chem. Mater. **20**, 6095–6106 (2008)
221. Kim, M.G., Jo, M., Hong, Y.S., Cho, J.: Template-free synthesis of Li $[Ni_{0.5}Li_{0.15}Mn_{0.6}]O_2$ nanowires for high performance lithium battery cathode. Chem. Commun. **1**, 218–220 (2009)
222. Ito, A., Li, D., Ohsawa, Y., Sato, Y.: A new approach to improve the high-voltage cyclic performance of Li-rich layered cathode material by electrochemical pretreatment. J. Power Sources **183**, 344–346 (2008)
223. Yabuuchi, N., Yoshii, K., Myung, S.T., Nakai, I., Komaba, S.: Detailed studies of a high-capacity electrode material for rechargeable batteries, $Li_2MnO_3 - LiCo_{1/3}Ni_{1/3}Mn_{1/3}O_2$. J. Am. Chem. Soc. **133**, 4404–4419 (2011)
224. Li, G.R., Feng, X., Ding, Y., Ye, S.H., Gao, X.P.: AlF_3-coated $Li(Li_{0.17}Ni_{0.25}Mn_{0.58})O_2$ as cathode material for Li-ion batteries. Electrochim. Acta **78**, 308–315 (2012)
225. Zhang, H.Z., Qiao, Q.Q., Li, G.R., Gao, X.P.: PO_4^{3-} polyanion-doping for stabilizing Li-rich layered oxides as cathode materials for advanced lithium-ion batteries. J. Mater. Chem. A **2**, 7454–7460 (2014)
226. Xu, K., Jie, Z., Li, R., Chen, Z., Wu, S., Gu, J., Chen, J.: Synthesis and electrochemical properties of CaF_2-coated for long-cycling $Li[Mn_{1/3}Co_{1/3}Ni_{1/3}]O_2$ cathode materials. Electrochim. Acta **60**, 130–133 (2012)
227. Iftekhar, M., Drewett, N.E., Armstrong, A.R., Hesp, D., Braga, F., Ahmed, S.A., Hardwick, L.J.: Characterization of aluminum doped lithium-manganese rich composites for higher rate lithium-ion cathodes. J. Electrochem Soc. **161**, A2109–A2116 (2014)
228. Knight, J.C., Nandakumar, P., Kan, W.H., Manthiram, A.: Effect of Ru substitution on the first charge–discharge cycle of lithium-rich layered oxides. J. Mater. Chem. A **3**, 2006–2011 (2015)
229. Min, J.W., Yim, C.J., Im, W.B.: Facile synthesis of electrospun $Li_{1.2}Ni_{0.17}Co_{0.17}Mn_{0.5}O_2$ nanofiber and its enhanced high-rate performance for lithium-ion battery applications. ACS Appl. Mater. Interfaces **5**, 7765–7769 (2013)
230. Shen, C.H., Wang, Q., Fu, F., Huang, L., Lin, Z., Shen, S.Y., Su, H., Zheng, X.M., Xu, B.B., Li, J.T., Sun, S.G.: Facile synthesis of the Li-rich layered oxide $Li_{1.23}Ni_{0.09}Co_{0.12}Mn_{0.56}O_2$ with superior lithium storage performance and new insights into structural transformation of the layered oxide material during charge-discharge cycle. In situ XRD characterization. ACS Appl. Mater. Interfaces **6**, 5516–5524 (2014)
231. Naoi, K., Yonekura, D., Moriyama, S., Goto, H., Iwama, E., Kubota, S., Ishimoto, S., Naoic, W.: Lithium-rich layered oxide nanoplate/carbon nanofiber composites exhibiting extremely large reversible lithium storage capacity. J. Alloys Compd. **605**, 137–141 (2014)
232. Jung, Y.S., Cavanagh, A.S., Yan, Y., George, S.M., Manthiram, A.: Effects of atomic layer deposition of Al_2O_3 on the $Li[Li_{0.20}Mn_{0.54}Ni_{0.13}Co_{0.13}]O_2$ cathode for lithium-ion batteries. J. Electrochem Soc. **158**, A1298–A1302 (2011)
233. Uzun, D., Doğrusöz, M., Mazman, M., Biçer, E., Avci, E., Şener, T., Kaypmaz, T.C., Demir-Cakan, R.: Effect of MnO_2 coating on layered $Li(Li_{0.1}Ni_{0.3}Mn_{0.5}Fe_{0.1})O_2$ cathode material for Li-ion batteries. Solid State Ionics **249**, 171–176 (2013)
234. Yuan, W., Zhang, H.Z., Liu, Q., Li, G.R., Gao, X.P.: Surface modification of Li $(Li_{0.17}Ni_{0.2}Co_{0.05}Mn_{0.58})O_2$ with CeO_2 as cathode material for Li-ion batteries. Electrochim. Acta **135**, 199–207 (2014)
235. Kim, I.T., Knight, J.C., Celio, H., Manthiram, A.: Enhanced electrochemical performances of Li-rich layered oxides by surface modification with reduced graphene oxide/$AlPO_4$ hybrid coating. J. Mater. Chem. A **2**, 8696–8704 (2014)

236. Qiao, Q.Q., Zhang, H.Z., Li, G.R., Ye, S.H., Wang, C.W., Gao, X.P.: Surface modification of Li-rich layered $Li(Li_{0.17}Ni_{0.25}Mn_{0.58})O_2$ oxide with Li–Mn–PO_4 as the cathode for lithium-ion batteries. J. Mater. Chem. A **1**, 5262–5268 (2013)
237. Liu, B., Zhang, Q., He, S., Sato, Y., Zheng, J., Li, D.: Improved electrochemical properties of $Li_{1.2}Ni_{0.18}Mn_{0.59}Co_{0.03}O_2$ by surface modification with $LiCoPO_4$. Electrochim. Acta **56**, 6748–6751 (2011)
238. Sun, Y.Y., Li, F., Qiao, Q.Q., Cao, J.S., Wang, Y.L., Ye, S.H.: Surface modification of Li $(Li_{0.17}Ni_{0.2}Co_{0.05}Mn_{0.58})O_2$ with $LiAlSiO_4$ fast ion conductor as cathode material for Li-ion batteries. Electrochim. Acta **176**, 1464–1475 (2015)
239. Fu, Q., Du, F., Bian, X., Wang, Y., Yan, X., Zhang, Y., Zhu, K., Chen, G., Wang, C., Wei, Y.: Electrochemical performance and thermal stability of $Li_{1.18}Co_{0.15}Ni_{0.15}Mn_{0.52}O_2$ surface coated with the ionic conductor Li_3VO_4. J. Mater. Chem. A **2**, 7555–7562 (2014)
240. Li, X., Liu, J., Banis, M.N., Lushington, A., Li, R., Cai, M., Sun, X.: Atomic layer deposition of solid-state electrolyte coated cathode materials with superior high-voltage cycling behavior for lithium ion battery application. Energy Environ. Sci. **7**, 768–778 (2014)
241. Welsch, A.M., Behrens, H., Horn, I., Roß, S., Heitjans, P.: Self-diffusion of lithium in $LiAlSi_2O_6$ glasses studied using mass spectrometry. J. Phys. Chem. A **116**, 309–318 (2012)
242. Shin-ichi, F., Satoshi, S., Kaduhiro, S., Hitoshi, T.: Preparation and ionic conductivity of β-$LiAlSiO_4$ thin film. Solid State Ionics **167**, 325–329 (2004)
243. Tarascon, J.M., McKinnon, W.R., Coowar, F., Bowmer, T.N., Amatucci, G., Guyomard, D.: Synthesis conditions and oxygen stoichiometry effects on Li insertion into the spinel $LiMn_2O_4$. J. Electrochem. Soc. **141**, 1421–1431 (1994)
244. Gummow, R.J., Kock, A.D., Thackeray, M.M.: Improved capacity retention in rechargeable 4 V lithium/lithium-manganese oxide (spinel) cells. Solid State Ionics **69**, 59–67 (1994)
245. Thackeray, M.M., Kock, A.D., Rossouw, M.H., Liles, D., Bittihn, R., Hoge, D.: Spinel electrodes from the Li-Mn-O system for rechargeable lithium battery applications. J. Electrochem Soc. **139**, 363–366 (1992)
246. Wan, C., Wu, M.: Synthesis of spherical $LiMn_2O_4$ cathode material by dynamic sintering of spray-dried precursors. Powder Technol. **199**, 154–158 (2010)
247. Shukla, A.K., Kumar, T.P.: Materials for next-generation lithium batteries. Curr. Sci. **94**, 314–331 (2008)
248. Kim, D.K., Muralidharan, P., Lee, H.W., Ruffo, R., Yang, Y., Chan, C.K., Peng, H., Huggins, R.A., Cui, Y.: Spinel $LiMn_2O_4$ nanorods as lithium ion battery cathodes. Nano Lett. **8**, 3948–3952 (2008)
249. Hosono, E., Kudo, T., Honma, I., Matsuda, H., Zhou, H.: Synthesis of single crystalline spinel $LiMn_2O_4$ nanowires for a lithium ion battery with high power density. Nano Lett. **9**, 1045–1051 (2009)
250. Xia, Y., Zhou, Y., Yoshio, M.: Capacity fading on cycling of 4 V Li/ $LiMn_2O_4$ cells. J. Electrochem. Soc. **144**, 2593–2600 (1997)
251. Pistoia, G., Antonini, A., Rosati, R., Zane, D.: Storage characteristics of cathodes for Li-ion batteries. Electrochim. Acta **41**, 2683–2689 (1996)
252. Jang, D.H., Shin, Y.J., Oh, S.M.: Dissolution of spinel oxides and capacity losses in 4 V Li/ $Li_xMn_2O_4$ cells. J. Electrochem. Soc. **143**, 2204–2211 (1996)
253. Yamada, A.: Lattice instability in $Li(Li_xMn_{2-x})O_4$. J. Solid State Chem. **122**, 160–165 (1996)
254. Song, D., Ikuta, H., Uchida, T., Wakihara, M.: The spinel phases $LiAl_yMn_{2-y}O_4$ ($y = 0$, 1/12, 1/9, 1/6, 1/3) and $Li(Al, M)_{1/6}Mn_{11/6}O_4$ (M = Cr, Co) as the cathode for rechargeable lithium batteries. Solid State Ionics **117**, 151–156 (1999)
255. Iqbal, M.J., Zahoor, S.: Synthesis and characterization of nanosized lithium manganate and its derivatives. J. Power Sources **165**, 393–397 (2007)
256. Eftekhari, A.: Electrochemical performance and cyclability of $LiFe_{0.5}Mn_{1.5}O_4$ as a 5 V cathode material for lithium batteries. J. Power Sources **124**, 182–190 (2003)

257. Gummow, R.J., Dekock, A., Thackeray, M.M.: Improved capacity retention in rechargeable 4 V lithium/lithium-manganese oxide (spinel) cells. Solid State Ionics **69**, 59–67 (1994)
258. Lee, J.H., Hong, J.K., Jang, D.H., Sun, Y.K., Oh, S.M.: Degradation mechanisms in doped spinels of $LiM_{0.05}Mn_{1.95}O_4$ (M=Li, B, Al Co, and Ni) for Li secondary batteries. J. Power Sources **89**, 7–14 (2000)
259. Park, S.H., Park, K.S., Sun, Y.K., Nahm, K.S.: Synthesis and characterization of a new spinel, $Li_{1.02}Al_{0.25}Mn_{1.75}O_{3.97}S_{0.03}$, operating at potentials between 4.3 and 2.4 V. J. Electrochem Soc. **147**, 2116–2121 (2000)
260. Amatucci, G.G., Perera, N., Zheng, T., Plitz, L., Tarascon, J.M.: Enhancement of the electrochemical properties of $LiMn_2O_4$ through chemical substitution. J. Power Sources **81**, 39–43 (1999)
261. Subramania, A., Angayarkanni, N., Vasudevan, T.: Synthesis of nano-crystalline $LiSr_xMn_{2-x}O_4$ powder by a novel sol-gel thermolysis process for Li-ion polymer battery. J. Power Sources **158**, 1410–1413 (2006)
262. Kovacheva, D., Gadjov, H., Petrov, K., Mandel, S.: Synthesizing nanocrystalline $LiMn_2O_4$ by a combustion route. J. Mater. Chem. **12**, 1184–1188 (2002)
263. Bach, S., Henry, M., Baffier, N., Livage, J.: Sol-gel synthesis of manganese oxides. J. Solid State Chem. **88**, 325–333 (1990)
264. Pereira-Ramos, J.P.: Electrochemical properties of cathodic materials synthesized by low-temperature techniques. J. Power Sources **54**, 120–126 (1995)
265. Barboux, P., Tarascon, J.M., Shokoohi, F.K.: The use of acetates as precursors for the low-temperature synthesis of $LiMn_2O_4$ and $LiCoO_2$ intercalation compounds. J. Solid State Chem. **94**, 185–196 (1991)
266. Liu, W., Farrington, G.C., Chaput, F., Dunn, B.: Synthesis and electrochemical studies of spinel phase $LiMn_2O_4$ cathode materials prepared by the pechini process. J. Electrochem. Soc. **143**, 879–884 (1996)
267. Thirunakaran, R., Ravikumar, R., Vanitha, S., Gopukumar, S., Sivashanmugam, A.: Glutamic acid-assisted sol-gel synthesis of multi-doped spinel lithium manganate as cathode materials for lithium rechargeable batteries. Electrochim. Acta **58**, 348–358 (2011)
268. Thirunakaran, R., Kalaiselvi, N., Periasamy, P., Babu, B.R., Renganathan, N.G., Muniyandi, N., Raghavan, M.: Significance of Mg doped $LiMn_2O_4$ spinels as attractive 4 V cathode materials for use in lithium batteries. Ionics **7**, 187–191 (2001)
269. Guo, S., Zhang, S., He, X., Pu, W., Jiang, C., Wan, C.: Synthesis and characterization of Sn-doped $LiMn_2O_4$ cathode materials for rechargeable Li-ion batteries. J. Electrochem. Soc. **155**, A760–A763 (2008)
270. Thirunakaran, R., Sivashanmugam, A., Gopukumar, S., Dunnill, C.W., Gregory, D.H.: Electrochemical behaviour of nano-sized spinel $LiMn_2O_4$ and $LiAl_xMn_{2-x}O_4$ (x = Al: 0.00–0.40) synthesized via fumaric acid-assisted sol-gel synthesis for use in lithium rechargeable batteries. J. Phys. Chem. Solids **69**, 2082–2090 (2008)
271. Thirunakaran, R., Sivashanmugam, A., Gopukumar, S., Dunnill, C.W., Gregory, D.H.: Phthalic acid assisted nano-sized spinel $LiMn_2O_4$ and $LiCr_xMn_{2x}O_4$ (x = 0.00–0.40) via sol-gel synthesis and its electrochemical behaviour for use in Li-ion-batteries. Mater. Res. Bull. **43**, 2119–2129 (2008)
272. Michalska, M., Lipińska, L., Mirkowska, M., Aksienionek, M., Diduszko, R., Wasiucionek, M.: Nanocrystalline lithium-manganese oxide spinels for Li-ion batteries—Sol-gel synthesis and characterization of their structure and selected physical properties. Solid State Ionics **188**, 160–164 (2011)
273. Wang, Z.J., Du, J.L., Duan, W., Niu, Y.Q., Wu, Z.: $LiMn_{2x}Ni_xO_4$ spinel oxides as high-temperature lithium battery cathode materials for borehole applications. Int. J. Electrochem. Sci. **8**, 6231–6242 (2013)

274. Wang, Z., Du, J., Li, Z., Wu, Z.: Sol-gel synthesis of Co-doped $LiMn_2O_4$ with improved high-rate properties for high temperature lithium batteries. Ceram. Int. **40**, 3527–3531 (2014)
275. Liu, W., Farrington, G.C., Chaput, F., Dunn, B.: Intercalation-induced stress and heat generation within single lithium-ion battery cathode particles. J. Electrochem. Soc. **143**, 879–884 (1996)
276. Guoping, W., Qingtang, Z., Zuolong, Y., MeiZheng, Q.: The effect of different kinds of nano-carbon conductive additives in lithium ion batteries on the resistance and electrochemical behavior of the $LiCoO_2$ composite cathodes. Solid State Ionics **179**, 263–268 (2008)
277. Sheem, K., Lee, Y.H., Lim, H.S.: High-density positive electrodes containing carbon nanotubes for use in Li-ion cells. J. Power Sources **158**, 1425–1430 (2006)
278. Wang, L., Huang, Y., Jiang, R., Jia, D.: Nano-$LiFePO_4$/ MWCNT cathode materials prepared by room-temperature solid-state reaction and microwave heating. J. Electrochem. Soc. **154**, A1015–A1019 (2007)
279. Li, J., Tang, S., Lu, L., Zeng, H.C.: Preparation of nanocomposites of metals, metal oxides, and carbon nanotubes via self assembly. J. Am. Chem. Soc. **129**, 9401–9409 (2007)
280. Chen, G., Wang, Z., Xia, D.: One-pot synthesis of carbon nanotube@SnO_2 − Au coaxial nanocable for lithium-ion batteries with high rate capability. Chem. Mater. **20**, 6951–6956 (2008)
281. Liu, X.M., Huang, Z.D., Oh, S., Ma, P.C., Chan, P.C.H., Vedam, G.K., Kang, K., Kim, J.K.: Sol-gel synthesis of multiwalled carbon nanotube-$LiMn_2O_4$ nanocomposites as cathode materials for Li-ion batteries. J. Power Sources **195**, 4290–4296 (2010)
282. Idemoto, Y., Narai, H., Koura, N.: Crystal structure and cathode performance dependence on oxygen content of $LiMn_{1.5}Ni_{0.5}O_4$ as a cathode material for secondary lithium batteries. J. Power Sources **119–121**, 125–129 (2003)
283. Sun, Y., Wang, Z., Huang, X., Chen, L.: Synthesis and electrochemical performance of spinel $LiMn_{2-x-y}Ni_xCr_yO_4$ as 5-V cathode materials for lithium ion batteries. J. Power Sources **132**, 161–165 (2004)
284. Talyosef, Y., Markovsky, B., Salitra, G., Aurbach, D., Kim, H.J., Choi, S.: The study of $LiNi_{0.5}Mn_{1.5}O_4$ 5-V cathodes for Li-ion batteries. J. Power Sources **146**, 664–669 (2005)
285. Ohzuku, T., Takeda, S., Iwanaga, M.: Solid-state redox potentials for $Li[Me_{1/2}Mn_{3/2}]O_4$ (Me: 3d-transition metal) having spinel framework structures: a series of 5 volt materials for advanced lithium-ion batteries. J. Power Sources **81–82**, 90–94 (1999)
286. Xu, X.X., Yang, J., Wang, Y.Q., NuLi, Y.N., Wang, J.L.: $LiNi_{0.5}Mn_{1.5}O_{3.975}F_{0.05}$ as novel 5 V cathode material. J. Power Sources **174**, 1113–1116 (2007)
287. Arunkumar, T.A., Manthiram, A.: Influence of lattice parameter differences on the electrochemical performance of the 5 V spinel $LiMn_{1.5-y}Ni_{0.5-z}M_{y+z}O_4$ (M=Li, Mg, Fe, Co, and Zn). Electrochem. Solid State Lett. **8**, A403–A405 (2005)
288. Lee, Y., Mun, J., Kim, D.W., Lee, J.K., Choi, W.: Surface modification of $LiNi_{0.5}Mn_{1.5}O_4$ cathodes with $ZnAl_2O_4$ by asol-gel method for lithium ion batteries. Electrochim. Acta **115**, 326–331 (2014)
289. Arrebola, J.C., Caballero, A., Hernán, L., Morales, J.: Re-examining the effect of ZnO on nanosized 5 V $LiNi_{0.5}Mn_{1.5}O_4$ spinel: An effective procedure for enhancing its rate capability at room and high temperatures. J. Power Sources **195**, 4278–4284 (2010)
290. Wu, H.M., Belharouak, I., Abouimrane, A., Sun, Y.K., Amine, K.: Surface modification of $LiNi_{0.5}Mn_{1.5}O_4$ by ZrP_2O_7 and ZrO_2 for lithium-ion batteries. J. Power Sources **195**, 2909–2913 (2010)
291. Fan, Y., Wang, J., Tang, Z., He, W., Zhang, J.: Effects of the nanostructured SiO_2 coating on the performance of $LiNi_{0.5}Mn_{1.5}O_4$ cathode materials for high-voltage Li-ion batteries. Electrochim. Acta **52**, 3870–3875 (2007)

292. Kang, H.B., Myung, S.T., Amine, K., Lee, S.M., Sun, Y.K.: Improved electrochemical properties of BiOF-coated 5 V spinel $Li[Ni_{0.5}Mn_{1.5}]O_4$ for rechargeable lithium batteries. J. Power Sources **195**, 2023–2028 (2010)
293. Zawadzki, M.: Synthesis of nanosized and microporous zinc aluminate spinel by microwave assisted hydrothermal method (microwave–hydrothermal synthesis of $ZnAl_2O_4$). Solid State Sci. **8**, 14–18 (2006)
294. Shen, S., Hidajat, K., Yu, L.E., Kawi, S.: Simple hydrothermal synthesis of nanostructured and nanorod Zn–Al complex oxides as novel nanocatalysts. Adv. Mater. **16**, 541–545 (2004)
295. Yan, J., Yuan, W., Xie, H., Tang, Z., Mao, W., Ma, L.: Novel self-catalyzed sol-gel synthesis of high-rate cathode $Li_3V_2(PO_4)_3/C$ for lithium ion batteries. Mater. Lett. **71**, 1–3 (2012)

Chapter 7
Sol-Gel Chemistry Engineering for Corrosion Protection

Killian Barton, Maikki Cullen and Brendan Duffy

7.1 Introduction

7.1.1 Sol-Gel Chemistry

The chemistry underlying the Sol-Gel process may be described as a bottom-up approach, whereby particles or interconnected polymeric strands are synthesised molecule by molecule and the precursors and reaction conditions determine the final product. The range of materials that can be produced is very wide including monodisperse nanoparticles, low impurity ceramics, coatings, fibres and materials with novel properties [1].

Precursors for sol-gel chemistry include inorganic alkoxides of the general form $M(OR)_x$, where M is a metal or metalloid and R is an alkyl. These alkoxides will undergo hydrolysis in the presence of water:

$$M(OR)_n + xH_2O \rightarrow M(OR)_{n-x}(OH)_x + xROH$$

The hydrolyzed product may then undergo condensation to form an inorganic polymer network. The reactions involved in condensation involve oxolation (where water is the leaving group) and alcoxolation (where the leaving group is ROH) [2];

$$2M(OH)_x \rightarrow M{-}O{-}M(OH)_{x-1} + H_2O$$

$$M(OR)_{n-x}(OH)_x + M(OR)_{n-x}(OH)_x \rightarrow M{-}O{-}M(OR)_{n-x-1}(OH)_{x-1} + ROH$$

K. Barton (✉) · M. Cullen · B. Duffy
Dublin Institute of Technology, Focas Institute, Camden Row, Dublin 8, Ireland
e-mail: killian.barton@dit.ie

S.C. Pillai and S. Hehir (eds.), *Sol-Gel Materials for Energy, Environment and Electronic Applications*, Advances in Sol-Gel Derived Materials and Technologies,
DOI 10.1007/978-3-319-50144-4_7

In the case of transition metal and main group based precursors, the electronegativity of the metallic element will dictate its reactivity. Metals with high electronegativity are seen to react very readily with water and usually require some chelating agent to protect the metal centre and hence reduce the rate of reaction. Si-based precursors are more hydrolytically stable and require catalysis to increase the rate of the reaction to an acceptable level. In each case, the catalyst or the chelate will have an impact on the physical and chemical characteristics of the final product [3].

Reaction conditions can control characteristics of the output of a sol-gel process such as morphology. Temperature and pH of a reaction influence the supersaturation and nucleation behaviour hence controlling particle size and distribution. Structure and phase changes of sol-gel prepared materials can be altered by the introduction of dopants. In the case of nanoparticle (NP) synthesis through the sol-gel technique, this may be accomplished by incorporation of the dopant material in the structure of the chelate [4].

Silica prepared using the hydrolysis and condensation of silicon alkoxide precursors will, during processing, transform from an initial sol state to a final gel state. The degree of cross-linking and the rate of hydrolysis will depend on the catalysis used to form the sol-gel. Slow hydrolysis and weak cross-linking result when low pH catalysis is used and highly cross-linked polymers result from high pH catalysis. This affects the structure of the resulting gel after solvent evaporation. Acid-catalysed systems proceed via an SN_1 nucleophilic substitution reaction (Fig. 7.1) and tend to form high density, low pore volume gels after solvent removal (Fig. 7.3). Whereas base-catalysed systems proceed via an SN_2 nucleophilic substitution reaction (Fig. 7.2) and form densely packed arrays of identifiable particles around which are large voids (Fig. 7.3). It is important to note that from this that *gels produced using silicon alkoxide precursors are generally porous* regardless of the catalysis used. This is especially important when considering the use of these materials as barriers to the ingress of corrosion initiating materials [5].

Sol-gel chemistry has seen much use in the last 50 years and its flexibility is still opening up new avenues in tailor made substances. Organofunctional alkoxysilane precursors have allowed for the devlopment of organically modified silicates (Ormosils) via the sol-gel route. Ormosils are classified as Class II inorganic-organic hybrid materials. The classification here refers to *covalent* linking of the organic and

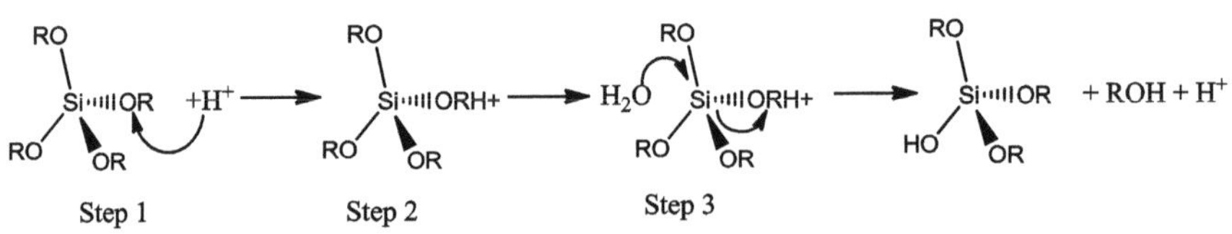

Fig. 7.1 Mechanism for S_N1 nucleophilic substitution

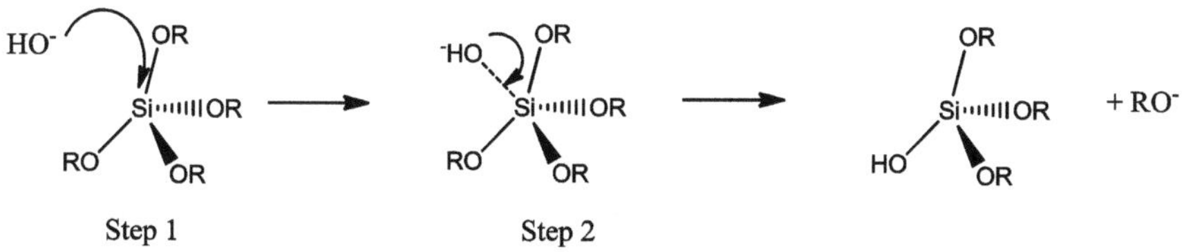

Fig. 7.2 Mechanism for S_N2 nucleophilic substitution

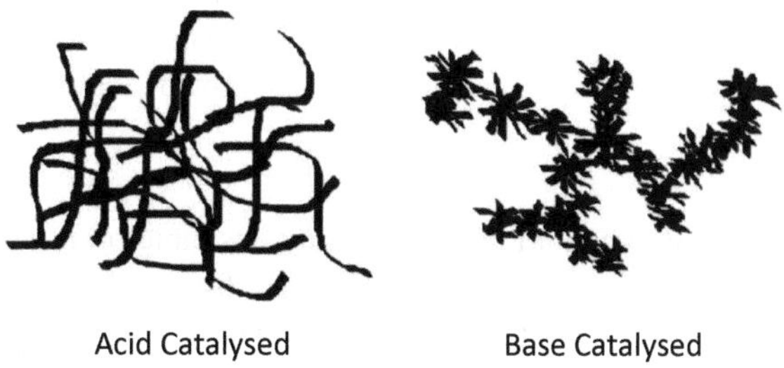

Fig. 7.3 Effects of pH on sol formation

inorganic moieties in the hybrid material. In the design of coatings, ormosils allow for the incorporation of attributes conferred by both inorganic alkoxides and organic based polymers. The former allows the coating to covalently bond to inorganic substrates providing a more strongly anchored coating. The organic moiety confers flexibility and low processing temperatures to the material. The presence of a mixture of inorganic and organic materials results in multi-compatibility with pure inorganic and organic materials [3].

The advantages of ormosil-based coating design goes further as it allows for the incorporation of moieties that can perform specific functions and which are not there merely to enhance compatibility and mechanical characteristics of the coating. Dyes, fluorescent materials, enzymes and various reagents have been incorporated into sol-gel coatings using the ormosil approach [6].

7.2 Corrosion Control for Aluminium Alloys

7.2.1 Overview of Aluminium Alloys

Aluminium displays excellent corrosion resistance due to its affinity for oxygen; a very thin oxide film is formed when the surface is exposed to the atmosphere. The application of aluminium and its alloys in industry is based on three main properties; its low density (approximately 2.7 g cm^{-3}), the relatively high corrosion resistance of the pure metal and the high mechanical strength that can be achieved by suitable alloying and heat treatments. Its strength to weight ratio makes it a very popular material for the automotive and aerospace industries where weight and CO_2 reduction are targets.

Table 7.1 Principal alloying elements of the aluminium series [8]

Alloy group	Principal alloying element	Applications
1xxx	Unalloyed aluminium	Chemical tanks and piping
2xxx	Copper	Aircraft engine parts
3xxx	Manganese	Forming applications (utility and van trailer sheet)
4xxx	Silicon	Filler metal
5xxx	Magnesium	Chemical storage tanks, pressure vessels and structural applications
6xxx	Magnesium and silicon	Automotive, pipe, railings, structural applications
7xxx	Zinc	Aircraft industry
8xxx	Other elements	Aerospace industry

Although aluminium has many desirable properties, unalloyed it is too soft and impractical for many applications. In order to enhance its properties, such as increased strength, it is alloyed with other metals. Aluminium alloys are designated by a four-digit numerical system which is an industry standard for wrought alloys. The first digit specifies the main alloying element and the series are shown in Table 7.1 [7].

The aluminium alloy 2024 is distinguished by its improved stength and toughness and is extensively used in the aerospace industry. The following sections will present the composition and corrosion mechanism of the 2024 alloy.

7.2.2 *Aluminium Alloy 2024 T3*

Introduced in the 1930s, the 2024 T3 was widely adopted in the aerospace industry due to its excellent toughness and strength-to-weight ratio. Today the alloy still sees extensive use as it continues to meet specification for application in aircraft manufacturing. The composition of the alloy is presented in Table 7.2. Of the alloyed metals, copper is included at highest percentage. Precipitation hardening is used to strengthen the alloy whereby careful control of the cooling temperature and rate, promotes dissolution of copper–aluminium particles which deposit at the grain boundary [9]. The sheet form is manufactured using cold rolling which results in elongation of grains in the plain of processing [10]. The 2024 alloy is used widely in the aviation industry. Statistics compiled by the International Civil Aviation Organization indicate that 3.1 billion passengers travelled by air in 2013, therefore it is essential that corrosion of the materials used in aircraft manufacturing is controlled effectively in order to guarantee safety [11]. The safety of the passengers and crew is dependent on the reliable performance of the aircraft components. As well as this, the aviation industry consumes a vast amount of this type of alloy and any gain in the lifetime of parts utilising 2024 will increase the efficiency of resource management in the long term.

Table 7.2 Elemental composition of Al 2024 alloy expressed as percentage [10]

Cu	Mg	Mn	Fe	Si	Zn	Cr	Ti	Al
3.8–4.9	1.2–1.8	0.3–0.9	0.5	0.5	0.3	0.1	0.1	Balance

7.2.3 Corrosion Mechanism of Al 2024 T3

The fabrication of Al 2024 T3 results in the formation subphase particles in the alloy matrix. Early studies concluded that approximately 4–5% of the surface area of the alloy is made up of these particles. They have a range of compositions with the most important, from a corrosion perspective, being the S phase (Al_2CuMg) and the AlCuFeMn particles. Of these, the S phase particles are the most common accounting for 60% of the overall subphase [10] (Fig. 7.4).

The role of S phase particles in the corrosion of the 2024 alloy has been studied by several groups and the following observations have been made [10, 12, 13].

- The S phase particles may be protected by a Mg hydroxide passivation layer that forms in air (Fig. 7.4a). Once this layer is breached the S phase particle is an anodic site with respect to the alloy matrix
- This results in rapid dissolution of Mg and Al due to water and oxygen reduction at the exposed aluminium surface accelerated by the large surface area of the aluminium relative to the S phase particle (Fig. 7.4b, c)
- This serves to dramatically increase the rate of water and oxygen reduction in the vicinity of the S phase particle
- Reduction reactions produce hydroxide locally resulting in dissolution of the Aluminium oxide and exposure of the underlying metal (Fig. 7.4c)
- Oxide formation occurs at the exposed metal followed by oxide dissolution due to the high pH. This process continues until the S phase particle is free of Mg and Al (Fig. 7.4c)
- The S phase particle following de-alloying consists of a porous Cu particle
- The Cu remnant is noble to the matrix, however, it constitutes too small an electrode area to facilitate any significant dissolution of the aluminium matrix
- The porous Cu breaks down in an effort to reduce its surface area and the Cu is redistributed by mechanical processes in the region of the newly formed pit (Fig. 7.4d)

7.2.4 Chromium Based Anticorrosion Strategies

The chief application of chromium in relation to the 2024 alloy is the use of chromate conversion coatings (CCCs). The process is simple and fast and involves the exposure of the sheet alloy piece to chromic acid in the presence of other compounds including ferricyanite and sodium fluoride in a chemical bath. The

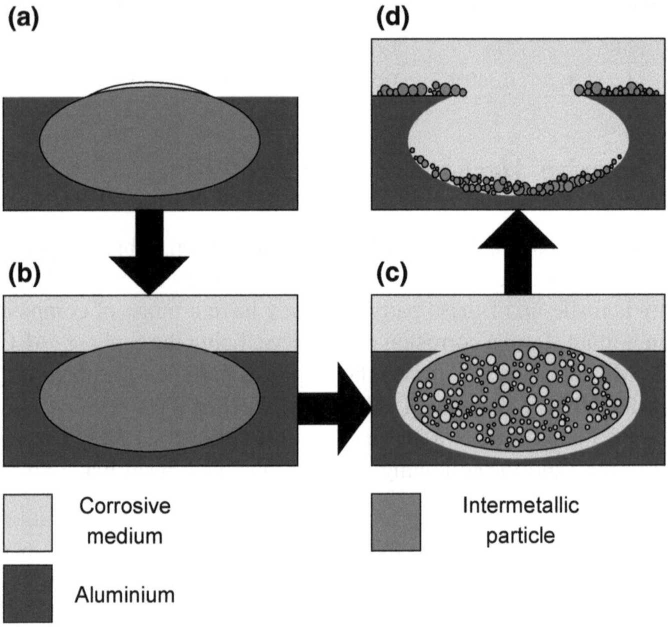

Fig. 7.4 Corrosion mechanism of Al 2024 T3 alloy. The intermetallic particle indicated here is an S phase particle

exposure is short (<1 min) and results in the formation of an oxide at the alloy surface containing chromium in a 3:1 mixture of trivalent and hexavalent oxidation states [14].

The other compounds in the CCC bath play important roles in the formation of the protective oxide. Fluoride is active at the substrate surface in the initial moments of the immersion in the solution. This delays the interaction of the chromate with the surface which otherwise would form a chromium hydroxide which does not have the strength or toughness possessed by the CCC [14]. During this delay in interaction between the Cr(VI) and the surface, ferricyanite catalyses the reduction of the hexavalent chromium to Cr(III).

$$[CrO_4]^{2-} + Al^0 \rightarrow Cr^{3+}_{(aq)} + Al^{3+}_{(aq)}$$

The metal ion adopts a hexahydrate form in solution that is then hydrolyzed through the following reaction:

$$\left[Cr(H_2O)_6\right]^{3+} \rightarrow \left[Cr(OH)(H_2O)_5\right]^{2+} + H^+$$

The resulting hydrolysed Cr(II) species may subsequently condense and form the CCC oxide [15] (Fig. 7.5).

The anticorrosion properties of the oxide are excellent. The primary features [16] of the CCC protection mechanism are:

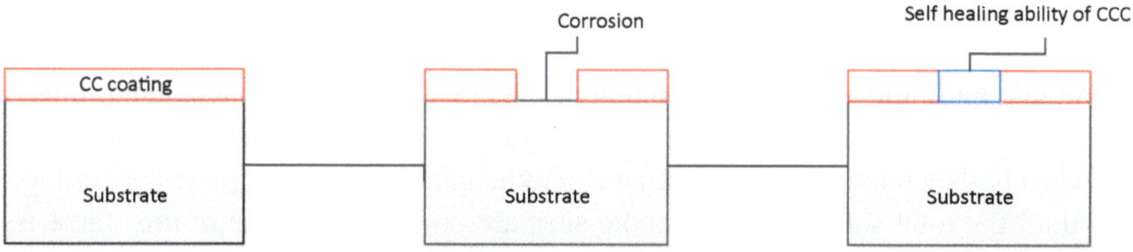

Fig. 7.5 CCC mechanism on an aluminium substrate

- pH increases at active corrosion sites ***trigger*** Cr(VI) release
- Solubility of Cr(VI) over a wide pH range facilitates ***diffusion*** to sites of attack
- Cr(VI) inhibits **anodic** corrosion processes via its high reduction potential
- Cr(VI) inhibits **cathodic** corrosion processes via the formation of a barrier layer

The most significant characteristic of the CCC is therefore its self-healing ability which is the result of the combination of the factors listed above. The Cr(VI) is contained in the coating until conditions produced by or favouring corrosion trigger its release. Its solubility allows it to diffuse easily to the site of action where it interacts with the surface and creates a new layer of oxide. Thus, the coating is repaired through processes controlled by its environment.

Environmental and health concerns have rendered the indefinite use of chromate based anticorrosion treatments unviable. Chromium compounds act both as carcinogens and mutagens. Cr(VI) exhibits solubility over a wide pH range thereby facilitating its proliferation once released into the environment. As a result, the elimination of the use of dangerous chromium containing compounds has been a target of environmental protection legislation. EU directives relating to the reduction of chromium usage include 'End of Life Vehicle', 'Waste Electrical and Electronic Equipment' and 'Restriction of Hazardous Substances' directives [17].

7.2.5 Corrosion Inhibition Chemistry

While sol-gel materials themselves may perform the role of the barrier in the corrosion protection, other materials may be introduced that will provide the self-healing characteristics typical of the chromium based systems. Corrosion inhibitors may be used as components in new film designs. These are compounds that interrupt the electrochemical reactions taking place at corrosion sites on a given metal substrate.

Corrosion inhibitors are split into two classes, environmental conditioners and interface inhibitors. Environmental conditioners act by removing an active corrosion species such as dissolved oxygen from the area of interest. This slows down reduction reactions and hence the rate of corrosion. The most important of the interfacial inhibitors are the liquid phase inhibitors which are extensively used in the protection of metals. There are three types of liquid phase inhibitor; anodic, cathodic and mixed.

Anodic inhibitors mitigate dissolution of the metal. They form passivating films on the active anodic sites on the substrate. Given that the rate of corrosion at an anodic site is inversely dependent on its surface area, anodic inhibitors will increase the corrosion rate until the total surface of the anodic sites are covered. Because of this, there is a critical concentration below which the inhibitor will accelerate corrosion.

Cathodic inhibitors act in two ways. They can decrease the rate of reduction or precipitate and form a barrier layer of insoluble compounds at cathodic sites.

The majority of inhibitors in the liquid phase class are mixed inhibitors. These are organic compounds that protect a metal surface through a mixture of anodic and cathodic inhibition. Adsorption of the mixed inhibitor determines its efficacy of protection. Adsorption is controlled by the structure of the compound, the surface charge and the electrolyte. Inhibitors that adsorb chemically rather than physically are generally seen to provide the best corrosion protection [18].

7.2.6 Corrosion Inhibition and the 2024 Alloy

The heterogeneous nature of the 2024 alloy means that an appropriate corrosion inhibitor must be selected. The S phase particles act as corrosion sites in the alloy matrix the stability of which is undermined by the much larger matrix surface area. The heterogeneous nature of the subphase particles also requires that the inhibitor chemistry being chosen addresses the corrosion behaviour specific to the element being targeted and its environment.

7.2.7 Organic Inhibitor Compounds

Benzotriazole (BTAH) has seen extensive use as an inhibitor for copper particularly in immersed environments. It protects the copper surface by the formation of a polymer comprised of itself and copper ions. This layer deposits on the polymer surface and acts as a barrier to ion interaction with the surface therefore inhibiting further reaction. At low pH, the BTAH interacts with the surface through Van der Waal's forces creating a barrier layer [19]. The efficacy of BTAH in protecting copper has led to several investigations of its use in inhibiting the breakdown of S phase particles at the 2024 alloy surface. It has been shown that BTAH acts to inhibit corrosion at copper containing intermetallic particles (IMPs). It has also been shown that octyl ester BTAH derivatives offer a better degree of corrosion inhibition to copper alloys which may translate to application with the 2024 alloy [20].

An SVET study of the 2024 alloy surface in the presence and absence of BTAH revealed that film formation reduced the oxygen reduction reaction at the surface and hence inhibited corrosion [21]. Another study showed that BTAH had no effect on

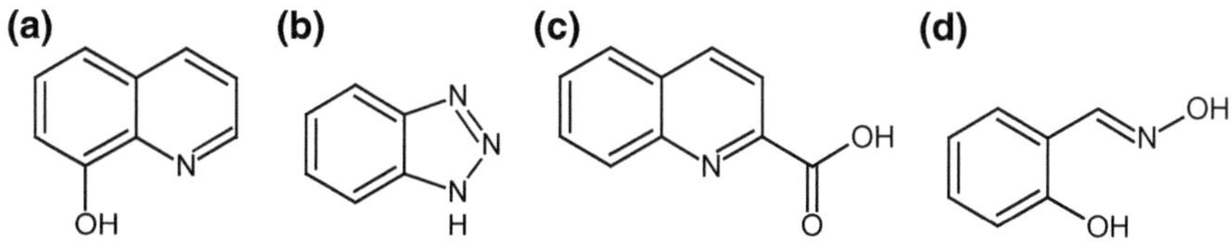

Fig. 7.6 Structures of organic inhibitor compounds: **a** 8-hydroxyquinoline, **b** benzotriazole, **c** quinaldic acid and **d** salicylaldoxine

the corrosion of pure Al in acidic chloride containing media. This further strengthens the position that BTAH corrosion inhibition is specific to its interaction with the copper rich S phase particles. Interestingly, this study showed that no attack had occurred at the S phase particle boundaries after a week immersion in the corrosion medium [22]. This would suggest that BTAH interacts rapidly with the S phase particles and prevents dissolution of the Mg and Al contained therein as these metals have been shown to dissolve rapidly in the initial stages of corrosion [10, 12, 13].

The dissolution of Mg and Al in the initial phase of S phase decomposition has been studied. Compounds used for the gravimetric analysis of Mg, Al and Cu were investigated. It was shown that 8-hydroxyquinoline, quinaldic acid and salicylaldoxime suppressed the dissolution of these elements at S phase sites in chloride media. As well as this, highly insoluble inhibitor–metal complexes were identified. The formation of these complexes was proposed to have resulted in the formation of a barrier layer at the S phase particles preventing adsorption of chloride ions at the surface and hence retarding the rate of corrosion [23] (Fig. 7.6).

7.2.8 Inorganic Inhibitor Compounds

The most promising inorganic corrosion inhibitors are vanadium based. Vanadium conversion coatings (VCCs) have been developed which are produced much in the same manner as CCCs. The 2024 substrate is exposed to sodium vanadate in the presence of ferricyanate and fluoric acid. The mechanism of film formation has been proposed to be driven by hydrolysis and condensation. The VCCs showed excellent corrosion protection under salt spray testing. The VCC protected the substrate by cathodic inhibition. It also exhibited self-healing qualities similar to that of CCCs [14].

The specific interaction of vanadates with the 2024 surface has been examined using a split cell technique. Here, anodic and cathodic processes may be studied separately. The vanadate solution was observed to reduce the cathodic reactions significantly. It had no impact on the anodic reactions. Some evidence was also found that dissolution of the S phase was suppressed, however, the impact of this on the corrosion protection of the alloy was determined to be much less than the cathodic inhibition [24].

In order to further elucidate the inhibition mechanism of vanadates, monovanadates and decavanadates and their respective interactions with the 2024 alloy surface were analysed separately. Monovanadates were seen to adsorb to the surface following a Langmuir isotherm, however, adsorption was dependent on a very slow reduction reaction. Once adsorbed, the monovanadate layer is assumed to inhibit Cl^- adsorption at the alloy surface as well as reduce the extent of oxygen reduction. Decavanadates were shown to offer little or no corrosion inhibition [25].

Other inorganic compounds that have been investigated include Mg^{2+} ions. It was shown that Mg^{2+} delayed the dealloying of Mg from intermetallics on the 2024 surface thus retarding the overall corrosion rate by controlling the initiation step [26]. ZnO has been applied as a corrosion inhibitor also, this paper is reviewed later in the text [27]. Sodium silicate was investigated as a corrosion inhibitor for 2024 by Garrity et al. It was observed that its inhibition performed best at high pH with protective film formation at the surface occurring involving aluminosilicate ions. At low pH, films formed from silica and silica-based derivates were porous and offered little corrosion protection [28].

7.2.9 Rare Earth Metals

Lanthanides have been the subject of a great deal of research in corrosion inhibition. Cerium and it's salts have shown the most promise to date. There is a number of ways to deploy cerium compounds as part of a corrosion protection system. These include:

- The formation of a cerium conversion coating [29, 30]
- The inclusion of CeO_2 nanoparticles in sol-gel coatings [31]
- The inclusion of $Ce(NO_3)_3$ in zirconia nanoparticles in sol-gel coatings [32]

The mechanism by which cerium protects the alloy depends on the precipitation of insoluble hydroxides of the element. This reaction is triggered by the sharp rise in pH that results at the surface during the initial de-alloying stage at the S phase particles. The high pH environment favours the formation of insoluble Ce(III) hydroxide. Tetravalent Ce may also be formed in these circumstances and it inhibits corrosion more efficiently than the Ce(III). The precipitated hydroxide acts in two ways to inhibit corrosion. It reduces the local pH leading to a reduction in the rate of the cathodic reaction. It also forms a barrier at the S phase particles that prevents further diffusion of corrosive species to the alloy surface [33]. Shi et al. have reported that cerium cinnamate offers improved corrosion protection on 2024 substrates when added to an epoxy coating. They compared the protection offered by cinnamate and cerium salts with that of the cerium cinnamate and found that cerium cinnamate gave improved protection suggesting a synergistic effect [34]. A similar result was reported by Catubig et al. on the corrosion protection offered by salts of Ce, Pr and mercaptoacetate. Comparison was then made with mercaptoacetates of the rare earth metals which exhibited altered electrochemical activity at the surface of 2024 [35].

7.2.10 Conducting Polymers

Electronically conducting organic polymers (CPs) provide suitable materials for corrosion control research. This is due to the physical characteristics of CPs which include electronic conductivity, redox activity and the presence of heteroatoms. It is assumed that advantage can be taken of these features in interrupting or altering the redox and galvanic processes that occur during corrosion.

Practical features of CP film deposition have been overcome by the use of catalysis. Electron transfer mediation has been used previously to catalyse the deposition of CPs at inert metal electrodes. The principle has since been transferred to use on active metals with excellent results producing a film that exhibits higher conductivity and better substrate film connectivity [36]. CP films deposited in this way were examined using SVET in a novel experimental set up that isolated the physical contact of the CP and the 2024 alloy yet maintained electrical contact. It was found that a significantly reduced current flowed at the entire CP film surface when both the 2024 alloy and the indium tin oxide (ITO) mounted CP film were exposed to dilute Harrison's solution. The author suggests that reduction of oxygen at the film surface plays a role in passivating the 2024 alloy [37].

Other forms of anticorrosion activity observed in CP films include sacrificial cathodic activity in polythiophenes and triggered corrosion inhibitor release. CP film anticorrosion behaviour is dependent on the polymer in use and its oxidation state. This was highlighted in a study on doped and undoped polypyrroles and polythiophenes where observed protection mechanisms included sacrificial cathodic activity, oxygen scavenging and oxidation promotion at defect sites [38].

7.2.11 Commercially Available Sol-Gel Pretreatments

Examples of current commercially available silane based pretreatments in the aerospace industry are based on patents published by Blohowiak and Osborne [39, 40]. Socomore and Advanced Chemistry and Technology (AC Tech) offer the Socogel and AC-130 systems respectively. The former is marketed as a chromium free conversion coating that may be applied to a range of metal substrates. The product may be applied by dipping, spray coating or brush and is distributed as a two part kit that must be used within 12 h of opening. There is an initial induction period of 30 min and a minimum drying period of 1 h [41]. The AC 130 product is marketed as a surface preparation for adhesive bonding. Application conditions and instructions for the AC 130 are similar to that of the socogel system (Table 7.3).

Table 7.3 A reference chart to the materials, processing conditions and analytical methods used in the research presented in Sect. 7.2

Film precursors	Particle precursors	Additive	Film precursor processing	Particle precursor processing	Deposition method	Control	Analysis	Ref
GPTMS, TEOS	–	–	HNO_3, H_2O	–	Dip, spray	Comparative study	NMR, NSS	[42]
$RSiOCH_3$, R = methyl, dimethyl, n-propyl, n-butyl, isobutyl, n-hexyl, n-octyl, isooctyl	–	–	HNO_3, H_2O	–	Spray	Comparative study	EIS, NSS, FTIR, contact angle	[43]
TEOS, VTMOS, MAPTMS	–	–	HNO_3	–	Spray	Comparative study	EIS, NSS, FTIR	[44]
BSS	–	–	H_2O, EtOH	–	Dip coating	AA 2024-T3	EIS, SEM/EDX, PDS	[45]
BSS, BAS	–	–	H_2O, EtOH, acetic acid (BAS only)	–	Dip coating	AA 2024-T3	EIS, PDS, contact angle, ellipsometry	[46]
BAS, VTAS	–	–	H_2O	–	Dip coating	AA 2024-T3, alodine	EIS, NSS, CASS, machu, scribe, seawater immersion, DC polarisation, FTIR, ellipsometry	[47]
BTSE, BTESPT, MAPTMS	–	–	H_2O, MeOH	–	Dip coating	AA 2024-T3, accelagold	AES, Auger mapping, XPS, EIS, cross hatch, fatigue	[48]
GPTMS, TEOS	–	–	H_2O, HNO_3, TEPA	–	Spin (1000 rpm 30 s)	AA 2024-T3	FTIR, ^{29}Si NMR, ^{13}C NMR, PDS	[49]
GPTMS, TEOS	–	–	H_2O, HNO_3, 2-propanol	–	Dip coating	AA 2024-T3	EIS, PDS, NSS, SEM/EDX	[50]
ZSM-5	–	–	Tetrapropylammonium	–	Immersion	AA 2024-T3, STOPSRUST (chromate)	EIS, PDS, SEM	[51]
GPTMS	–	–	H_2O, KF, AP4IB4 POSS	–	Dip coating	AA 2024-T3	NSS, PDS, FTIR, ^{29}Si NMR	[52]

(continued)

Table 7.3 (continued)

Film precursors	Particle precursors	Additive	Film precursor processing	Particle precursor processing	Deposition method	Control	Analysis	Ref
AniSi, OTES	–	–	H_2O, MeOH, acetic acid	–	Immersion	AA 2024-T3	RAIR, XPS, SEM/EDX	[6]
TMOS, GPTMS	–	–	*p*, *o* &, *m*-pheneylenediamine, MeOH, H_2O, surfactant	–	Immersion	Comparative study	EIS, SVET, UATR-IR, Contact angle, UV-Vis	[53]
GPTMS	–	–	H_2O, EtOH, acetic acid	–	Cathodic polarisation/dip coating	AA 2024-T3	EIS, SEM/EDX	[54]
MTEOS, APTEOS, GPTMS	–	LUDOX, $Ce(NO_3)_3$	H_2O, EtOH, acetic acid	–	Dip coating	AA 2024-T3	NSS, cross hatch, pencil hardness, SEM/EDX, PDS	[55]
TEOS	–	–	H_2O, EtOH, HCl, pluronic P123	–	Dip coating	AA 2024-T3	Hardness, thickness, TGA, GDOES, FTIR, EIS, contact angle	[56]
3-Glycidoxypropyl terminated PDMS, APTES	–	–	H_2O, EtOH, acetic acid	–	Dip coating	AA 2024-T3	RAIR, SEM, EIS, Raman, Thickness	[57]
APTES	–	Epoxy resin	*p*-phenylenediamine	–	Dip coating	Comparative study	SEM/EDX, BET, EIS, SVET, nanoindentation	[58, 59]
GPTMS, TEOS	TPOZ	–	2-propanol,	EtAcAc	Dip coating	AA 2024-T3	EIS, SKPFM, SEM, XRD, XPS, TEM	[60]
GPTMS	TPOZ	–	Acetic acid	Acetic acid	Dip coating	AA 2024-T3, Alodine 1200	AFM, NSS, EIS, cyclic salt spray,	[62]
MAPTMS	TPOZ	–	HNO_3,	MAAH	Spin coating	N/A	TEM, FTIR, UV-Vis	[63]

(continued)

Table 7.3 (continued)

Film precursors	Particle precursors	Additive	Film precursor processing	Particle precursor processing	Deposition method	Control	Analysis	Ref
GPTMS	TPOZ, TAPO	–	HNO_3,	EtAcAc, AcAc	Dip coating	Unidentified chromate conversion coating	SEM, EIS, TGA, DSC, NSS	[64]
MAPTMS	TPOZ	–	HNO_3, H_2O	AcOH, IBA, MAAH, AcAc, Bipy	Spin coating	Comparative study	DLS, DSC, EIS, PDS, FTIR, NSS	[65]
MAPTMS	TPOZ	–	HNO_3	3,4-DABA	Spin coating 1000 rpm	Comparative study	EIS, PDS, ^{29}Si NMR, DSC, AFM, FTIR, NSS	[66]
GPTMS, TEOS	–	Boehmite	HNO_3, IPA, aluminium isopropoxide	–	Dip coating	Comparative study	SEM, AFM, EIS, PDS	[67]
MAPTMS	TZPO, AZPO	Aluminium	HNO_3, H_2O	MAAH	Spin coating	AA 2024-T3	DLS, SEM/EDX, PDS, XPS	[68]
MAPTMS	NbOEt	Niobium	0.01 N HNO_3, H_2O	MAAH	Spin coating	AA 2024-T3	DLS, DSC, SEM, XPS, AFM, PDS, NSS, thermal stress testing	[69]
TEOS, TEOCS	ZPO	$Ce(NO_3)_3$ CeO_2	10 v% HNO_3	Ethylaceto-acetate	Dip coating	Comparative study	DLS, XPS, SEM, EDX, AFM, PDS	[70]
BSS	–	Silica NPs	H_2O ethanol	–	Dip coating	Comparative study	PDS, MTS, nanoindentation, XPS, CSM, ellipsometry	[71]
GPTMS	TPOZ	Benzotriazole	2-propanol acidified water	Ethylaceto-acetate	Dip coating	Comparative study	DLS, SEM, SVET, EIS, AFM, viscosity	[72]
TEOS, MTEOS, SiO_2, GPTMS	–	hydroxyphenylpropane 1-methylimidazole, CeO_2	HCl	–	Dip coating	Comparative study	TEM, EIS, NSS, scratch and abrasion resistance	[31]
VTMS, TEOS	–	Sodium montmorillonite	H_2O, Ethanol	–	Dip coating	Comparative study	FTIR-RA, SEM, EDX, EIS	[73, 74]

(continued)

Table 7.3 (continued)

Film precursors	Particle precursors	Additive	Film precursor processing	Particle precursor processing	Deposition method	Control	Analysis	Ref
GPTMS	TPOZ	cerium molybdate nanowires	2-propanol HNO_3	Ethylaceto-acetate	Dip coating	Comparative study	EIS, SEM/EDX, TEM, XRD, SVET	[75]
GPTMS, TEOS	–	PDMS	H_2O		Spin coating	Comparative study	FTIR, ^{29}Si NMR, ^{13}C NMR, TGA/DTG, PDS	[76]
GPTS, ECET	–	Chitosan	H_2O, acetic acid	–	Brush coating	Comparative study	Viscosity, EIS, adhesion, NSS, UV/Vis, FTIR	[77]
GPTMS, DTMS, TEOS, VTES	Epoxy resin (type E-44)	Organotin compound	Xylene		Spin coating	Comparative study	EIS, Machu Test, FTIR	[78]
GPTMS, TEOS	TPOZ	MTB MBI DETA	H_2O, Acetic acid 2-propanol ethylacetoate n-propanol		Dip coating	Comparative study	PDS, SEM, EIS, UATR-IR	[79]
APTMS, TEOS	–	PACM Ancamide −2050, −2353 Epikure −3164	BGC resins	HDB Glycidol	Draw Bar	Comparative study	FTIR, GPC, ^{13}C NMR, TGA/DSC, hardness, crosshatch, contact angle	[80]
TEOS, MAPTMS	LUDOX AS-40 2,2′-Azobis (isobutyronitrile)	EGDMA	HNO_3		Dip coating	Comparative sudy	FTIR, UATR-IR, viscosity,EIS, PDS, rheometry, laser diffraction, contact angle, profilometry, DLS	[81]
TEOS, GPTMS		EHA, GMA, AIBN	HCl, H_2O	2M THF	Spin coating	Comparative study	AFM, FTIR, DSC/TGA, EIS	[82]
TEOS, APTES	–	Epikure F205, DGEBA	THF, Ethanol, HCl	–	Draw Bar	Comparative study	FTIR, ^{29}Si NMR, ^{1}H NMR, SEM, EIS, TGA	[83]

(continued)

Table 7.3 (continued)

Film precursors	Particle precursors	Additive	Film precursor processing	Particle precursor processing	Deposition method	Control	Analysis	Ref
TMOS, GPTMS, EAPTES, DETA	–		0.05 M acetic acid DI water pH 7, DETA	Surfactant (3 M FC-430, 0.04 wt./v% and FC-171, 0.01 wt./v%)	Dip coating	Comparative study	PDS, EIS, SVET, Raman, AFM	[84]
TMOS, GPTMS	–	β-cyclodextrin, MBT, MBI, MBISA, TSA	0.05 M Acetic acid DI water pH 7, DETA	–	Dip coating	Comparative study	PDS, EIS, SVET	[85]
BSS, BAS VTAS	–	TTA, BTAH, $Ce(NO_3)_3$	DI water pH 7 ethanol	–	Dip coating	Comparative study	PDS, TOF-SIMS, SEM, EDX, ICP/MS, Scratch cell test	[86]
GPTMS, TEOS	TPOZ	$Ce(NO_3)_3$	Acidified water, pH ~ 1 2-propanol	Ethylaceto-acetate	Dip coating	Comparative study	EIS, SEM, EDS, AFM	[87]
GPTMS	TPOZ	Benzotriazole	2-propanol DI water, pH 5	Ethylacetoacetate	Dip coating	Comparative study	EIS, AFM, SEM/EDX, SVET	[88]
GPTMS	TPOZ	$Ce(NO_3)_3$ BTAH	Isopropanol pH 0.5, HNO_3	Ethanol pH ~ 1, HNO_3	Dip coating	Comparative study	FTIR, Raman, SEM, TEM, EIS, SVET, XRD	[89]
MTEOS	–	$Mg(NO_3)_2$	Dilute HNO_3 EtOH	–	Flood coating, spin coating	Alodine 1200	EIS, PDS, NSS, AFM, SEM	[90]
GPTMS, TEOS	Aluminium isopropoxide	Boehmite $Ce(NO_3)_3$	Isopropanol, HNO_3	$Al(OC_3H_7)_3$ DI water, pH 7	Dip coating	Comparative study	EIS, SEM/EDX, FTIR	[91]
GPTMS	TPOZ	$Ce(NO_3)_3$, Na_2MoO_4, $NaVO_3$	FC-171 fluorochemical surfactant	Acetic acid DI water, pH 7	Dip coating	Alodine 1200	PDS	[92]
TEOS, GPTMS	$AlCl_3 . 6H_2O$	$Ce(NO_3)_3$	EtOH glacial acetic acid	–	Dip coating	Comparative study	PDS, EIS, SEM, exfoliation testing	[93]
GPTMS, TEOS	–	Alumina sol (formulation withheld)	EtOH, H_2O, acetic acid	–	Dip coating	AA 2024 T3	EIS, SEM, PDS	[94]
GPTMS, TMOS	–	Cerium salts	Glacial acetic acid, DETA	–	Dip coating	Comparative study	ESEM, XRD, XPS, EIS	[95]

(continued)

Table 7.3 (continued)

Film precursors	Particle precursors	Additive	Film precursor processing	Particle precursor processing	Deposition method	Control	Analysis	Ref
MAPTMS, TEOS	–	SiO_2 NP, $CeNO_3$	EGDMA, GMA, AIBN	–	Dip coating	Comparative study	EIS, GDOES	[96]
EAPTES modified epoxy	–	BTAH, 8HQ, PVP, $CoNO_3$, Ce NO_3, Mn NO_3	Ethanol, acetone, 2,2′-diaminoethylamine	–	Dip coating	Comparative study	UATR-IR, SEM/EDX, EIS	[97]
GPTMS	–	Ce NO_3	propanol, $Al(O^sBu)_3$	–	Dip coating	AA 2024-T3	SEM, EIS, hardness, ^{29}Si, ^{13}C &^{27}Al NMR	[97]
GPTMS	TPOZ	Calcined hydrotalcite	HNO_3,	EtOAc	Dip coating	Comparative study	TGA, DSC, XRD, FTIR, EIS, NSS	[98]
GPTMS	TPOZ	Ce NO_3	HNO_3, EtAcAc	–	Dip coating	Comparative study	EIS, SEM/EDX	[99]
GPTMS, APTES	–	CeO_2 NP, ZnO NP	HCl	–	Dip coating	Comparative study	Thickness, ESEM, NSS, EIS	[27]

Both products are based on zirconia and organic hybrid silanes. The patent states that the coating acts as an effective oxygen barrier which may go some way in explaining the corrosion inhibition offered by the coating. Research into the corrosion protection mechanism of the Si/Zr system has been conducted since the date of patenting of the zirconia based sol-gel coating technology described here and this will be presented in the following sections.

7.3 Sol-Gel Chemistry Engineering for Corrosion Coatings

7.3.1 Sol-Gel Film Design

Metroke et al. performed a multifaceted characterisation of GPTMS-TEOS coatings on AA 2024 T3. Both dip and spray deposition were used and the organic and water content were varied. The samples were subsequently tested using salt spray exposure testing (SST) for 168 h. Spray coated samples performed well with the use of about 10% organic content and high water content. Dip-coated samples were seen to perform well with higher (67%) organic content and lower water content. The impact of water content on the extent of epoxy ring opening was identified using ^{1}H-^{13}C NMR. At high organic content, the epoxy ring opening reaction required more hydrolysis water content in order to go to completion. The group proposed that the hydrophobicity and density of the polymer network dictated the performance of the coatings [42].

In a subsequent study, spray coated substrates were prepared using TEOS and a range of ormosils. Alkyl moieties of varying geometry and carbon content were employed. The organic concentration was varied from 0 to 16% in line with the optimum concentration from the previous work. Straight chain alkyls with 4 or more carbon atoms were seen to give the best corrosion protection. Hexyl modified silane (HTMOS) based samples gave the best performance in SST, maintaining satisfactory film integrity after exposure of up to 672 h. EIS analysis showed that the pore resistance of the films increased with increasing organic concentration. Long-term immersion studies indicated that, while the HTMOS performed well in SST, the coating capacitance increased dramatically after exposure times longer than 4 weeks. Hydrophobicity of the coatings was measured through contact angle measurements and results indicated that improved performance was related to the degree of hydrophobicity exhibited by the coat. The group proposed that the better performance of the long straight chain alkyls was a result of shielding of silanol end groups from interaction with adsorbed water. This in turn reduced the rate of ingress of water into the film and improved its protection qualities [43].

Further investigation of the scope for optimisation of hybrid inorganic–organic coatings was conducted by Metroke and coworkers. A paper published in 2004 looked at the impact of solvent dilution immediately prior to spray application of a

TEOS/VTMOS/MAPTMS coating. A range of solvents and concentrations were tested. Results showed that the use of low concentrations of small alcohols and high concentrations of aprotic solvents yielded coatings that withstood 168 h SST with low pitting observed. All solvent additions yielded thinner coatings as their concentration increased. EIS results cast doubt over the stability of the coatings system using the aprotic and small alcohol solvents. Coating capacitance for these systems was seen to undergo significant increase after prolonged exposure. This was assumed to be the result of increased hydrophilicity due to the inorganic content of the coating. Alcohols with larger organic character were seen to inhibit the condensation reaction of the silane through the production of reesterification products [44].

Van Ooij et al. have conducted extensive research into the efficacy of silane based coatings that are based on bis-silane compounds. The formation of a dense interfacial layer by bis-sulphur silane (BSS) was proposed based on PDS and EIS results published in 2003. The PDS results indicated that the coating inhibited corrosion through a barrier mechanism. EIS results were interpreted as indicating the presence of an interfacial layer primarily responsible for retarding the corrosion rate at pits on the surface. Cracking and delamination were observed to be the main processes occurring during failure of the coating. It was proposed that curing at a higher temperature would lower the likelihood of cracking. The increased adhesive strength such as that offered by the BSS would impede the onset of delamination [45].

In another study by Van Ooij et al., a coating chemistry applicable to a range of substrates (hot dipped galvanised steel (HDG) and Al 2024) was developed. Bis-amino silane (BAS) and BSS were applied to galvanised steel and 2024 substrates both separately and as part of a mixture. Coatings prepared using just the BAS exhibited poor corrosion protection. BSS based coatings were seen to perform well on 2024 and poorly on HDG steel. Mixing the two yielded a coating that worked well as corrosion protection on both substrates. It was proposed that the low concentrations of BAS improved wetting of the coating and the large concentration of the BSS renders the coating hydrophobic enough to offset the hydrophilicity of the BAS [46].

A waterborne coating formula which uses BAS and VTAS was developed by Zhu et al. The pH of the sol, a function of the ratio of BAS to VTAS content determined the interaction of the sol with the substrate surface. Galvanised steel and aluminium 2024 alloy were both examined as well as the 6061 and 3003 alloys. The pH at which the metal oxide of the substrate exhibits passivity was shown to be the optimum pH that the BAS: VTAS film must be adjusted to for satisfactory adhesion and film formation to occur. Stability of the coatings in solution form was also shown to be dependent on coating pH with improved stability seen for coatings with higher pH. Thickness was shown to be a function of overall silane content in the aqueous solution, increasing as the silane content increased. The coating thickness was also compared to that of chromated systems and shown to be considerably lower. Neutral SST showed that the waterborne systems performed just as well as commercially available CCC systems [47].

Research reported by Cabral focussed on silane pretreatments using bis-silanes, mercapto-monosilanes and BSS. Auger electron analysis (AES) of the surface showed films rich in Si and O were formed using the bis-silanes. These films were dense and thick as a result. BSS films were shown to deposit preferentially on Cu rich IMPs which was demonstrated by the use of AES mapping of the pretreated surface. Adhesion for both the bis-silane and BSS films was shown to be satisfactory when judged using the ASTM D 33598 standard [48].

Amine cured GPTMS-TEOS sol-gel coatings were prepared by Wu et al. FT-IR indicated that the addition of tetraethylenepentamine (TEPA) in order to enhance the formation of the organic network was effective. It showed the disappearance of absorption peaks associated with the Glycidoxy ring as the N-H and C-N absorption increases. Also, the silica absorption bands were seen to shift to a higher wave number as the organic network became denser. NMR analysis allowed for the determination of the degree of condensation which, for the TEPA-cured coatings was 94% larger than that of the coatings which were untreated. The spin diffusion path length was also derived from NMR analysis and decreased with increasing TEPA concentration indicating a denser network in the cured coating [49].

Tavandashti et al. prepared a detailed study of the coatings that may be produced from mixtures of GPTMS and TEOS sols. Hydrolysis water and GPTMS content were varied and the coatings were characterised using SST, PDS, SEM and EDX. The results showed that hydrolysis water content was at an optimum when at the stoichiometric amount. High GPTMS content provided the best protection of the 2024 alloy at the optimum water content. This result was further reinforced by the PDS results which showed the lowest corrosion current density for this system [50].

Nonporous zeolites have been applied to 2024 by Cheng et al. through an in situ crystallisation technique. Structure directing agents such as tetrapropylammonium reduced intra-crystal porosity and aided the formation of an excellent barrier film. The formulation has low viscosity and crystal formation proceeds at the surface of the substrate therefore accommodating for substrates that possess complex geometry. The relative concentrations of the formulation components control the thickness of the resulting coatings. The only disadvantage as expressed by the author of the work lies in the need for a long high temperature curing step [51].

The enhancement of the GPTMS-based coatings by the addition of amino-isobutyl functionalised POSS has been shown by Jerman and coworkers. Raman and IR results indicated that water content of the GPTMS formulation should be such that one water molecule is present for each hydrolysable methoxy group. Also, the amino content should be in a 1:1 ratio to that of the glycidoxy groups available for reaction. Any higher concentration leads to the formation of pockets of unreacted amine which adversely affects the integrity of the coating. Those samples that exhibited the best protection in SST possessed the lowest POSS content. PDS measurements also showed that the system containing the least POSS had the least corrosion current. This was attributed to the formation of a denser film through cross-linking of the POSS amine groups and the glycidoxy groups of the silane [52].

The inclusion of corrosion inhibiting moieties, in particular monomers of the conducting polymer polyaniline has been explored by Trueba et al. The performance of films formed from N-(3-trimethoxysilylpropyl) aniline was compared with coatings formed using octylsilane. Reflection absorption IR spectroscopy and XPS both provided data that indicated the interaction of the conducting polymer with the S phase particle. It was shown that Cu ions released from the S phase formed emeraldine and perigraniline salts in the coating matrix. SEM imaging of coated S phase particles which had undergone extended (30 days) exposure showed dendritic copper structures. This was interpreted as having resulted from the controlled extraction of Cu ions from the S phase particles by an interconnected aniline phase in the coating and furthermore that the galvanic coupling of the Cu and Al had been reduced in the presence of the film [6].

TMOS-GPTMS coatings were prepared by Croes et al. and the impact of applying isomers of phenylenediamine as the cross-linking agent was examined. After extended periods of immersion (ca. 2 months) the coatings showed colouration which was dependent on the isomer used. Colouration was ruled out as a substrate induced effect as coatings deposited on glass and 2024 alloy were subjected to immersion. EIS analysis of the coatings showed extended durability of the coatings when compared to coatings cured using aliphatic amines. The group hypothesised that oligomeric species formed through oxidation of the PDA in corrosive media contributed to the enhanced protection offered by the coatings through two mechanisms. The first is that the oxygen reduction reaction is concentrated in the coating as opposed to the metal surface. The second being the scavenging of Cl^- ions through the formation of positively charged species as a result of the oxidation processes [53].

Sol-gel film formation via electrodeposition was investigated by Mrad et al. The coatings were deposited at varying potentials and in the presence of $NaNO_3$. The use of $NaNO_3$ was justified by the ability of nitrates to promote the formation of metallic oxides and presumably to improve the formation of the Si–O–Al network at the substrate coating interface. The potential applied to the coating solution dictated the morphology of the coatings produced. At very negative potentials (ca. −1.3 V) electrochemically induced hydrogen evolution interrupts the film formation and results in a high degree of porosity. EIS results confirmed the reduced barrier properties of coatings formed at the most negative potentials [54].

Lakshmi et al. prepared sol-gel coatings using MTEOS, GPTMs and colloidal silica. Although they did apply corrosion inhibitors during this study, it is included in this section due to the groups application of a silane coupling agent in one of the group's sample sets. APTS was used to pretreat the Al2024 prior to coating application and produced a coating that outperformed all others in accelerated corrosion testing [55].

Recloux et al. explored the use of templating agents in forming mesoporous silica coatings. More specifically, the subsequent removal of the templating agent was investigated. They compared the use of UV/Ozone treatment with calcination and found that the damage caused to the 2024 substrate by calcination could be avoided by the use of UV/Ozone curing. The removal of the templating agent was detected using FTIR. The resultant film was shown to be porous with interconnected pathways with diameters between 5 and 7 nm. They proposed that this diameter was suitable for the containment of organic corrosion inhibitors which share a similar average diameter [56].

Rodosek et al. prepared coatings containing hydrolysed APTES, epoxy terminated polydimethylsiloxane and polyhedral oligomeric silsesquioxanes (POSS). They reported that the coatings were electrochemically active at the surface, greatly reducing the cathodic current density [57].

Kartsonakis et al. conducted a study of coatings based on epoxy and hydrolysed APTES mixtures. These were then loaded with a range of different corrosion control additives, most of which are described in the body of this paper [58, 59].

7.3.2 Introduction of Nanoparticles into the Protective Coating Via the Sol-Gel Route

Zheludkevich et al. have investigated the enhancement of corrosion protection offered by silane films through the incorporation of zirconia nanoparticles (NPs). These NPs may be produced using a sol-gel technique which allows for their addition by mixing with a hybrid organic–inorganic based sol-gel coating solution (GPTMS-TEOS). Hydrolysis time dictates the concentration of NPs in the resulting film and hence its corrosion protection performance. The group proposed that a pore blocking mechanism was responsible for the improved anticorrosion properties of the Zr NP films under immersion conditions [60].

TPOZ and GPTMS coatings were synthesised by Voevodin's group. The use of GPTMS allows for the incorporation of epoxy functionality in the coating that overcomes issues with dispersion that arise from the direct addition of epoxy to silica based sol-gel reaction mixtures [61]. Glacial acetic acid used in the preparation of the TPOZ sol was shown to act as a chelating agent and its presence in the coating was confirmed despite curing temperatures up to 150 °C being applied. The zirconia XPS spectra indicated preferential deposition at the substrate coating interface while silicon was seen to be present at higher concentrations at the outer surface of the coating (see Fig. 7.7) [62].

The structure of the MAPTMS/ZPO system has been captured using TEM by Hicks et al. in work investigating the effect of the condensation of hybrid organic–inorganic sol-gel materials on the optical properties of trypan blue. The size and distribution of the Zr NPs in the coating is clearly seen [63] (Fig. 7.8).

The influence of chelating agents and the structure of the precursor for titania and zirconia containing coatings was investigated by Poznyak et al. The stability of the titania containing coatings during EIS testing was strongly dependent on the nature of the precursor and the chelating agent employed. Acetylacetonate (Acac) performed better than ethylacetoacetate (Etacac) as a chelating agent presumably benefiting from the absence of the ethyl group which may be a source of steric

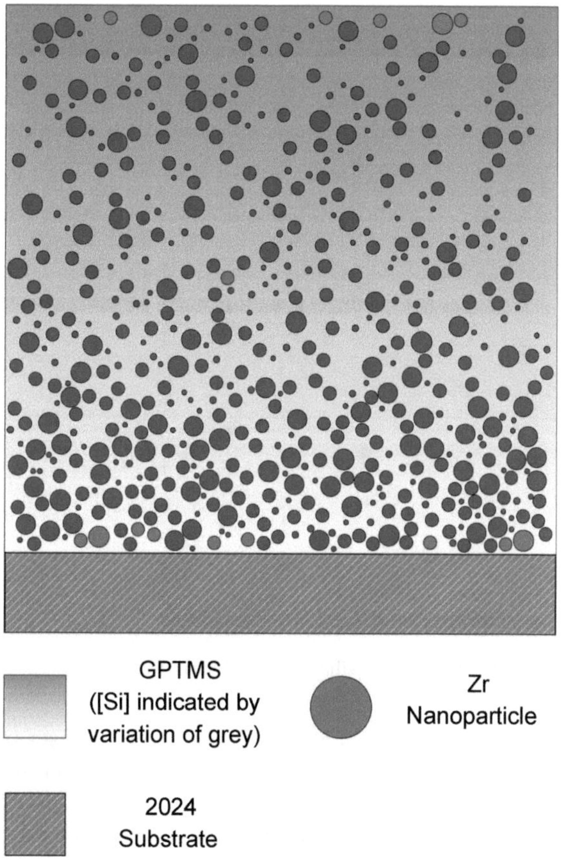

Fig. 7.7 Diagram of coating system

hindrance during condensation. Titanium isopropoxide was seen to perform better than titanium(IV)-bis(acetylacetonate)diisopropoxide (TAPO) as the precursor. Here again steric factors are assumed to play a role.

Zirconia sols were prepared using both Etacac and Acac as chelating agents, however, the difference in efficacy of the chelating agents was seen to be negligent. It is known that zirconia precursors are more reactive than titania precursors and therefore the titania precursors are more likely to show greater sensitivity to the chemical environment during sol preparation [64].

Findings relating to the impact of ligands on anticorrosion coatings prepared using zirconia sols have been published. Acetic, isobutyric and methacrylic acid were used as well as acetylacetone and bipy to prepare zirconia sols. The zirconia sols were then mixed with a MAPTMS based sol and coatings deposited on 2024 alloy. TGA analysis of the sols exhibited two glass transition temperatures (T_g) one for the MAPTMS sol and one for the TPOZ sol. The thermal stability of the TPOZ sol produced using the bipy ligand is high and may result from coordination bonding between the bipy and the zirconia as indicated by the FTIR data reported.

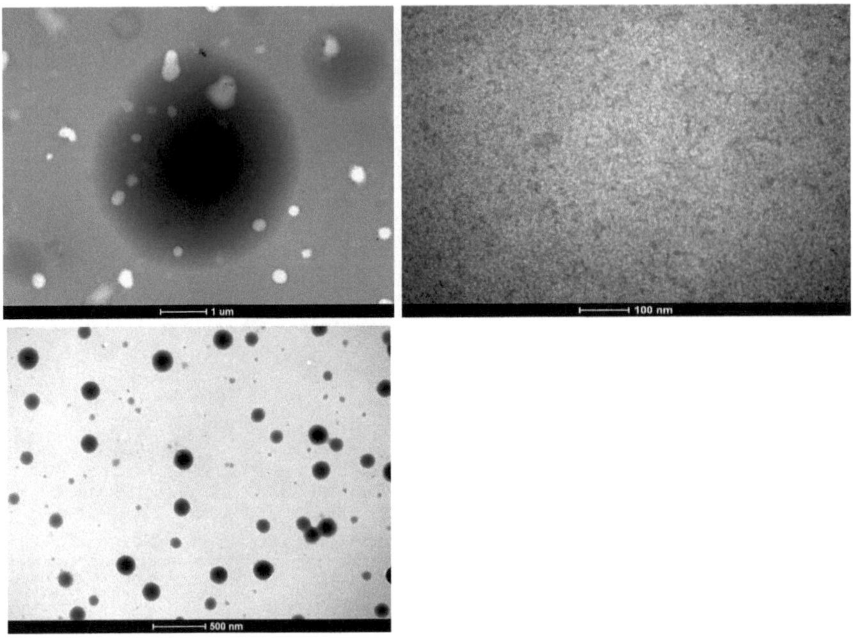

Fig. 7.8 Zr NPs in a silica coating Matrix [62]

For the carboxylic acids and AcAc, the FTIR data suggest that the interactions are not as strong as the characteristic absorption bands as both ligands remain unchanged relative to those of the bipy ligand. FTIR data indicates that the AcAc exist in an equilibrium of resonance states that may result in chelation with more than one zirconia atom. SST results showed that the methacrylic and acetic acids produced coatings with the best corrosion protection. It was proposed that the small molecule ligands reduced the size of the zirconia particles formed by the TPOZ thus forming a denser coating. Pitting was greatly reduced in these systems also suggesting that passivation of the copper IMPs had occurred to some extent [65].

In a follow up study, Varma et al. investigated the efficacy of diaminobenzoic acid (DABA) as a ligand in the MAPTMS–TPOZ system. The study aimed to incorporate the chelation of carboxylic acids on TPOZ and the corrosion inhibition qualities of the compound. DSC results showed that the DABA conferred improved stability to the MAPTMS–TPOZ mixture. FTIR confirmed that the DABA retained its structure and hence acted as a bidentate ligand in the TPOZ sol. EIS confirmed that the interfacial oxide layer resistance was also better in the DABA coating and this result was further confirmed by the performance of the MAPTMS/TPOZ/DABA system in SST. ^{29}Si NMR indicated that the mixture containing DABA proceeded to a more complete condensation than that of the AcAc [66].

Figure 7.9 shows the ^{29}Si NMR spectra of the MAPTMS/Zr/acac at different hydrolysis stages, pre-hydrolysed MAPTMS results in peaks at −49.5 (T_1^2), −59.6 (T_2^0) and a small amount of (T_0^0), characteristic of the precursor. When the Zr/acac

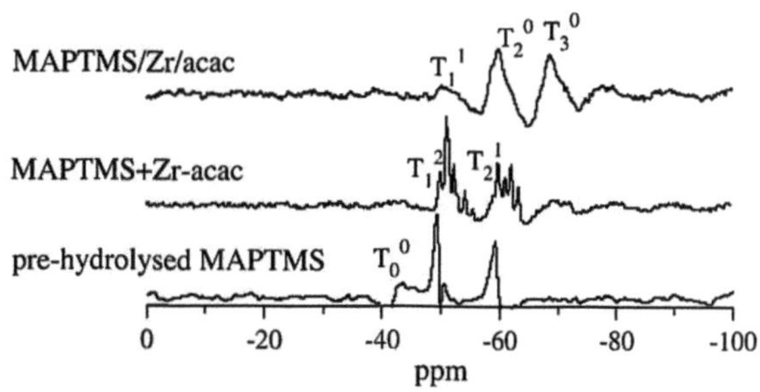

Fig. 7.9 ^{29}Si NMR of pre-hydrolysed MAPTMS, pre-hydrolysed MAPTMS + Zr–acac and MAPTMS/Zr/acac

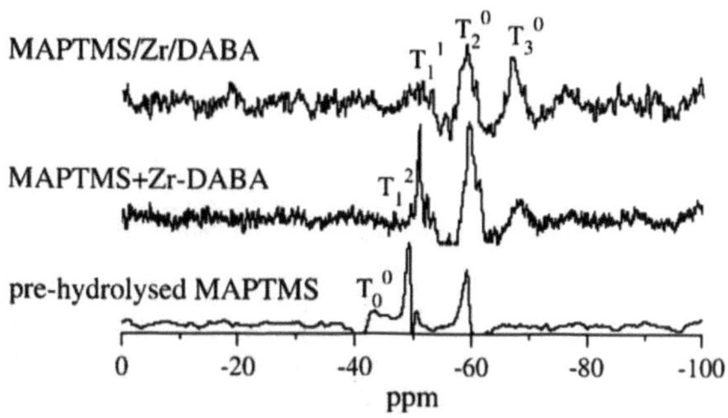

Fig. 7.10 ^{29}Si NMR of pre-hydrolysed MAPTMS, pre-hydrolysed MAPTMS + Zr–DABA and MAPTMS/Zr//DABA

was added the (T_1^2) species disappear indicating the self-condensation reactions catalysed by the introduction of the Zr/acac. The formation of new species at −50.8 (T_1^1), −58.7 (T_1^1) and a broad peak at −68.6 (T_3^0) is also observed. An increase in the condensation, shown by an increase in ($T^0{}_3$) species and the disappearance of (T_2^1) and (T_1^1) species occurs during the final step when the sol-gel is formed. The ^{29}Si NMR spectra of MAPTMS/Zr/DABA (Fig. 7.10) shows similar peaks to that of MAPTMS/Zr/acac, however, the formation of (T_2^1) species was not observed in the second step. Although they are similar it is observed that the (T_2^0) and the (T_3^0) signals were more intense which verifies that the level of condensed siloxane bonds is greater in MAPTMS/Zr/DABA than in MAPTMS/Zr/acac.

Boehmite NPs were incorporated in a GPTMS–TEOS matrix by Tavandashti et al. and the resulting coatings were characterised. The NPs were prepared using the sol-gel method with aluminium isopropoxide as precursor. Ultrasonic agitation and reduction of the hydrolysis and condensation time both aided in reducing the size of the NPs. EIS spectra indicated that a coating containing an optimised concentration of the boehmite formed an excellent intermediate or oxide layer at the surface of the 2024 substrate. The oxide capacitance and resistance behaviour showed no sign of ingress of water due to the formation of pores after extended immersion [67].

Oubaha et al. reported similar results when studying coatings prepared form a mixture of MAPTMS, TPOZ and aluminium–zirconium propoxide (AZPO). DSL results showed that the resulting formulations contained a mixture of silica and zirconia–alumina NPs each with distinct size distributions. Increase of the AZPO

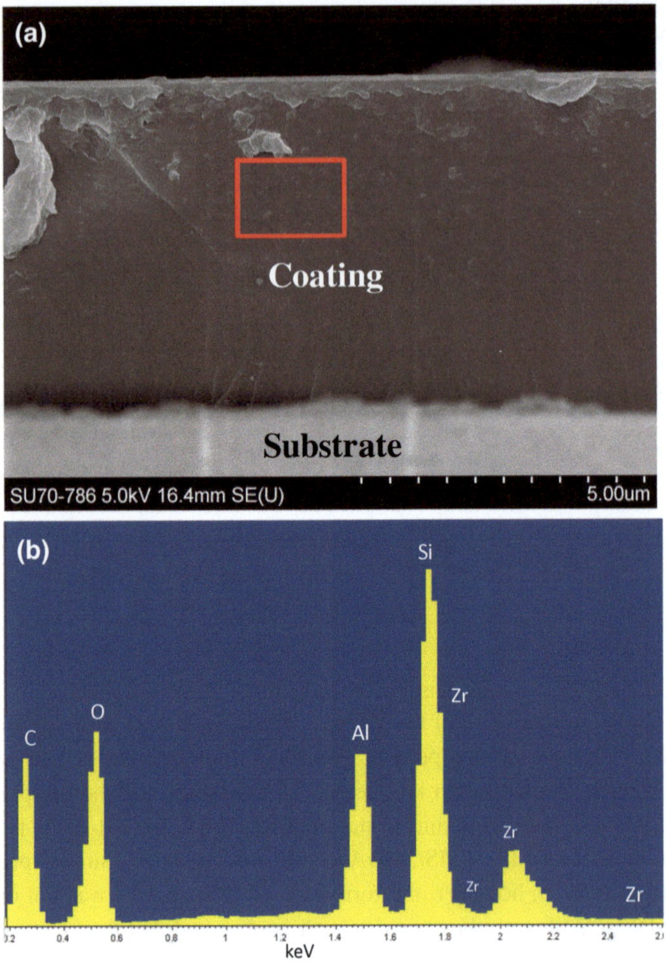

Fig. 7.11 **a** SEM image of 15% Al coating showing the coating and substrate; **b** EDS of the 15% Al coating showing the presence of Si, Al, Zr and O (*Red rectangle* indicates the area of EDS study)

concentration resulted in a change in the size distribution of the silica NPs which is thought to result from the ability of the Si and Al to form oxo-bridges. XPS detected no Al signal at the surface of the film, however, SEM EDX detected Al in the body of the coating suggesting that there is preferential deposition of Al towards the coating–substrate interface (see Figs. 7.11 and 7.12). PDS results indicated that the film was denser owing to its reduced corrosion current. No electrochemical interaction of the film was observed given that very slight changes to the E_{corr} were recorded [68].

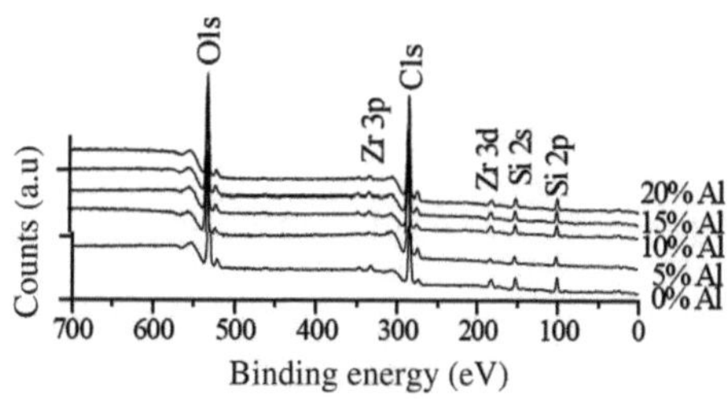

Fig. 7.12 XPS characterisation of Sol-Gel coating with different concentration of Al

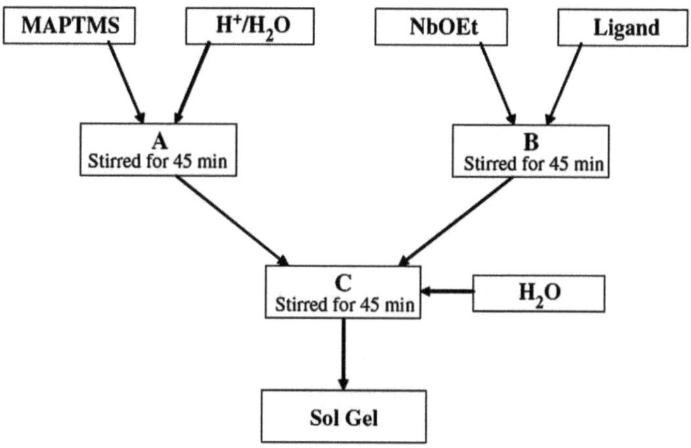

Fig. 7.13 Flow chart for the preparation of hybrid sols

Varma et al. examined the effects of the introduction of Niobium NPs into an MAPTMS coating through a sol-gel process (see Fig. 12.13). The corrosion performance after thermal cycling was tested. Samples were treated to ten cycles of −45 to 245 °C and then characterised using microscopy and salt spray testing. The corrosion performance of the doped MAPTMS coating was excellent in comparison to that of the undoped system (see Table 7.3). This improvement resulted from the formation of a denser Si–O network in the presence of the niobium. FE-SEM images of the MAPTMS/Ni coating after thermal cycling revealed the increase in the presence of Si/Ni particles [69] (Table 7.4).

Gonzales et al. prepared sol-gel coatings using TEOS, octyltriethoxysilane (OTES) which were then mixed with TPOZ prepared zirconia NPs. To these systems, the effect of adding $CeNo_3$ was investigated. They found that the best protection was offered by a coating containing ceria and zirconia NPs. They proposed that the zirconia NPs prevented aggregation of the ceria NPs. This led to stabilisation of the ceria which could then be released in the environment of any corrosion sites on the surface [70].

Table 7.4 Corrosion parameters estimated from potentiodynamic scan for bare AA 2024-T3 and sol-gel coatings

Sol-Gel coatings	I_{corr} (A/cm^2)	E_{corr} (V)	|βa| (V/decade)	|βc| (V/decade)	R_{pol} (Ω. cm^2)
Bare AA 2024-T3	3.91×10^{-7}	−0.580	0.037	0.086	1.33×10^3
MAPTMS	3.60×10^{-7}	−0.510	0.018	0.082	7.70×10^3
MAPTMS/Nb	1.54×10^{-9}	−0.50418.	0.231	0.297	2.83×10^6

7.3.3 *Direct Introduction of Commercially Available Nanoparticles*

In order to improve the mechanical characteristics of BSS coatings, Palanivel et al. incorporated silica nanoparticles (NPs) into a coating formulation. The effects of the concentration of the NPs were examined. PDS data showed that the NPs interacted with the cathodic reactions at the 2024 surface:

$$SiO_2 + 2OH^- \rightarrow SiO_3^{2-} + H_2O$$

The SiO_3^{2-} reacts with Al^{3+} leading to the formation of Al silicate compounds and strengthening of the interfacial layer. The optimum concentration of NPs was found to be between 5 and 15 ppm with film integrity being severely compromised by overloading of NPs. 50 ppm of silica in corrosion protection coatings was observed to lead to coating porosity which was detected by poor performance by the coating in EIS testing [71].

In an effort to emulate the self-healing nature of CCC, the following system was developed by Zheludkevich et al. Silica NPs were coated with a polyelectrolyte shell consisting of poly(ethylene imine) (PEI) and poly(styrene sulfonate) (PSS) through the layer by layer (LbL) deposition technique. During the coating process, BTAH was incorporated as a positive layer sandwiched between layers of PSS. Artificial defects were created in the film exposing the metal to chloride solution. Scanning vibrating electrode technique was then applied to the area to map current flow with time. The BTAH-NP containing silane film showed inhibition of corrosion as indicated by the disappearance of ionic current flow at the defect recorded using the SVET. Changes in pH changes induced by the corrosion event were proposed to be responsible for altering the surface permeability of the polyelectrolyte shell and hence releasing the BTAH into the environment of the defect [72].

Both silica and ceria nanoparticles were included in an inorganic–organic hybrid matrix by Schem et al. It may be assumed that one of the principle aims of this work was to capitalise on the corrosion inhibition properties of cerium containing compounds. The coating matrix consisted of GPTMS-TEOS-MTEOS and bisphenol which was polymerised in the coating mixture prior to deposition. Lower concentrations of ceria NPs in the coating produced the best results in terms of coatings protection and the abrasion resistance was also seen to benefit from their inclusion. By examining creepage of corrosion product on scribed samples subject to SST, it was concluded that the ceria NPs were active as corrosion inhibitors [31].

Balaskas et al. prepared epoxy-silane-based films that incorporated TiO_2 nanocontainers loaded with the corrosion inhibitor 8-hydroxyquinoline.

The nanocontainers were prepared via a templating route using polystyrene. TGA of the loaded containers indicated that the 8-HQ was retained and UV absorption analysis was used to provide evidence of release of the inhibitor.

EIS results indicated:

- Direct addition of the 8-HQ provided slight improvement of the epoxy silane coating
- The addition of unloaded nanocontainers improved barrier properties
- Nanocontainers loaded with 8-HQ showed improvement barrier performance and polarisation resistance under continued exposure to corrosive medium suggesting self-healing capability of the coating

SEM imaging of the coating showed that the incorporation of the TiO_2 nanocontainers had no impact on the surface morphology of the film [59].

Dalmoro et al. prepared sol-gel coatings using VTMS and TEOS precursors and introduced montmorillonite clay as a NP additive. Sodium and quaternary ammonium salts of montmorrillonite were applied with the latter providing the best corrosion protection as shown by EIS results. Low concentrations of the montmorillonite additive were beneficial providing more uniform coatings [73]. A study of a similar coating and its protection of HDG mild steel was carried out by Fedel et al. They concluded that the montmorrilonite clay undergoes a structural change when it is in the sol-gel matrix [74].

Cerium molybdite nanowires were incorporated in TPOZ/GPTMS coatings by Yasakua et al. They observed that the coatings containing the CeMo nanowires suppressed corrosion at defects created in the coating. Release studies showed that Ce was more likely to be released than Mo. The interfacial resistance of the coatings containing the nanowires was also seen to be enhanced [75].

7.3.4 Application of Organic Polymers

Voevodin et al. have studied the effect of mixing TEOS with a vinyl precursor to form a hybrid coating. The corrosion prevention of the resulting coatings was tested and showed that the mixture provided reasonable corrosion protection. The films exhibited cracking and aggressive pitting corrosion was seen to occur at these sites. The poor film integrity indicates that the dispersion of the inorganic and organic phases was limited and the film properties were dictated more so by the silane content [61].

GPTMS TEOS sols were treated with polydimethylsiloxane (PDMS) as a curing agent in a follow on work from the previously mentioned TEPA cure paper [49]. TEPA polymers with number average molar mass of 4400, 7800 and 27,000 were used in order to determine the impact of the chain length on the coating structure. The findings of the study were similar to those for the TEPA study, however, it was shown that the denser configuration of the polymer resulting from incorporation of the PDMS resulted in a lowering of the thermal stability. It was proposed that this was due to the increased thermal conductivity of the highly crosslinked polymer. The chain length of the PDMS was seen to interfere with silanol group condensation in the polymer matrix as it increased [76].

Chitosan film forming capacity has been modified by the introduction of organically modified silane and vanadate corrosion inhibitors by Kumar and Buchheit. The resulting coatings exhibited reduced hydrophilicity and optimum adsorption in the pH range 3–5. FTIR analysis of the coatings revealed that the vanadate adsorbs at the amine groups of the chitosan. UV-Vis spectroscopy showed that the vanadate may undergo pH controlled adsorption and desorption. Viscosity of the coating formulation was seen to control adhesion strength and also corrosion resistance. Formulations with high viscosity were seen to outperform their counterparts that exhibited low viscosity [77].

Silanes have been incorporated in an organic matrix by Ji et al. with a view to enhancing corrosion resistance by condensation in the presence of corrosive media. Epoxy resin was reacted with GPTMS, DTMS, TEOS and VTES in the presence of an organotin catalyst. IR analysis indicated that silane monomers bonded to the alcohol groups on the polymer backbone. EIS showed that the silane modified epoxy coatings acted as a better barrier than their unmodified counterparts. EIS also showed that charge transfer resistance and double layer capacitance fluctuated under exposure to corrosive media and maintained values consistent with improved corrosion resistance [78].

The introduction of the hyperbranched polymer PEI into GPTMS-TEOS coatings has been investigated by Roussi et al. PEI of different number average molar mass (5000 and 25,000) were employed in a dual role acting as both curing agent and organic corrosion inhibitor reservoir. The corrosion inhibitors used were imidazole and triazole based. It was found that coatings prepared using PEI performed better than Si–Zr based systems and that the partially oxidised PEI was able to respond to oxidative stresses in the corrosion environment by initiating further cross-linking hence maintaining the integrity of the barrier protection (see Fig. 7.14) [79].

Hybrid coatings were formed from a mixture of silyl functionalised glycidyl carbamate resin and TEOS by Chattopadhyay et al. The study found that hydrogen bonding in the resin changes with temperature. Hardness testing showed that TEOS concentration and the amine crosslinking agent dictated the hardness of the resulting coatings [80].

Ethylene glycol dimethacrylate (EGDMA) was incorporated in a TEOS-MAPTMS formulation by Roserro-Navarro et al. The effect of the polymerization processes for both the inorganic and organic components of the coating were examined. Hydrolysis and condensation were seen to proceed rapidly (ca. 17 min) for the inorganic components when this reaction was initiated with 0.6wt% HNO_3. The amine cross-linker 2,2′-azobis(isobutyronitrile) (AIBN) was employed and allowed for control of the degree of polymerisation of the organic component.

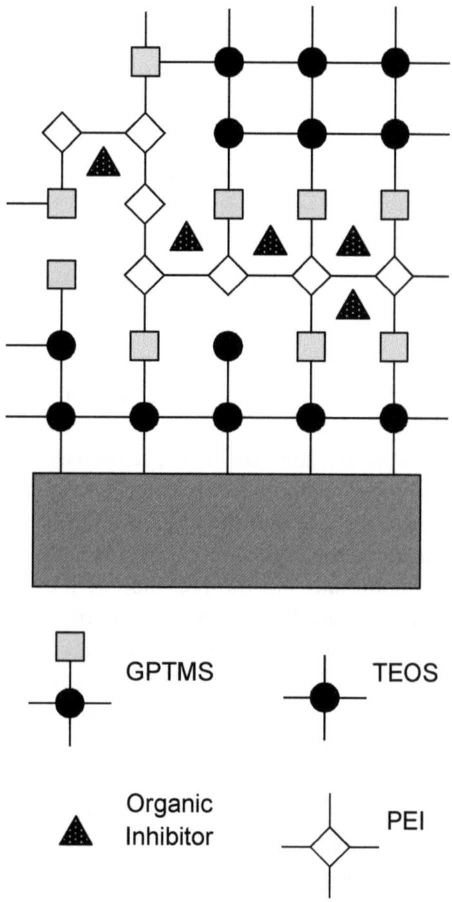

Fig. 7.14 Diagram of coating system

Silica nanoparticles incorporated in the formulation were seen to agglomerate as the degree of polymerisation increased. This behaviour was detected as a change in particle size and was determined using DLS. Thinner coatings were formed on 2024 using formulations where the degree of polymerisation was kept low. Those formed from more highly polymerised formulations exhibited higher defect densities. This indicated that protective coatings in which organic components are included are sensitive to the degree of polymerization of the organic phase [81].

Hybrid films were developed by Khelifa et al. whereby an organic matrix was modified by the inclusion of silica NPs prepared using a sol-gel process. The organic coating was prepared using 2-ethylhexyacrylate and glycidylmethacrylate as monomers. Silica NPs were prepared using acid catalysed TEOS which was then modified using GPTMS as a coupling agent. It was found that silica NP concentration, morphology and the ratio of TEOS to GPTMS were factors that controlled the efficacy of the resulting hybrid film [82].

Bakhshandeh et al. prepared coatings of varying inorganic content using a mixture of APTES and bisphenol A which was subsequently added to varying amounts of hydrolysed TEOS. Attributes such as hardness and adhesion were seen to depend on the TEOS concentration. The group proposed that an improvement in the corrosion protection offered by the coatings was attributable to the formation of a silica layer at the coating substrate interface [83].

7.3.5 Application of Corrosion Inhibitors

Epoxy coatings were prepared by Khramov et al. which utilised amino silanes rather than amine curing agents for the cross-linking of the epoxy groups [84]. Organic corrosion inhibitors were incorporated in this coating system with and without the addition of β-cyclodextrin, which is known to enhance solubility of nonpolar compounds in polar media. The organic inhibitors used fell into two categories; ionisable and non-ionisable. The latter were seen to outperform the former as their weak interactions with the coating allowed for diffusion to corrosion initiation sites. In the presence of β-cyclodextrin, both kinds of organic inhibitor were seen to deliver improved corrosion protection [85].

The incorporation of inorganic and organic inhibitors was examined by Palinivel et al. Tolytriazole(TTA), benzotriazole(BTAH) and cerium nitrate were included in waterborne silane coatings consisting of BAS and VTAS. The behaviour of the inhibitors was examined separately in NaCl solutions in order to see how their incorporation in the silane films affected their corrosion protection mechanism. The organic examples were seen to enhance the barrier properties of the film but showed no indication of self-healing. Cerium nitrate loaded films however, did show self-healing behaviour. This was confirmed by the observation of scribes on samples of cerium nitrate–silane coated 2024 samples that showed very little corrosion in comparison to coatings without the inhibitor after several days' immersion. SEM EDX of the scribes detected the presence of cerium indicating that the inhibitor was mobile in the film under corrosion conditions. The pH dependence of the release of cerium was also examined and confirmed using ICP-MS. It was shown that cerium concentrations in solution rose by two orders of magnitude at pH 11. It was proposed that high pH conditions caused the oxidation of Ce^{3+} to Ce^{4+}, the latter being more soluble in solution and thus able to migrate to areas where corrosion reduction reactions are happening at the coating–surface interface. Once there, the Ce^{4+} is reduced and forms insoluble cerium oxides which precipitate at the reduction site and inhibit the reduction reactions driving corrosion [86].

Zirconia NPs may act as reservoirs for inhibitors as demonstrated by Zheludkevich et al. EIS studies showed that silane films with Zr NP and $Ce(NO_3)_3$ incorporated in the Zr sol exhibited excellent corrosion protection characteristics. The films exhibited very stable coating capacitance (C_{coat}) as well as stable oxide and coating resistances (R_{oxide} and R_{Coat}). The polarisation resistance of these coatings was also seen to be high. Cerium nitrate concentration had an adverse impact on the coating performance as it increased and also in coatings where the inhibitor was added to the silicon sol as opposed to the Zr sol [32, 87].

A study by Lamaka et al. presented a bilayer anticorrosion system (see Fig. 7.15). The first layer was prepared using a novel process whereby titanium isopropoxide sol was mixed with a block copolymer dissolved in ethanol. This was deposited on the 2024 substrate and shown to produce a highly porous TiO_2 layer, the morphology of which depended strongly on the morphology of the substrate. The block copolymer was assumed to play a templating role during the hydrolysis and condensation of the TiO_2 producing a well-defined network of the metal oxide [88] The TiO_2 coated substrates were immersed in solutions containing Ce $(NO_3)_3$ and BTAH. TPOZ-GPTMS was then coated on some substrates to act as an extra barrier layer. The substrates coated with the inhibitor doped TiO_2 layer and GPTMS-TPOZ sol-gel exhibited excellent oxide resistance. EIS results showed that the oxide resistance fluctuated with time as active corrosion protection proceeded. R_{ox} maintained a high value after 300 h of ca. $10^8 \Omega$ [89].

Fig. 7.15 Diagram of coating system

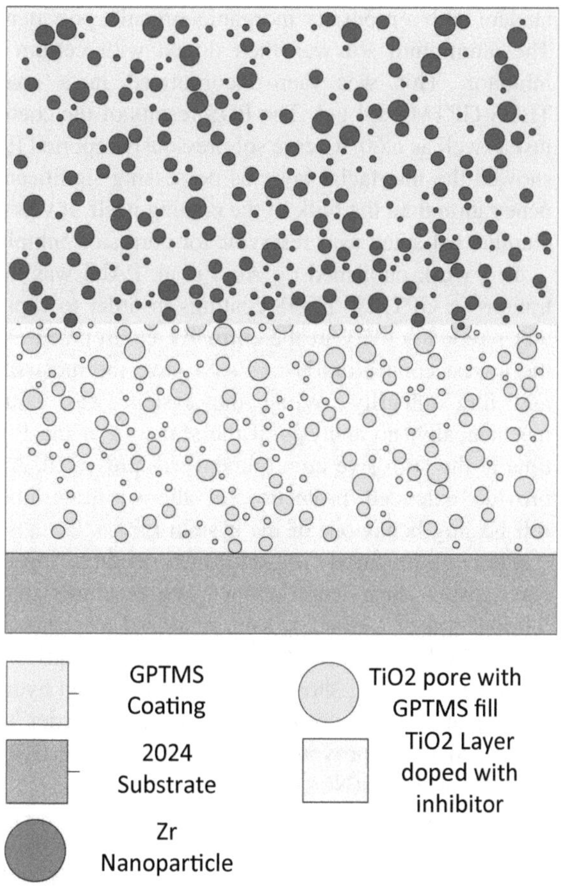

The introduction of inorganic inhibitors into MTEOS coatings and its effect on the corrosion resistance of the 2024 alloy was investigated by Varma et al. An optimum concentration of 0.7%w/w $Mg(NO_3)_2$ was found at which a pore free coating which protected the substrate for 120 h under SST was prepared. The mechanism of inhibition by the $Mg(NO_3)_2$ involves the creation of insoluble hydroxides which act as pore blockers and help to maintain the barrier properties of the coating. EIS results showed that the interfacial resistance of the films with varying concentration of $Mg(NO_3)_2$ did not exhibit better performance as this parameter increased which supports the theory that the corrosion protection occurs in the body of the sol-gel film [90].

The addition of $Ce(NO_3)_3$ to the Boehmite–GPTMS–TEOS coating was the subject of a publication by the CREST group. The group reported the diffusion of the cerium nitrate to sites that had been scratched from the coating. Also, the coating had an improved and more durable oxide resistance as the cerium nitrate acted to inhibit corrosion reactions at weak spots in the coating [91].

Room temperature cure anticorrosion sol-gel coatings were the focus of a publication by Wang et al. Aluminium salts have been used as a precursor in tandem with a propriety inorganic–organic, polymer-like solvent to form a sol-gel. The aluminium sol was then doped with cerium nitrate acting as a corrosion inhibitor. This was then incorporated in a coating at a 1:1 ratio with a TEOS-GPTMS sol-gel. The PDS results of the coating indicated that it performed just as well as a 60 °C cure sol previously reported [92]. EIS analysis of the coating showed the interfacial layer as possessing significantly more resistance to solvent penetration than the bulk of the coating itself. It was proposed by the group that the alumina sol acted as a reservoir for corrosion inhibitor [93].

In a work published by Akid et al. PANI was incorporated at various concentrations in GPTMS–TEOS coatings in order to capitalise on the conducting polymer's redox activity in the corrosive environment. Of three concentrations tested, the lowest concentration of PANI gave the most satisfactory corrosion inhibition and film integrity. While the system was examined electrochemically and mechanically, no analysis of the structure of the PANI as a function of exposure time to the corrosive environment was presented. A scribed coating did, however, provide extended protection of the substrate under SST, indicating possible self-healing behaviour of the system [53].

Shi et al. prepared GPTMS/TEOS coatings doped with Ce based inhibitors and cured using diethylenetriamine. The resulting characterisation showed that the introduction of $Ce(NO_3)_3$ salts produced the best corrosion inhibition. XRD data confirmed the presence of $Ce(OH)_3$ in the sols prior to their application to the 2024 substrate. XPS data showed that Ce oxides and hydroxides were released from the coating at sites of corrosion acting as a barrier to further corrosion. EIS data confirmed the improvement in corrosion protection of the substrate provided the addition of the $Ce(NO_3)_3$ [94].

Multilayer coatings were designed by Paussa et al. in an attempt to avoid the porosity that is inherent in sol-gel based coatings where corrosion inhibitors have been introduced directly into the preparation. $Ce(NO_3)_3$ was incorporated into a hybrid organic–inorganic system comprising TEOS, MAPTMS and polymerized EGDMA and GMA. The doped sols were introduced as an intermediate layer between undoped sols and deposited on 2024. GDOES was used to examine the elemental composition of the coatings before and after immersion and the results were interpreted as indicating that Ce in the coating that was dissolved in the electrolyte initially subsequently deposited at the coating–metal interface. These results were supplemented by EIS information indicating high R_{CT} in systems containing Ce [95].

The impact of direct addition of inorganic and organic corrosion inhibitors to an AETMS-epoxy based coating has been examined by Balaskas et al. $Co(NO_3)_3$, Ce $(NO_3)_3$ and $Mg(NO_3)_3$ were applied as inorganic inhibitors with the Ce and Mg salts providing the best protection of the coating formulations prepared. 8-hydroxyquinolone, benzotriazole and mercaptobenzothiazole served as the organic inhibitors, however, their ability to prevent corrosion was less than that of the inorganic salts. EIS results showed that the most effective inhibitors, $Ce(NO_3)_3$ and $Mg(NO_3)_3$ both enhanced the oxide resistance of the coatings [96].

The structural changes that occur in an inorganic–organic hybrid coating as the content of corrosion inhibitor $Ce(NO_3)_3$ increases have been studied by Cambon et al. using NMR. ^{29}Si, ^{13}C and ^{27}Al NMR were used as well as EIS and nanoindentation to characterise an aluminosilicate based coating. GPTMS was used as the silane and aluminium tri-sec butoxide as the alumina source. Results showed that an optimum concentration was found for $Ce(NO_3)_3$ concentration above which, the barrier properties of the film were compromised and the protection offered by the coating reduced. The NMR analysis showed that high Ce concentration lead to the preferential organic polymerisation and silanol cage formation. This leads to segregation of the organic and inorganic phases of the coating and leads to loss of the beneficial characteristics that a more homogeneous mixture exhibit [97].

Collazo et al. have investigated the benefits of adding thermally activated hydrotalcite (HT) to GPTMS-TPOZ based sol-gel films. The HT acts by scavenging aggressive ions in corrosive media and its ability to do this can be dramatically improved by subjecting the HT to calcination at 450 °C for 24 h prior to introduction to the formulation. EIS data revealed that the calcined HT had a detrimental effect on the barrier properties of the TPOZ-GPTMS film. However, SST showed that this system outperformed all others in the study. EIS results showed that the main benefit of the calcined HT addition was to greatly improve polarisation resistance at the alloy surface. Also, redox reactions involving Cu ions on the alloy surface during corrosion were impeded by HT containing coatings. It was stressed in the reporting of this study that the HT acted as an absorptive additive and that the benefits seen here did not result from the ion exchange behaviour that HT is mainly known for [98].

Dias et al. prepared ZrO_2/SiO_2 sol-gel coatings loaded with cerium enriched zeolites. The Ce enriched zeolites improved coating performance in accelerated corrosion testing when compared to the film on its own and films impregnated with cerium free zeolites. They proposed that the dissolution of copper from 2024 intermetallic particles facilitated the release of Ce(III) ions from the zeolites through an ion exchange mechanism with Cu^{2+} [99].

Wittmar et al. investigated the use of CeO_2 and ZnO NPs as corrosion inhibitors in a GPTMS hybrid film. The organic component of the film was provided by including 2, 2′-bis(hydroxyphenol)-propane (BPA). Cross-linking of the GPTMS was facilitated by including APTES. They concluded that improvements observed in the protection provided by coatings loaded with NPs were due to the NPs performing a dual role of pore fillers and corrosion inhibitors [27].

Perspectives

The process of reviewing the published research on the area of sol-gel based corrosion protection for the AA 2024 T3 alloy has allowed for the following trends and opportunities to be identified:

The followings trends and opportunities in the area of sol-gel based corrosion protection have been identified:

With regard to **sol-gel film design**, alkoxysilanes such as TEOS, GPTMS and BAS have been studied extensively. For example, 31 of the 64 papers reviewed herein used GPTMS as a primary sol-gel coating component. Overall, 11 different alkoxysilane precursors were investigated. This appears to be a small amount when a quick search on a research chemical supplier's online catalogue indicates that there are in excess of 60 different alkoxysilane precursors available (Fig. 7.16).

The area of sol-gel film design has shown the impact of the precursor structure on the coating produced. The precursor allows for tuning of the adhesion, wettability and hydrophobicity of the resulting coating. Functional moieties incorporated in the precursor can contribute to a self-healing effect on the anticorrosive coating. It is reasonable to assume that there are many more avenues for research given the wide range of sol-gel precursors available that have not yet been studied.

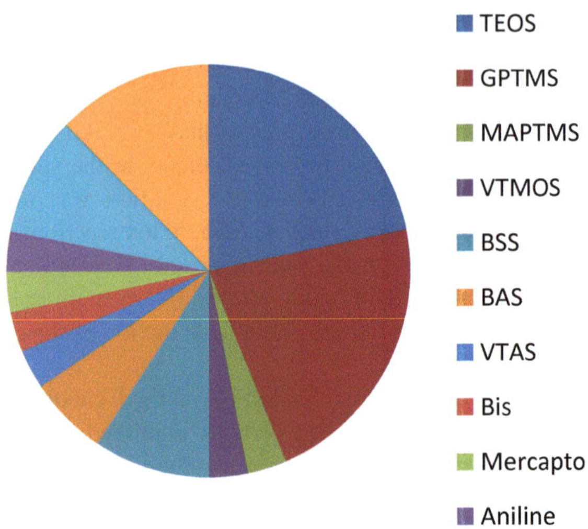

Fig. 7.16 Alkoxysilane precursors applied in 2024 corrosion protection. The trends noted are presented section by section

Two different areas of **nanoparticle inclusion** have been studied; preparation and inclusion of NPs using the sol-gel route and the direct addition of commercially available NPs. The first strategy has its advantages in that the chelating agents required to prepare transition metal based sol-gel materials may be chosen to perform a dual role; both chelating the transition metal precursor and performing as active compounds within the coating. These nanoparticles may also act as reservoirs for corrosion inhibition compounds. The second strategy allows the researcher to select and incorporate NPs that may be composed of corrosion inhibition compounds as in the case of ceria. It also allows for the preparation of complex NP systems that may be specifically tailored to contain and release active coating compounds.

Organic polymers have been used to perform tasks similar to NPs. Examples include the adjustment of mechanical properties and aiding the inclusion of otherwise incompatible corrosion inhibitors. In the course of this research additional benefits conferred by the inclusion of organic polymers have been observed such as the ability of polymers that have undergone incomplete cross-linking to mitigate the stresses on the coating in corrosive environments.

The multi-compatibility of the ormosil coatings allows for the incorporation of organic polymers in that they may be chemically grafted to the sol-gel matrix using cross-linkers. This allows for the design of coating systems that take advantage of the best aspects of both inorganic and organic coating systems. In the context of corrosion control for the AA 2024 T3 alloy, epoxy-based organic polymers have been the subject of the majority of the research and it is likely that further success may achieved by analysing coatings prepared from alternative organic precursors.

The use of **corrosion inhibitors** in the sol-gel processing of coatings has been investigated. These compounds offer the researcher the option of preparing active coatings that react to chemical stimuli in the corrosion environment such as pH or the presence of corrosion initiating materials. $CeNO_3$ is the corrosion inhibiting compound that has dominated this field of research.

The reviewed sol-gel systems offer many different and unique protection mechanisms, which can be split into the following categories (Fig. 7.17):

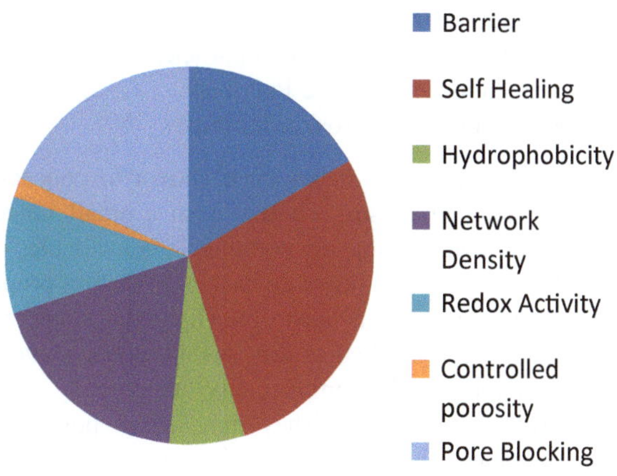

Fig. 7.17 Strategies for protection of 2024 explored in the text

- Barrier: while all the coating systems are a barrier to some extent those included in this category carry out that role specifically
- Hydrophobicity
- Network density: network density has been modified by altering the condensation of alkoxysilane based systems or those that incorporate organic polymers
- Redox activity: systems that exhibit electrochemical reactivity in the body of the coating
- Pore blocking: systems that improve the barrier by blocking the pores, usually by inclusion of a sol-gel produced nanoparticulate system

The self-healing systems may be seen to be the focus of the majority of the research presented in this review and this may reflect a dominating trend. This may be indicative of an over-reliance on the belief that the final protective coating system prepared using transition metal and silicon alkoxysilanes must resemble the chromate system that they are sought to replace. While this assumption is not a bad one, it may preclude the investigation of different materials and strategies that could serve to solve this problem in an equally effective yet novel manner.

The use of an acceptable control in the research is a matter for concern. In this instance it is proposed that an acceptable control would be the chromate treated AA2024 T3 system. This is a rational choice as control for this research as the systems that have been prepared and analysed here are all intended as replacements for this system.

7.4 Conclusions

The research related to the application of sol-gel based materials as corrosion protection for aluminium alloys has been analysed in the previous sections including:

- Sol-Gel Film Design
- Introduction of Nanoparticles (via the sol-gel route)
- Direct Introduction of Nanoparticles
- Application of Organic Polymers
- Application of Corrosion Inhibitors

Sol-Gel film design has been shown to concentrate on films prepared using alkoxysilane precursors chosen from a small range considering the number of precursor materials that are commercially available. Within those systems studied it has been shown that novel corrosion protection properties (such as those observed in the aniline functionalised ormosil coating) can be achieved.

Nanoparticles, whether they are introduced into the coating using the sol-gel route or are bought commercially and directly introduced, have been shown to perform multiple roles. They may act as pore blockers, vehicles for corrosion inhibitors and also as agents for improving the mechanical features of the coating.

Research into the incorporation of organic polymers into sol-gel based systems has been focussed on epoxy based compounds. There is a wide range of organic monomers available and it is reasonable to assume that further exploration of these materials is worthwhile.

A large part of the research relating to the incorporation of corrosion inhibiting compounds into sol-gel based coating systems has focussed on $Ce(NO_3)_3$. Again, this is a field of research that may be expanded upon.

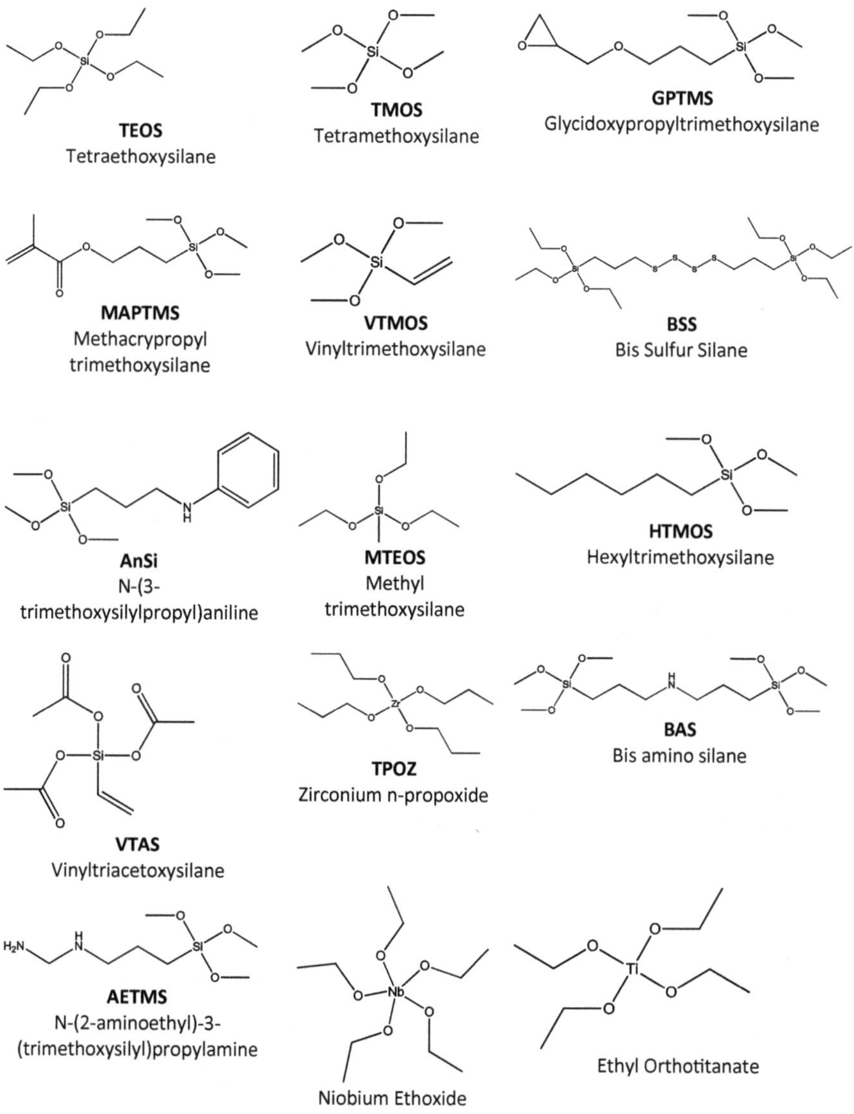

Fig. 7.18 Structures of Sol-Gel precursors used in Al 2024 corrosion control

The identification and use of an appropriate control is necessary in order to allow for the direct comparison and evaluation of the corrosion protection systems that are being investigated.

The application of sol-gel technology in the effort to develop an alternative to environmentally harmful chromium based corrosion control materials, is a demonstration of the flexibility offered by sol-gel chemistry when designing novel materials. The benefits of both inorganic and organic compounds may be combined in an unlimited variety of ways to provide an ever developing and improving range of materials (Fig. 7.18).

References

1. Hench, L.L., West, J.K.: The Sol-Gel process. Chem. Rev. **90**, 33–72 (1990). doi:10.1021/cr00099a003
2. Celzard, A., Marêché, J.F., De Chimie, L., Henri, U., Nancy, P.: Advanced chemistry classroom and laboratory applications of the sol—gel process using well-tested recipes **79** (2002)
3. Judeinstein, P., Sanchez, C.: Hybrid organic-inorganic materials: a land of multidisciplinarity chemistry: synthesis of hybrid materials **6**, 511–525 (1996)
4. Nolan, N.T., Synnott, D.W., Seery, M.K., Hinder, S.J., Van Wassenhoven, A., Pillai, S.C.: Effect of N-doping on the photocatalytic activity of Sol-Gel TiO_2. J. Hazard. Mater. **211–212**, 88–94 (2012). doi:10.1016/j.jhazmat.2011.08.074
5. Buckley, A., Greenblatt, M.: The Sol-Gel preparation of silica gels. J. Chem. Educ. http://pubs.acs.org/doi/abs/10.1021/ed071p599 (1994). Accessed 27 June 2014
6. Trueba, M., Trasatti, S.P., Flamini, D.O.: The effect of aluminium alloy secondary phases on aniline-based silane protection capacity. Corros. Sci. **63**, 59–70 (2012). doi:10.1016/j.corsci.2012.05.028
7. The Aluminum Association Inc.: International alloy designations and chemical composition limits for wrought aluminum and wrought aluminum alloys. Alum. Assoc. Arlington, Virginia. **28** (2006)
8. Welding Science & Technology, (n.d.). http://www.weldinginfocenter.com/sci_tech/st_03.html. Accessed 13 April 2016
9. Starke, E.A., Staley, J.T.: Application of modern aluminium alloys to aircraft. Prog. Aerosp. Sci. **32**, 131–172 (1996). http://static.gest.unipd.it/esercizi/IIP_pmmi/Al_aerospace_03.pdf
10. Buchheit, R.G.: Local dissolution phenomena associated with S phase (Al[sub 2]CuMg) particles in aluminum alloy 2024-T3. J. Electrochem. Soc. **144**, 2621 (1997). doi:10.1149/1.1837874
11. International Civil Aviation Organization.: International Civil Aviation Organization Safety Report 2014, Montreal, 2014. http://www.icao.int/safety/Documents/ICAO_2014 Safety Report_final_02042014_web.pdf
12. Sinclair, J.D., Soc, J.E., Martin, Y., Abraham, D., Wickramasinghe, H.: Corrosion study of AA2024-T3 by scanning kelvin probe force microscopy and in situ atomic force microscopy scratching. **145** (1998)
13. Zhu, D., van Ooij, W.J.: Corrosion protection of AA 2024-T3 by bis-[3-(triethoxysilyl) propyl]tetrasulfide in neutral sodium chloride solution. Part 1: corrosion of AA 2024-T3. Corros. Sci. **45**, 2163–2175 (2003). doi:10.1016/S0010-938X(03)00060-X
14. Guan, H., Buchheit, R.G.: Corrosion protection of aluminum alloy 2024-T3 by vanadate conversion coatings, 284–296 (2004)

15. Osborne, J.H.: Observations on chromate conversion coatings from a sol-gel perspective. Prog. Org. Coatings. **41**, 280–286 (2001). doi:10.1016/S0300-9440(01)00143-6
16. Frankel, G.S., Mccreery, R.L.: Inhibition of Al alloy corrosion by chromates. Interface, 34–38 (2001)
17. Sonntag, B.: Substitution of Cr (VI)—containing chromate coatings by the European automobile industry, (n.d.). http://www.atotech.com/fileadmin/pdf/substitution_of_Cr_VI_.pdf. Accessed 27 Feb 2013
18. Winston Revie, R.: Uhlig's Corrosion Handbook, 2nd ed., Wiley, New York (2000)
19. Brusic, V., Frisch, M.A., Eldridge, B.N., Novak, F.P., Kaufman, F.B., Rush, B.M., et al.: Copper Corrosion With and Without Inhibitors. **138** (1991)
20. Huynh, N., Bottle, S.E., Notoya, T., Schweinsberg, D.P.: Inhibitive action of the octyl esters of 4- and 5-carboxybenzotriazole for copper corrosion in sulphate solutions. Corros. Sci. **42**, 259–274 (2000). doi:10.1016/S0010-938X(99)00075-X
21. Williams, G., Coleman, A.J., McMurray, H.N.: Inhibition of aluminium alloy AA2024-T3 pitting corrosion by copper complexing compounds. Electrochim. Acta. **55**, 5947–5958 (2010). doi:10.1016/j.electacta.2010.05.049
22. Qafsaoui, W., Huet, F., Takenouti, H.: Analysis of the inhibitive effect of BTAH on localized corrosion of Al 2024 from electrochemical noise measurements. J. Electrochem. Soc. **156**, C67 (2009). doi:10.1149/1.3040281
23. Lamaka, S.V., Zheludkevich, M.L., Yasakau, K.a., Montemor, M.F., Ferreira, M.G.S.: High effective organic corrosion inhibitors for 2024 aluminium alloy. Electrochim. Acta. **52**, 7231–7247 (2007). doi:10.1016/j.electacta.2007.05.058
24. Iannuzzi, M., Kovac, J., Frankel, G.S.: A study of the mechanisms of corrosion inhibition of AA2024-T3 by vanadates using the split cell technique. Electrochim. Acta. **52**, 4032–4042 (2007). doi:10.1016/j.electacta.2006.11.019
25. Iannuzzi, M., Frankel, G.S.: Mechanisms of corrosion inhibition of AA2024-T3 by vanadates. Corros. Sci. **49**, 2371–2391 (2007). doi:10.1016/j.corsci.2006.10.027
26. Collazo, A., Nóvoa, X.R., Pérez, C.: The role of Mg^{2+} ions in the corrosion behaviour of AA2024-T3 aluminium alloys immersed in chloride-containing environments. Electrochim. Acta. **124**, 17–26 (2014). doi:10.1016/j.electacta.2013.10.130
27. Wittmar, A., Caparrotti, H., Wittmar, M., Veith, M.: Simple preparation routes for corrosion protection hybrid Sol-Gel coatings on AA 2024. Surf. Interface Anal. **44**, 70–77 (2012). doi:10.1002/sia.3771
28. Lopez-Garrity, O., Frankel, G.S.: Corrosion inhibition of AA2024-T3 by sodium silicate. Electrochim. Acta. **130**, 9–21 (2014). doi:10.1016/j.electacta.2014.02.117
29. Campestrini, P., Terryn, H., Hovestad, A., De Wit, J.H.W.: Formation of a cerium-based conversion coating on AA2024 : relationship with the microstructure. **176**, 365–381 (2004). doi:10.1016/S0257-8972
30. Fahrenholtz, W.G., O'Keefe, M.J., Zhou, H., Grant, J.T.: Characterization of cerium-based conversion coatings for corrosion protection of aluminum alloys. Surf. Coatings Technol. **155**, 208–213 (2002). doi:10.1016/S0257-8972(02)00062-2
31. Schem, M., Schmidt, T., Gerwann, J., Wittmar, M., Veith, M., Thompson, G.E., et al.: CeO_2-filled sol-gel coatings for corrosion protection of AA2024-T3 aluminium alloy. Corros. Sci. **51**, 2304–2315 (2009). doi:10.1016/j.corsci.2009.06.007
32. Zheludkevich, M.L., Serra, R., Montemor, M.F., Yasakau, K.A., Salvado, I.M.M., Ferreira, M.G.S.: Nanostructured sol-gel coatings doped with cerium nitrate as pre-treatments for AA2024-T3. Electrochim. Acta. **51**, 208–217 (2005). doi:10.1016/j.electacta.2005.04.021
33. Yasakau, K.A., Zheludkevich, M.L., Lamaka, S.V., Ferreira, M.G.S.: Mechanism of corrosion inhibition of AA2024 by rare-earth compounds. J. Phys. Chem. B. **110**, 5515–5528 (2006). doi:10.1021/jp0560664
34. Shi, H., Han, E.-H., Lamaka, S.V., Zheludkevich, M.L., Liu, F., Ferreira, M.G.S.: Cerium cinnamate as an environmentally benign inhibitor pigment for epoxy coatings on AA 2024-T3. Prog. Org. Coatings. **77**, 765–773 (2014). doi:10.1016/j.porgcoat.2014.01.003

35. Catubig, R., Hughes, A.E., Cole, I.S., Hinton, B.R.W., Forsyth, M.: The use of cerium and praseodymium mercaptoacetate as thiol-containing inhibitors for AA2024-T3. Corros. Sci. **81**, 45–53 (2014). doi:10.1016/j.corsci.2013.12.001
36. Tallman, D.E., Vang, C., Wallace, G.G., Bierwagen, G.P.: Direct electrodeposition of polypyrrole on aluminum and aluminum alloy by electron transfer mediation. J. Electrochem. Soc. **149**, C173 (2002). doi:10.1149/1.1448820
37. He, J., Tallman, D.E., Bierwagen, G.P.: Conjugated polymers for corrosion control: scanning vibrating electrode studies of polypyrrole-aluminum alloy interactions. J. Electrochem. Soc. **151**, B644 (2004). doi:10.1149/1.1819632
38. Yan, M.C., Tallman, D.E., Rasmussen, S.C., Bierwagen, G.P.: Corrosion control coatings for aluminum alloys based on neutral and n-doped conjugated polymers. J. Electrochem. Soc. **156**, C360 (2009). doi:10.1149/1.3190162
39. PP423 Patent 6037060 Osborne Zr Silane 2000.pdf, (n.d.)
40. PP422 Patent 5939197 Osborne Zr Silane 1999.pdf, (n.d.)
41. Steel, S.: Chromium free conversion before painting. **33**, 3–5 (2012)
42. Metroke, T.L., Kachurina, O., Knobbe, E.T.: Spectroscopic and corrosion resistance characterization of GLYMO–TEOS Ormosil coatings for aluminum alloy corrosion inhibition. Prog. Org. Coatings **44**, 295–305 (2002). doi:10.1016/S0300-9440(02)00063-2
43. Metroke, T.L., Gandhi, J.S., Apblett, A.: Corrosion resistance properties of Ormosil coatings on 2024-T3 aluminum alloy. Prog. Org. Coatings **50**, 231–246 (2004). doi:10.1016/j.porgcoat.2004.03.001
44. Metroke, T.L., Apblett, A.: Effect of solvent dilution on corrosion protective properties of Ormosil coatings on 2024-T3 aluminum alloy. Prog. Org. Coatings **51**, 36–46 (2004). doi:10.1016/j.porgcoat.2004.06.002
45. Zhu, D., van Ooij, W.J.: Corrosion protection of AA 2024-T3 by bis-[3-(triethoxysilyl) propyl]tetrasulfide in sodium chloride solution. Corros. Sci. **45**, 2177–2197 (2003). doi:10.1016/S0010-938X(03)00061-1
46. Zhu, D., va. Ooij, W.J.: Enhanced corrosion resistance of AA 2024-T3 and hot-dip galvanized steel using a mixture of bis-[triethoxysilylpropyl]tetrasulfide and bis-[trimethoxysilylpropyl] amine. Electrochim. Acta. **49**, 1113–1125 (2004). doi:10.1016/j.electacta.2003.10.023
47. Zhu, D., van Ooij, W.J.: Corrosion protection of metals by water-based silane mixtures of bis-[trimethoxysilylpropyl]amine and vinyltriacetoxysilane. Prog. Org. Coatings **49**, 42–53 (2004). doi:10.1016/j.porgcoat.2003.08.009
48. Cabral, A.M., Duarte, R.G., Montemor, M.F., Ferreira, M.G.S.: A comparative study on the corrosion resistance of AA2024-T3 substrates pre-treated with different silane solutions. Prog. Org. Coatings. **54**, 322–331 (2005). doi:10.1016/j.porgcoat.2005.08.001
49. Wu, K.H., Chang, T.C., Yang, C.C., Wang, G.P.: Dynamics and corrosion resistance of amine-cured organically modified silicate coatings on aluminum alloys. Thin Solid Films **513**, 84–89 (2006). doi:10.1016/j.tsf.2006.01.022
50. Pirhady Tavandashti, N., Sanjabi, S., Shahrabi, T.: Corrosion protection evaluation of silica/epoxy hybrid nanocomposite coatings to AA2024. Prog. Org. Coatings **65**, 182–186 (2009). doi:10.1016/j.porgcoat.2008.10.010
51. Cheng, X., Wang, Z., Yan, Y.: Corrosion-resistant zeolite coatings by in situ crystallization. Electrochem. Solid-State Lett. **4**, B23 (2001). doi:10.1149/1.1359396
52. Jerman, I., Šurca Vuk, A., Koželj, M., Švegl, F., Orel, B.: Influence of amino functionalised POSS additive on the corrosion properties of (3-glycidoxypropyl)trimethoxysilane coatings on AA 2024 alloy. Prog. Org. Coatings. **72**, 334–342 (2011). doi:10.1016/j.porgcoat.2011.05.005
53. Akid, R., Gobara, M., Wang, H.: Corrosion protection performance of novel hybrid/Sol-Gel coatings on an aluminium 2024 alloy in neutral, alkaline and acidic solutions. Electrochim. Acta. **56**, 2483–2492 (2011). doi:10.1016/j.electacta.2011.06.046
54. Mrad, M., Montemor, M.F., Dhouibi, L., Triki, E.: Deposition of hybrid 3-GPTMS's film on AA2024-T3: dependence of film morphology and protectiveness performance on coating conditions. Prog. Org. Coatings. **73**, 264–271 (2012). doi:10.1016/j.porgcoat.2011.11.019

55. Lakshmi, R.V., Yoganandan, G., Mohan, A.V.N., Basu, B.J.: Effect of surface pre-treatment by silanization on corrosion protection of AA2024-T3 alloy by sol-gel nanocomposite coatings. Surf. Coatings Technol. **240**, 353–360 (2014). doi:10.1016/j.surfcoat.2013.12.051
56. Recloux, I., Debliquy, M., Baroni, A., Paint, Y., Lanzutti, A., Fedrizzi, L., et al.: Optimization of synthesis parameters of mesoporous silica sol-gel thin films for application on 2024 aluminum alloy substrates. Appl. Surf. Sci. **277**, 201–210 (2013). doi:10.1016/j.apsusc.2013.04.026
57. Rodošek, M., Rauter, A., Slemenik Perše, L., Merl Kek, D., Šurca Vuk, A.: Vibrational and corrosion properties of poly(dimethylsiloxane)-based protective coatings for AA 2024 modified with nanosized polyhedral oligomeric silsesquioxane. Corros. Sci. **85**, 193–203 (2014). doi:10.1016/j.corsci.2014.04.019
58. Kartsonakis, I.A., Athanasopoulou, E., Snihirova, D., Martins, B., Koklioti, M.A., Montemor, M.F., et al.: Multifunctional epoxy coatings combining a mixture of traps and inhibitor loaded nanocontainers for corrosion protection of AA2024-T3. Corros. Sci. **85** 147–159 (2014). doi:10.1016/j.corsci.2014.04.009
59. Balaskas, A.C., Kartsonakis, I.A., Tziveleka, L.-A., Kordas, G.C.: Improvement of anti-corrosive properties of epoxy-coated AA 2024-T3 with TiO_2 nanocontainers loaded with 8-hydroxyquinoline. Prog. Org. Coatings. **74** 418–426 (2012). doi:10.1016/j.porgcoat.2012.01.005
60. Zheludkevich, M.L., Serra, R., Montemor, M.F., Salvado, I.M.M., Ferreira, M.G.S.: Corrosion protective properties of nanostructured sol-gel hybrid coatings to AA2024-T3. Surf. Coatings Technol. **200**, 3084–3094 (2006). doi:10.1016/j.surfcoat.2004.09.007
61. Voevodin, N., Jeffcoate, C., Simon, L., Khobaib, M., Donley, M.: Characterization of pitting corrosion in bare and sol gel coated aluminum 2024-T3 alloy, 29–34 (2001)
62. Voevodin, N.N., Balbyshev, V.N., Donley, M.S.: Investigation of corrosion protection performance of sol-gel coatings on AA2024-T3. Prog. Org. Coatings. **52**, 28–33 (2005). doi:10.1016/j.porgcoat.2004.05.006
63. Lamaka, S.V., Montemor, M.F., Galio, A.F., Zheludkevich, M.L., Trindade, C., Dick, L.F., et al.: Novel hybrid sol-gel coatings for corrosion protection of AZ31B magnesium alloy, Electrochim. Acta. **53** 4773–4783 (2008). doi:10.1016/j.electacta.2008.02.015
64. Poznyak, S.K., Zheludkevich, M.L., Raps, D., Gammel, F., Yasakau, K.A., Ferreira, M.G.S.: Preparation and corrosion protective properties of nanostructured titania-containing hybrid sol-gel coatings on AA2024. Prog. Org. Coatings. **62** 226–235 (2008). doi:10.1016/j.porgcoat.2007.12.004
65. Varma, R., Colreavy, J., Cassidy, J., Mcdonagh, C.: Effect of organic chelates on the performance of hybrid sol-gel coated AA2024-T3 aluminium alloys (2009)
66. Varma, P.C.R., Colreavy, J., Cassidy, J., Oubaha, M., McDonagh, C., Duffy, B.: Corrosion protection of AA 2024-T3 aluminium alloys using 3, 4-diaminobenzoic acid chelated zirconium–silane hybrid sol-gels. Thin Solid Films **518**, 5753–5761 (2010). doi:10.1016/j.tsf.2010.05.088
67. PirhadyTavandashti, N., Sanjabi, S., Shahrabi, T.: Preparation and characterisation of silica/epoxy hybrid nanocomposite coatings containing boehmite nanoparticles for corrosion protection. Corros. Eng., Sci. Technol. **46**, 661–666 (2011). doi:10.1179/147842210X12680457338059
68. Oubaha, M., Varma, P.C.R., Duffy, B., Gasem, Z.M., Hinder, S.J.: Development of a novel hybrid aluminum-based Sol-Gel materials: application to the protection of AA2024-T3 alloys in alkaline environment. Adv. Mater. Phys. Chem. **04**, 75–84 (2014). doi:10.4236/ampc.2014.45010
69. Varma, P.C.R., Periyat, P., Oubaha, M., McDonagh, C., Duffy, B.: Application of niobium enriched ormosils as thermally stable coatings for aerospace aluminium alloys. Surf. Coatings Technol. **205**, 3992–3998 (2011). doi:10.1016/j.surfcoat.2011.02.023
70. Gonzalez, E., Pavez, J., Azocar, I., Zagal, J.H., Zhou, X., Melo, F., et al.: A silanol-based nanocomposite coating for protection of AA-2024 aluminium alloy. Electrochim. Acta. **56**, 7586–7595 (2011). doi:10.1016/j.electacta.2011.06.082

71. Palanivel, V., Zhu, D., van Ooij, W.J.: Nanoparticle-filled silane films as chromate replacements for aluminum alloys. Prog. Org. Coatings. **47**, 384–392 (2003). doi:10.1016/j.porgcoat.2003.08.015
72. Zheludkevich, M.L., Shchukin, D.G., Yasakau, K.A., Ferreira, M.G.S.: Anticorrosion coatings with self-healing effect based on nanocontainers impregnated with corrosion inhibitor. Chem. Mater. **100**, 402–411 (2007)
73. Dalmoro, V., Dos Santos, J.H.Z., Armelin, E., Alemán, C., Azambuja, D.S.: Sol-Gel hybrid films based on organosilane and montmorillonite for corrosion inhibition of AA2024. J. Colloid Interface Sci. **426**, 308–313 (2014). doi:10.1016/j.jcis.2014.04.021
74. Fedel, M., Callone, E., Diré, S., Deflorian, F., Olivier, M.-G., Poelman, M.: Effect of Na-montmorillonite sonication on the protective properties of hybrid silica coatings. Electrochim. Acta. **124**, 90–99 (2014). doi:10.1016/j.electacta.2013.11.006
75. Yasakau, K.A., Kallip, S., Zheludkevich, M.L., Ferreira, M.G.S.: Active corrosion protection of AA2024 by sol-gel coatings with cerium molybdate nanowires, Electrochim. Acta. **112**, 236–246 (2013). doi:10.1016/j.electacta.2013.08.126
76. Wu, K.H., Chao, C.M., Yang, C.J., Chang, T.C.: Synthesis and characterization of polydimethylsiloxane-cured organically modified silicate hybrid coatings. Polym. Degrad. Stab. **91**, 2917–2923 (2006). doi:10.1016/j.polymdegradstab.2006.08.016
77. Kumar, G., Buchheit, R.G.: Development and characterization of corrosion resistant coatings using natural biopolymer Chitosan. ECS Trans. **1**, 101–117 (2006)
78. Ji, W.-G., Hu, J.-M., Liu, L., Zhang, J.-Q., Cao, C.-N.: Improving the corrosion performance of epoxy coatings by chemical modification with silane monomers. Surf. Coatings Technol. **201**, 4789–4795 (2007). doi:10.1016/j.surfcoat.2006.09.100
79. Roussi, E., Tsetsekou, A., Tsiourvas, D., Karantonis, A.: Novel hybrid organo-silicate corrosion resistant coatings based on hyperbranched polymers. Surf. Coatings Technol. **205** 3235–3244 (2011). doi:10.1016/j.surfcoat.2010.11.037
80. Chattopadhyay, D., Zakula, A., Webster, D.: Organic–inorganic hybrid coatings prepared from glycidyl carbamate resin, 3-aminopropyl trimethoxy silane and tetraethoxyorthosilicate. Prog. Org. Coatings. **64**, 128–137 (2009). doi:10.1016/j.porgcoat.2008.09.008
81. Rosero-Navarro, N.C., Pellice, S.A., Castro, Y., Aparicio, M., Durán, A.: Improved corrosion resistance of AA2024 alloys through hybrid organic–inorganic sol-gel coatings produced from sols with controlled polymerisation. Surf. Coatings Technol. **203**, 1897–1903 (2009). doi:10.1016/j.surfcoat.2009.01.019
82. Khelifa, F., Druart, M.-E., Habibi, Y., Bénard, F., Leclère, P., Olivier, M., et al.: Sol-gel incorporation of silica nanofillers for tuning the anti-corrosion protection of acrylate-based coatings. Prog. Org. Coatings. **76**, 900–911 (2013). doi:10.1016/j.porgcoat.2013.02.005
83. Bakhshandeh, E., Jannesari, A., Ranjbar, Z., Sobhani, S., Saeb, M.R.: Anti-corrosion hybrid coatings based on epoxy–silica nano-composites: toward relationship between the morphology and EIS data. Prog. Org. Coatings. **77**, 1169–1183 (2014). doi:10.1016/j.porgcoat.2014.04.005
84. Khramov, A., Balbyshev, V., Voevodin, N., Donley, M.: Nanostructured sol-gel derived conversion coatings based on epoxy- and amino-silanes. Prog. Org. Coatings. **47**, 207–213 (2003). doi:10.1016/S0300-9440(03)00140-1
85. Khramov, A.N., Voevodin, N.N., Balbyshev, V.N., Mantz, R.A.: Sol-gel-derived corrosion-protective coatings with controllable release of incorporated organic corrosion inhibitors. Thin Solid Films. **483**, 191–196 (2005). doi:10.1016/j.tsf.2004.12.021
86. Palanivel, V., Huang, Y., van Ooij, W.J.: Effects of addition of corrosion inhibitors to silane films on the performance of AA2024-T3 in a 0.5 M NaCl solution. Prog. Org. Coatings. **53** 153–168 (2005). doi:10.1016/j.porgcoat.2003.07.008
87. Zheludkevich, M., Serra, R., Montemor, M., Ferreira, M.: Oxide nanoparticle reservoirs for storage and prolonged release of the corrosion inhibitors. Electrochem. Commun. **7**, 836–840 (2005). doi:10.1016/j.elecom.2005.04.039
88. Lamaka, S.V., Zheludkevich, M.L., Yasakau, K.A., Montemor, M.F., Cecílio, P., Ferreira, M.G.S.: TiOx self-assembled networks prepared by templating approach as nanostructured

reservoirs for self-healing anticorrosion pre-treatments. Electrochem. Commun. **8**, 421–428 (2006). doi:10.1016/j.elecom.2005.12.019

89. Lamaka, S.V., Zheludkevich, M.L., Yasakau, K.A., Serra, R., Poznyak, S.K., Ferreira, M.G.S.: Nanoporous titania interlayer as reservoir of corrosion inhibitors for coatings with self-healing ability, Prog. Org. Coatings. **58** 127–135 (2007). doi:10.1016/j.porgcoat.2006.08.029
90. Varma, P.C.R., Duffy, B., Cassidy, J.: Influence of magnesium nitrate on the corrosion performance of sol-gel coated AA2024-T3 aluminium alloy. Surf. Coatings Technol. **204**, 277–284 (2009). doi:10.1016/j.surfcoat.2009.07.024
91. Tavandashti, N.P., Sanjabi, S.: Corrosion study of hybrid sol-gel coatings containing boehmite nanoparticles loaded with cerium nitrate corrosion inhibitor. Prog. Org. Coatings. **69**, 384–391 (2010). doi:10.1016/j.porgcoat.2010.07.012
92. Voevodin, N., Grebasch, N., Soto, W.: Potentiodynamic evaluation of sol-gel coatings with inorganic inhibitors. Surf. Coatings .... 24–28 (2001). http://www.sciencedirect.com/science/article/pii/S0257897201009999. Accessed 25 June 2014
93. Wang, H., Akid, R.: A room temperature cured sol-gel anticorrosion pre-treatment for Al 2024-T3 alloys. Corros. Sci. **49**, 4491–4503 (2007). doi:10.1016/j.corsci.2007.04.015
94. Shi, H., Liu, F., Han, E.: Corrosion behaviour of sol-gel coatings doped with cerium salts on 2024-T3 aluminum alloy. Mater. Chem. Phys. **124**, 291–297 (2010). doi:10.1016/j.matchemphys.2010.06.035
95. Paussa, L., Rosero-Navarro, N.C., Andreatta, F., Castro, Y., Duran, A., Aparicio, M., et al.: Inhibition effect of cerium in hybrid Sol-Gel films on aluminium alloy AA2024. Surf. Interface Anal. **42**, 299–305 (2010). doi:10.1002/sia.3198
96. Balaskas, A.C., Kartsonakis, I. A., Snihirova, D., Montemor, M.F., Kordas, G.: Improving the corrosion protection properties of organically modified silicate–epoxy coatings by incorporation of organic and inorganic inhibitors. Prog. Org. Coatings. **72** 653–662 (2011). doi:10.1016/j.porgcoat.2011.07.008
97. Cambon, J.-B., Esteban, J., Ansart, F., Bonino, J.-P., Turq, V., Santagneli, S.H., et al.: Effect of cerium on structure modifications of a hybrid sol-gel coating, its mechanical properties and anti-corrosion behavior. Mater. Res. Bull. **47**, 3170–3176 (2012). doi:10.1016/j.materresbull.2012.08.034
98. Collazo, A., Hernández, M., Nóvoa, X.R., Pérez, C.: Effect of the addition of thermally activated hydrotalcite on the protective features of sol-gel coatings applied on AA2024 aluminium alloys, Electrochim. Acta. **56**, 7805–7814 (2011). doi:10.1016/j.electacta.2011.03.067
99. Dias, S.A.S., Marques, A., Lamaka, S.V., Simões, A., Diamantino, T.C., Ferreira, M.G.S.: The role of Ce(III)-enriched zeolites on the corrosion protection of AA2024-T3, Electrochim. Acta. **112**, 549–556 (2013). doi:10.1016/j.electacta.2013.09.026

Chapter 8
Graphene-Incorporated Sol-Gel Materials for Energy Applications

Honey John and Maheswary Kavirajan Kavitha

8.1 Introduction

International Energy Outlook [IEO] predicts that global energy consumption is expected to rise by 37–53% from 2013 to 2035 or by an average of 1.4% per year. There is an on-going search for clean and renewable energies due to fast depletion, and dangerous characteristics of fossil and nuclear fuels, which presently form the base for the commercial energies. During the last few decades, the scientists and technologists are in search of highly efficient energy conversion technologies, high capacitive energy storage devices, solar-based renewable energy scavenging techniques, etc. In this scenario, fuel cells, rechargeable batteries, electrochemical double-layer capacitors (EDLCs), hybrid capacitors, and solar cells are the best energy harvesting and sustaining technologies [1]. In all these technologies, the prime factor is their electrode materials which are responsible for the performance, durability, efficiencies, and cost of the devices. The major focus of this chapter is on the energy storage applications such as supercapacitor applications and solar cell applications of graphene-based sol-gel materials.

After the single-layer mechanical exfoliation of graphene from graphite, the monolayer carbon structured material exhibits extraordinary physiochemical properties like high carrier mobility at room temperature [2], good optical transparency [3], excellent thermal conductivity [4], good mechanical properties, and an

H. John (✉)
Department of Polymer Science and Rubber Technology,
Cochin University of Science and Technology, Kochi, India
e-mail: honey@cusat.ac.in

M.K. Kavitha
Department of Physics, Indian Institute of Technology, Chennai, India

S.C. Pillai and S. Hehir (eds.), *Sol-Gel Materials for Energy, Environment and Electronic Applications*, Advances in Sol-Gel Derived Materials and Technologies,
DOI 10.1007/978-3-319-50144-4_8

Fig. 8.1 Schematic representation of graphene

unusual quantum hall effect [5]. Figure 8.1 shows the typical structure of graphene sheet. Graphene also displays excellent optical properties. It can absorb entire UV to near IR [6] and the optical absorption of a single layer graphene is about 2.3% of the incident light [3]. Even though graphene is almost transparent, it is so dense that even the smallest helium atom cannot pass through it.

Graphene is considered to be an appropriate electrode material for electrochemical energy applications because of its extraordinary electronic and optical properties, high carrier mobility, large specific surface area ($\sim$2630 m^2/g), superior mechanical strength (Young's modulus of $\sim$1 TPa), and excellent electrical (0.96×10^6 S/cm) and thermal conductivity ($\sim$5000 W/m K). They can behave like relativistic particles and can be described by Dirac's equation for massless fermions [7]. Graphene-based nanohybrids or nanocomposites with polymer, metal, or metal oxides have unique mechanical, electronic, and electrochemical properties, which can be utilized for energy storage applications. Among various graphene-based supercapacitors, focus has been on the nanocomposites based on graphene/metal oxides such as RuO_2, MnO_2, SnO_2, TiO_2, ZnO, and SiO_2 which provide excellent electrochemical energy storage.

8.2 Graphene-Based Supercapacitors

The growth of electricity-demanding portable electronic devices and vehicles prompted the researchers to develop various energy devices such as batteries, fuel cells, and capacitors. Recently, work has been carried out on electrochemical capacitors also called as supercapacitors or ultracapacitors [8–10]. In electrochemical supercapacitors (ECs), the energy can be stored either by ion adsorption (electric double-layer capacitors (EDLCs) or by fast surface redox reactions (pseudocapacitors). Figure 8.2 shows the working of a typical EDLC. They possess higher energy than dielectric capacitors and higher power densities than rechargeable batteries with a long cycle life of the order of 10^5 and rapid, i.e., within seconds, charging and

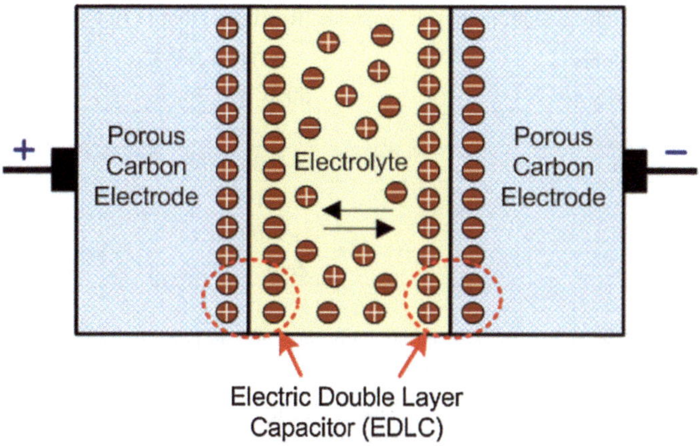

Fig. 8.2 Working of a Electric Double-Layer Capacitor {Originally published in book edited by Alisson V. Brito, ISBN 978-953-7619-68-8, Published: January 1, 2010 under CC BY-NC-SA 3.0 license. (Marcelo Gustavo Molina, (2010). Dynamic modeling and control design of advanced energy storage for power system applications, dynamic modeling, Alisson V. Brito (Ed.), InTech, doi:10.5772/7092} Available from: {doi:10.5772/7092OI})

discharging. So the property of an electrochemical capacitor is intermediate between traditional electrostatic capacitors and batteries. The main advantage of supercapacitors over batteries is their ability to provide energy in a very short time scale

The mechanisms associated with energy storage in supercapacitors are namely electrical double-layer charge discharge and pseudo-process charge discharge. The capacitance of the former is due to the charge accumulation at the electrode/electrolyte interface, and therefore highly dependent on the pore size and accessible surface area of the electrode. The capacitance of the latter comes from the fast reversible faradic transitions of electroactive species of the electrode such as surface functional groups, transition metal oxides, and conducting polymers. Now-a-days, a growing demand exists for hybrid type of supercapacitors to overcome the energy density limitation. Among hybrid supercapacitor materials, metal oxides find first place because of their high pseudo-capacitance originated from their redox reactions. Recent efforts on electrode materials have been devoted to designing advanced material with well-defined physical and chemical properties to achieve an enhanced energy density and great power density. Graphene, one of the allotropes of carbon, consists of a flat monolayer of sp^2 bonded carbon atoms with a 2D honeycomb lattice structure and excellent physical properties. Table 8.1 shows the important physical properties of graphene [11].

Among the various allotropes of carbon, graphene is considered to be an excellent electrode material for supercapacitor application because of its excellent electrical conductivity caused by the electronic properties such as massless Dirac fermion, ambipolar electric field effect, fast transport properties, high carrier mobility, and various physical properties such as high surface area, very high theoretical double-layer capacitance (21 mF/cm^2) [12, 13]. The large surface area

Table 8.1 Unique physical properties of graphene [Reprinted from [37], Copyright © 2013]

Charge carrier mobility	~200,000 cm^2/V s
Thermal conductivity	5000 W/m.K
Transparency	~97.4%
Specific surface area	~2630 m^2/g
Young's modulus	~1 TPa
Tensile strength	~1100 GPa
Bandgap	Zero

Table 8.2 Overview of graphene materials with capacitor properties [Reprinted from Ref. [1], copyright permission from Elsevier © July 2012]

Materials	Power density (kW/kg)	Energy density (W h/kg)	Specific capacitance (F/g)	References
Graphene	–	–	80 F/cm^2	[96]
rGO Reduced graphene	9.8	85.6	250	[18]
Functionalized GO	–	28.5	230	[97]
N-doped graphene	800	48	282	[98]
Carbon black graphene	–	–	175	[16]
CNT-graphene	–	–	385	[99]
Carbon sphere-GNS	15.4	–	198	[100]
PANi-graphene	0.14	37.9	1126	[14]
PEDOT-graphene	0.038	12	304	[101]
MnO_2-exfolitated graphite	110	12.5	315	[15]
RuO_2-graphene	0.05	20.1	570	[17]
NiO-graphene	–	–	400	[102]
Fe_3O_4-graphene	2.4	85	326	[103]

and high electrical conductivity reduce the resistance of transportation of electrons which favors capacitance properties of graphene. Also the composites of graphene with other materials such as activated metal oxide, carbon, and conducting polymers [14–17] will enhance the supercapacitor properties of graphene through synergetic effect. Table 8.2 shows the capacitance properties of graphene composites with various materials [1].

The most attractive feature of graphene-based supercapacitors is their high storage capacity. The upper theoretical limit of graphene-based supercapacitors is 550 F/g which is much higher than the current technology. Liu et al. [18] developed a supercapacitor with graphene-based electrodes with a specific energy density of 85.6 Wh/Kg at room temperature. The main advantage of this supercapacitor is its ultrafast charge–discharge capacity (i.e., within seconds to minute).

The incorporation metal oxides will act like a spacer for improving the surface area and also by the interaction with graphene; it will provide high capacitance through redox reactions. Among the metal oxides, manganese oxide which possesses high capacitive properties is considered to be an excellent environmental friendly electrode material for supercapacitive applications. The theoretical specific capacitance of MnOx is found to be as high as 1370 F/g [19]. Chen et al. [20] synthesized graphene oxide/MnO_2 hybrid by solution process and the supercapacitance performance of the composite is superior compared to the individual materials. They are reporting a specific capacitance of 197.2 F/g in 1 molar sodium hydroxide solution and it shows a high retention capacity of 84% even after 1000 cycles. MnO_2 nanomaterial is electrodeposited on a textile containing solution exfoliated graphene nanosheets for improving the capacitive property of graphene [15]. The hybrid graphene/MnO_2-based textile yields high-capacitance performance with specific capacitance of 315 F/g. They have successfully fabricated asymmetric electrochemical capacitors with graphene/MnO_2-textile as the positive electrode and single-walled carbon nanotubes (SWNTs) textile as the negative electrode in an aqueous Na_2SO_4 electrolyte solution. These devices exhibit promising characteristics with a maximum power density of 110 kW/kg, an energy density of 12.5 Wh/kg, and excellent cycling performance of ~95% capacitance retention over 5000 cycles. The main disadvantage of MnO_2 electrodes is its poor electronic and ionic conductivities which will lead to poor power density and cycling. To overcome these draw backs, Yu et al. [15] prepared graphene/MnO_2 based supercapacitor electrodes using conductive wrapping technique. When 3D wrapping of graphene/MnO_2 was done with carbon nanotubes or conducting polymers as shown in Fig. 8.3, the specific capacitance can be increased by ~20 or ~45%, respectively.

Fig. 8.3 Schematic illustration of conductive wrapping of Graphene and MnO_2; [Reprinted with permission from [104]. Permission from American Chemical Society Copyright © 2011]

8.2.1 Graphene-Based Supercapacitors from Sol-Gel Processing

The combination of graphene and sol-gel processing has been extensively studied. Wu et al. [17] anchored RuO_2 on graphene for high-performance electrochemical capacitor property. The exfoliated graphene sheets prepared by the chemical exfoliation of natural graphite powder are mixed with different loadings of Ru by combining sol-gel process and low-temperature annealing [21]. Typically, $RuCl_3$ is added to GS dispersion in ethanol:water (1:1) mixture at room temperature. Then RuO_2 is precipitated using 1 M NaOH solution. NaOH solution is slowly added till the pH of the solution becomes 7. The loading of RuO_2 is adjusted by adjusting the mass of $RuCl_3$ with respect to graphene sheets. The various analyses of RuO2 anchored graphene nanosheets reveals that the graphene sheets are well separated by fine RuO_2 particles of particle size ranging between 5 and 20 nm. The supercapacitor electrode developed using this composite shows a high specific capacitance of ~ 570 F g^{-1} at around 38.3 wt% Ru loading. It also shows a high energy density of 20.1 Wh kg^{-1} and around 97.9% retention after 1000 cycles and high power density (10,000 W kg^{-1}) [17].

The supercapacitance property of an electrode material can be enhanced by increasing the surface area of the electrode material. A new class of aerogel materials, graphene aerogel, is prepared through sol-gel synthesis route by Wang et al. [19]. It possesses high surface area [793 m^2/g], large pore volume [3 cm^3/g], and large pore size [17 nm] and enhances the supercapacitance performance of MnO_2 electrode material. They synthesized graphene aerogels for MnO_2 in supercapacitor applications by sol-gel chemistry using graphene oxide suspensions mixed with organic carbonaceous precursors like resorcinol and formaldehyde [22, 23]. The graphene aerogel thus prepared has large pore sizes, high specific surface areas, high pore volumes, and high electrical conductivities. Manganese oxide is electrochemically deposited on the graphene aerogel and around 61% of manganese oxide loading shows highest specific capacitance [410 F/g] at 2 mV/s. Another great advantage of this graphene aerogel/MnO_2 based supercapacitor is that the specific capacitance of this electrode material decayed by only 5% even after 50,000 cycles at 1000 mV/s. This may be due to the excellent electrochemical reversibility, cycling stability, columbic efficiency, and negligible iR drops of the electrode material. Figure 8.4 shows the CV and galvanostatic charge/discharge measurements for MnO_2-coated graphite substrate as the working electrode of a supercapacitor.

ZnO is considered to be a versatile material for optoelectronic applications. This is because of the progress in the development of crystalline ZnO with unique optical and electrical properties. Controlling the conductivity is an important challenge in ZnO research, because a small variation in the concentration and type of native defects can significantly affect the electrical and optical properties [24–26]. Combining ZnO with other functional materials like carbon nanotubes, graphene, noble metal nanoparticles, semiconductor quantum dots, etc., are the recent

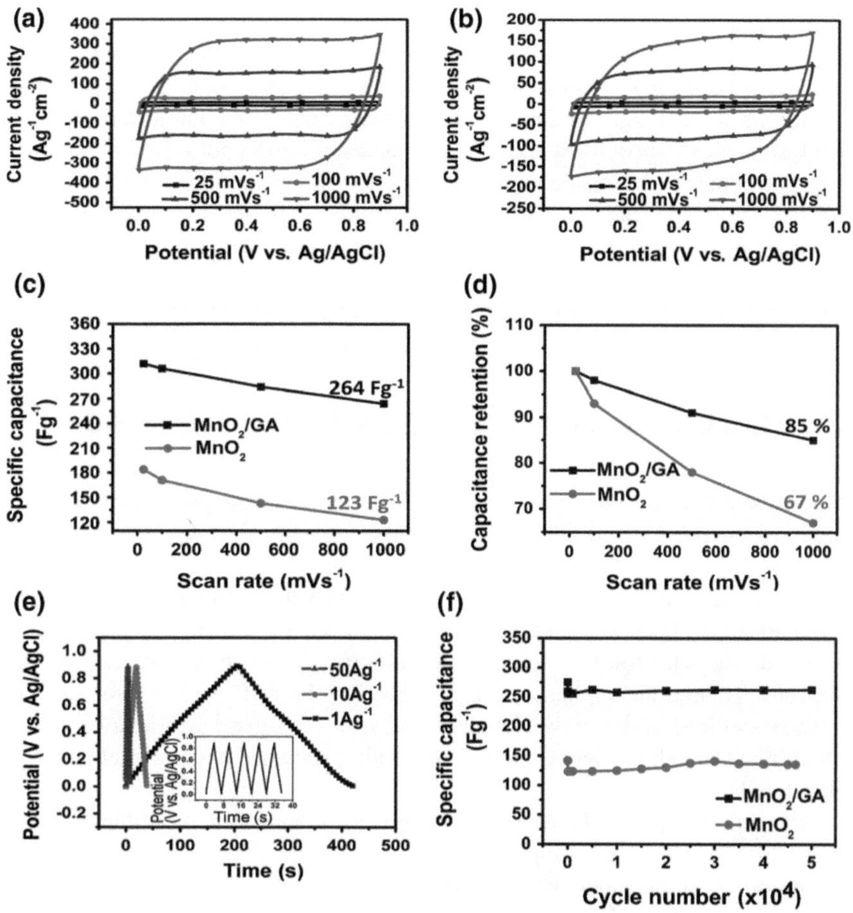

Fig. 8.4 CV loops for **a** MnO_2/GA composite electrode and **b** plain MnO_2 electrode at increasing scan rates from 25 to 1000 mVs^{-1} in 0.5 m Na_2SO_4; **c** Specific capacitance versus scan rate plot for MnO_2/GA composite and plain MnO_2 electrodes; **d** Capacitance retention of MnO_2/GA composite and plain MnO_2 electrodes as functions of scan rate; **e** Galvanostatic charging/discharging curves of MnO2/GA composite electrode at three different current densities, 1, 10, and 50 Ag^{-1}. Inset shows locally enlarged charging/discharging curves for the 50 Ag^{-1} case; **f** Specific capacitance as a function of cycle number for plain MnO_2 and MnO_2/GA composite electrodes at 1000 mVs^{-1} [Reprinted with permission from [19], Copyright © 2014 WILEY-VCH Verlag GmbH & Co. KGaA, Weinheim]

approaches to improve the optical and electrical performance of ZnO [27–35]. ZnO/graphene hybrids are one of the remarkable candidates for optoelectronic applications. ZnO/graphene hybrids can be synthesized by non-covalent interaction [36] or by in situ technique in which ZnO is directly grown on the surface of graphene sheets [28, 30, 37, 38].

Jun Wang et al. synthesized graphene nanosheets/ZnO composites via an in situ crystallization technique for supercapacitor applications. They have measured the power densities of various compositions of graphene/ZnO by varying the percentage of ZnO. In all the compositions, the composite shows excellent recyclability. Graphene/ZnO composite of 1:1 shows a capacitance of 62.2 F/g and power density of 8.1 kW/kg. Similarly, the graphene/ZnO films fabricated by spray pyrolysis also show excellent supercapacitor behavior [39]. Similarly, graphene/cobalt based nanocomposite is another promising electrode material for supercapacitor applications. Chen et al. [20, 40] synthesized graphene-$Co(OH)_2$ nanocomposite electrode material through a facile soft chemical route. They have reported the specific capacitance of the nanocomposite is as high as 972.5 F/g, which is higher than $Co(OH)_2$ [726.1 F g^{-1}] and graphene [137.6 F/g].

8.3 Graphene Nanocomposites for Solar Cell Applications

Photovoltaic devices convert light energy directly into electric power and have great impact on energy demand of mankind with negligible effect on environment. F.H.L. Koppens and coworkers from the Institute of Photonic Sciences (ICFO) have shown that graphene could prove to be far more efficient when it comes to transforming light into energy [41]. The study found that unlike silicon, which generates only one current-driving electron for each photon it absorbs, graphene can produce multiple electrons. Though the application of graphene in solar cells is only theoretical, the potential outlined in the study is remarkable; solar cells made with graphene could offer 60% solar cell efficiency—double the widely regarded maximum efficiency of silicon cells.

Tremendous interest is devoted to fabricate graphene-semiconductor hybrid materials for light harvesting applications [42–44]. Strong and tunable light absorption by semiconductor nanoparticles creates charge carriers. These charge carriers are readily accepted and shuttled through the 2D graphene sheet, due to its high charge mobility [45]. These electron accepting and shuttling mechanism is demonstrated in several graphene hybrids like TiO_2, ZnO, CdSe, and PdS [34, 46, 47]. Willaims et al. [34] demonstrated systematically the excited state interaction between ZnO and graphene oxide. The excited electrons in the conduction band of ZnO will migrate to rGO. This electron transfer mechanism will suppress the recombination of photo-induced electron–hole pairs and improve the conductivity [28]. This efficient charge separation, mobility, and broadband absorption of graphene-semiconductor materials point out the possibility of using this hybrid materials for various light harvesting applications like photovoltaics [48–50], photocatalysis [51, 52], and photodetectors [44, 53].

Sol-gel method is a widely used wet chemical method for the synthesis of graphene-semiconductor hybrid systems for light harvesting applications. This process involves the phase transformation of sol obtained from metal alkoxides or other organometallic precursors in alcoholic or aqueous solutions. For the sol-gel method, graphene oxide (GO) is well suited to combine with alkoxide precursors because of its water solubility and hydrogen bonding. The oxygen functionalities on

GO can form oxo and hydroxo bridges with metal centers. During the hydrolysis of alkoxides, interconnecting metal centers through oxo and hydroxo bridges occurs and these can be probably shared with hydroxo bridges of GO to form a sol followed by condensation. Gel formation will be enhanced by increasing the concentration of GO. These metal alkoxide/GO gels can be converted and reduced to metal oxide/reduced graphene oxide composite by hydrothermal or thermal treatment. For example reduced graphene oxide/TiO_2 microspheres are prepared by the non-hydrolytic sol-gel reaction of tetrabutyl orthotitanate and acetone followed by hydrothermal treatment [51]. Manga et al. [54] attempted sol-gel process to make TiO_2/GO composite using different alkoxides such that titanium alkoxides with longer side chain will slow down the rate of hydrolysis. They suggested an ionic salt titanium (IV) bis(ammonium lactate) dihydroxide (TBA) as titania precursor in order to overcome poor dispersion of GO in alcohol based sol and control the rapid hydrolysis of titanate sol. The precursor solution of titania and GO has viscosity suitable for inkjet printing proposing the ease of fabrication of transparent conductor and photovoltaic devices. Sol-gel method is one of the best methods to obtain uniform distribution of nanoparticles on graphene sheets. ZnO nanoparticles/rGO nanocomposite are prepared by sol-gel method with starch as a polymerising agent [55]. The nanoparticles with approximate size 25 $\pm$ 10 nm are uniformly decorated on graphene sheets as shown in Fig. 8.5.

The photoanode in DSSCs will perform dual functions as scaffold for dye molecules and electron transport from dye to external circuit. Large surface area and fast electron extraction rate will ensure better device performance. Graphene oxide is the ideal derivative of graphene material to make composite with metal oxide because the oxygen functional groups provide tremendous possibility of functionalizing GO surface with nanoparticles. This will tremendously improve the surface area and facilitate the transport of photoexcited electrons. Graphene/TiO_2 composite is one of the extensively studied photoanode in DSSCs since this combination compensates the intrinsic transport limitation of TiO_2 [48, 49, 56–58]. Figure 8.6 shows schematic of the functioning of DSSCs containing rGO/TiO_2 as photoanode [59]. When the device is illuminated, the dye molecules absorb light and get excited. The excited electrons from the dye molecules are injected to the conduction band of TiO_2, which is then transferred to ITO through rGO and compact TiO_2 blocking layer. The rGO in the photoanode will act as a bridge to improve the interfacial electron transfer process. This will increase the life time of holes and reduce the electron–hole recombination. The DSSC fabricated with rGO/TiO_2 photoanode shows an efficiency of 4.24% compared with that of unmodified TiO_2 (2.19%) as a result of fast injection and collection of elctrons by rGO.

Sun et al. [57] prepared TiO_2-graphene composite by heterogeneous coagulation to ensure a tight interfacial interaction between them. This graphene-incorporated device exhibits increased dye loading and longer electron lifetime with power conversion efficiency of 4.3% [57]. Graphene-incorporated nitrogen-doped TiO_2 prepared by sol-gel method is demonstrated as an efficient photoelectrode and the maximum power conversion efficiency obtained is 9.3% [60]. The interaction between graphene and TiO_2 can be physisorption, electrostatic binding or charge

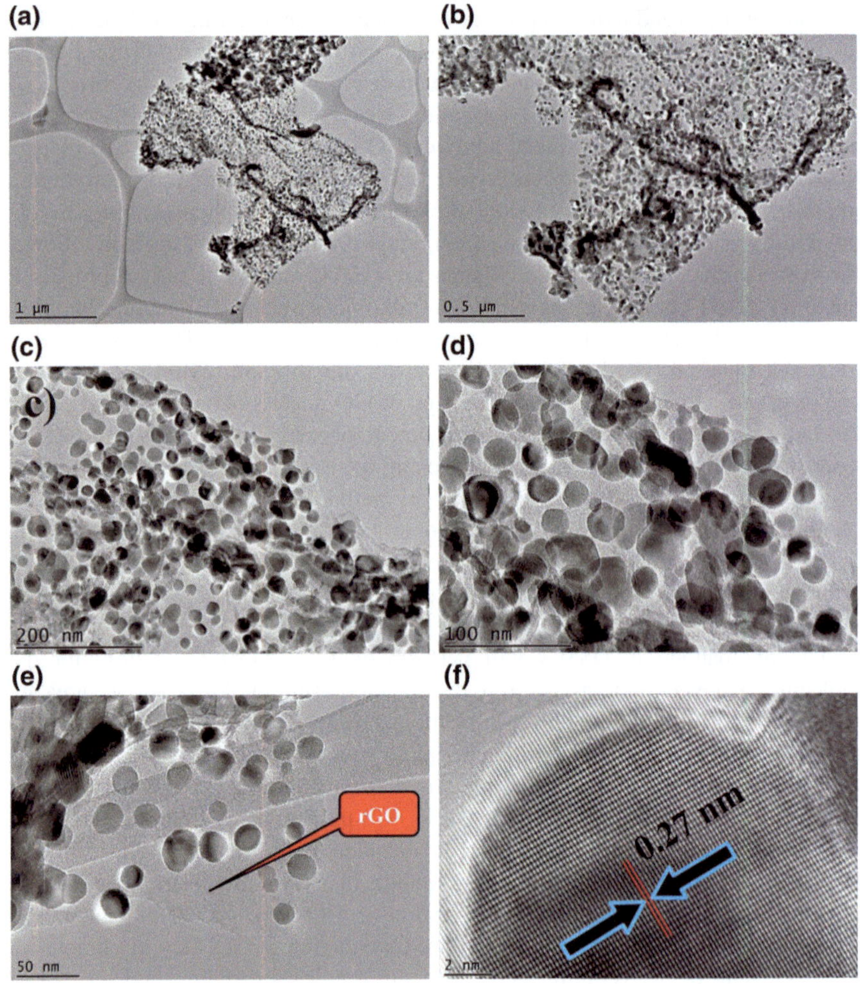

Fig. 8.5 Schematic illustration of the formation mechanism for ZnO nanoparticles/rGO composite via sol-gel method

transfer interactions. The addition of graphene will improve the dye absorptivity and decrease the electron transport time. Cheng et al. developed a two-step wet chemical method to synthesize rGO/TiO_2 composite for application in solar cell. The titanium glycolate precursor is prepared by sol-gel method using tetrabutoxy titanium and ethylene glycol in acetone water mixture. rGO/TiO_2 composite is synthesized by the hydrothermal treatment of titanium glycolate and GO. Figure 8.7 shows the proposed mechanism in which nanoparticles are crystallized from the glycolate precursor through a dissolution and crystallization process during hydrolysis. The titanium glycolate precursor nanoparticles are amorphous in nature. Formation of rGO/TiO_2 involves the hydrolysis of TiO_2 nanoporous

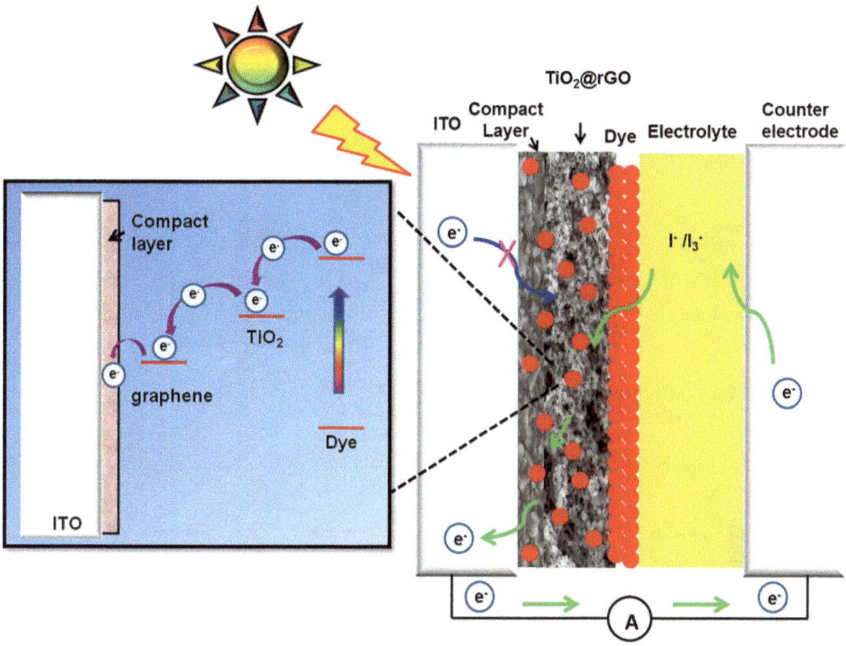

Fig. 8.6 Schematic of working of DSSC containing rGO/TiO_2 as photoanode

structure and reduction of graphene oxide. The prepared rGO/TiO_2 photoanode shows an efficiency of 7.68% compared to pure TiO_2 (4.78%) [61]. Another important parameter which determines the device performance is the amount of graphene added to the composite [39, 62]. Power conversion efficiency and short circuit current decrease with an increase in graphene content due to near-infrared absorption by graphene, which hinders the light absorption by dye molecules and reduces the photogenerated electrons [56]. Literature suggests that moderate amount of graphene enhances the DSSCs performance [61]. The conversion efficiencies of DSSCs with the GT0, GT0.8, GT1.6, GT4, and GT16 rGO/TiO_2 photoanodes are 4.78, 5.34, 7.68, 4.24, and 3.54%, respectively (Fig. 8.8a, b). Compared to the DSSC with the GT0 (pure TiO_2 nanoparticles) photoanode, the conversion efficiency and J_{SC} with GT1.6 are significantly improved by ~60 and ~38%, respectively [61]. These improvements clearly demonstrate that a small content of rGO in the hybrid photoanode is sufficient to enhance the electron injection and electron transfer in the device. The higher content of graphene in the hybrid phototoanodes may increase the resistance of the electrode and decrease the anodic reaction rate of the dye on the TiO_2 matrix, which may hinder the electron transportation upon illumination. This increase in graphene loading increases interfacial resistance and reduces the crystallinity of TiO_2 which in turn affect electron diffusion and transport processes [48, 62].

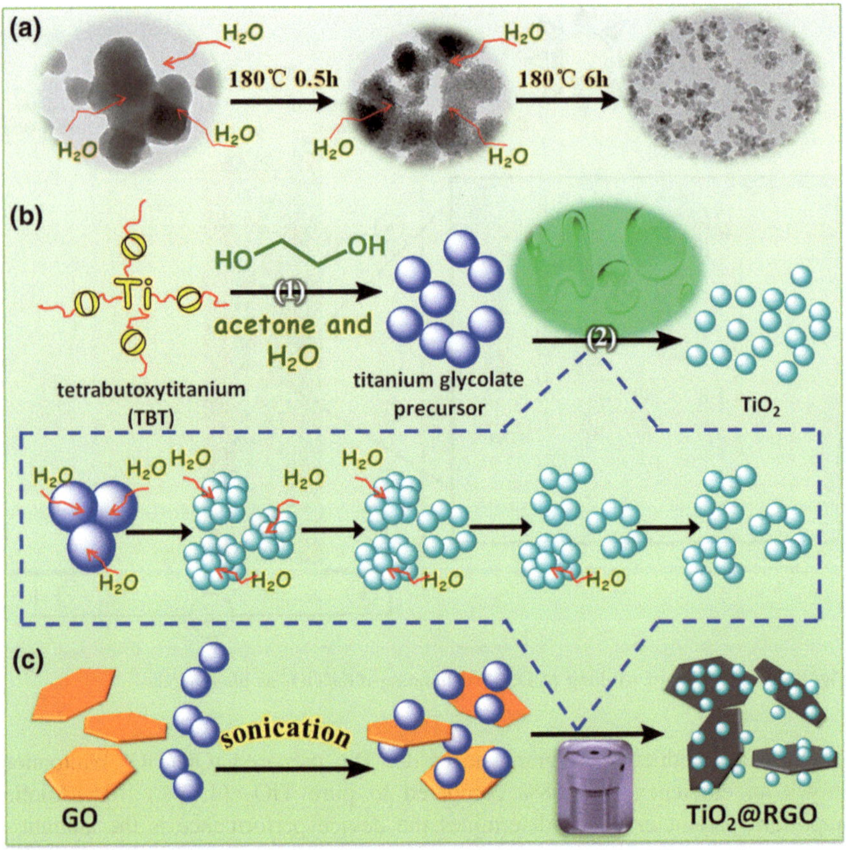

Fig. 8.7 **a** Illustration of the Preparation of Nanoparticulate TiO_2 **b** Possible Formation Process of TiO_2 Nanoparticles from the Titanium Glycolate Precursor: *1* Formation of titanium glycolate precursor by pouring the mixture of tetrabutoxytitanium and ethylene glycol into acetone and deionized water and *2* formation of TiO_2 nanoparticles in water from the titanium glycolate precursor via hydrothermal treatment: from nanoporous to nanoparticulate structures and **c** Possible Formation Process of the rGO/TiO_2 [61]

Sol-gel modified electrospinning is a versatile method to produce TiO_2 nanofibers with high surface to volume ratio. TiO_2/graphene nanocomposite with high surface porosity is prepared by electrospinning a carrier solution followed by calcination [63]. Electrospinning is a simple and cost-effective technique to produce fibrous nanostructures. The polymeric solution of poly vinyl acetate (PVAc) dispersed with graphene and titania precursor is subjected to electrospinning (Fig. 8.9). The prepared polymer/TiO_2/graphene composite fibers are sintered at 450 °C which causes disruptions in fibrous morphology resulting in a single rice-like mesostructure [63]. Polyvinyl pyrrolidone (PVP) and PVAc are the two polymers used to obtain electrospun metal oxide nanofibers. The PVP woven fibers have the advantage of retaining the fibrous mesostructure after sintering. Employing

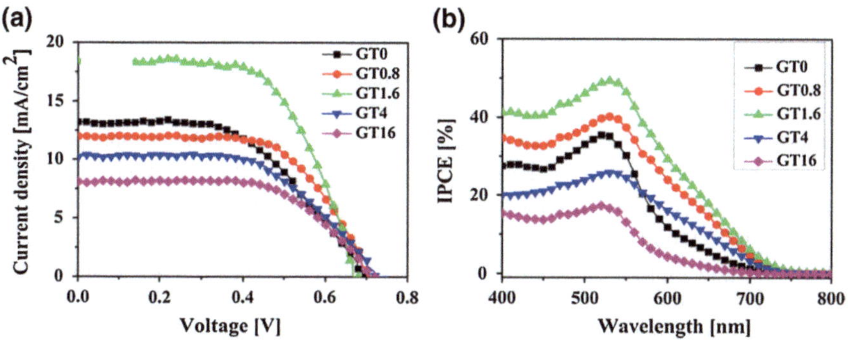

Fig. 8.8 **a** J–V characteristics and **b** IPCE spectra of the five cells with different electrodes prepared with TiO2@RGO with different amounts of reduced graphene oxide

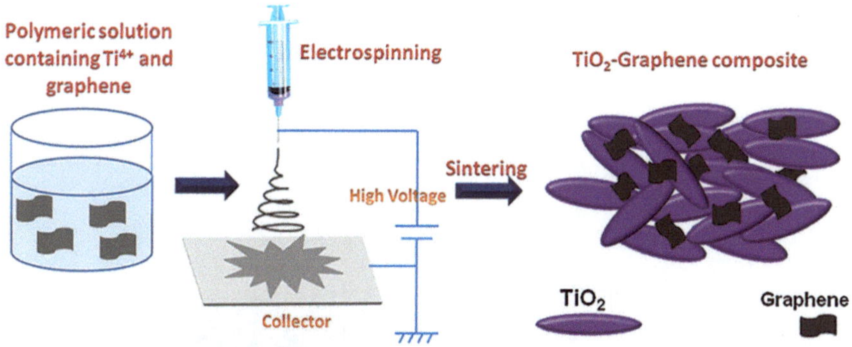

Fig. 8.9 Preparation of rGO/TiO_2 by electrospinning

this nanofibrous composite as the photoanode in DSSC, significantly reduced the recombination of charge carriers with increase in short circuit current with a power conversion efficiency of 7.6% [64]. ZnO is the best alternative to TiO_2 due to the large variety of nanostructures with different morphologies available by the sol-gel method. [65]. Graphene/ZnO nanocomposite photoelectrodes effectively enhance the photocurrent, open-circuit voltage, and the fill factor, thus the overall conversion efficiency could be improved due to the formation of rGO-ZnO Schottky barrier junction, by forming Schottky barrier junction between them [50], resulting in the improvement of electron transport/injection and the improved electrons moving from conduction band (CB) of ZnO to rGO.

8.3.1 Counter Electrodes from Sol-Gel Processing

In dye sensitized solar cells (DSSC) the catalytic counter electrode will be the cathode, its function is to collect electrons from the photoanode and act as an electrocatalyst for the regeneration of the redox couple of the electrolyte, which in turn results in the restoration of the ground state of dye molecules. For a high-performance device, materials with high electrical conductivity and high catalytic activity toward the reduction of electrolyte are chosen for counter electrode fabrication. Platinum (Pt) is one such ideal material, but graphene materials have attracted much attention as a replacement for Pt due to its high conductivity, good catalytic activity toward triiodide redox couple, low cost, and abundance [66]. Cheng et al. applied graphene aerogel as counter electrode, the aerogel is prepared by the organic sol-gel process involving graphene oxide [67]. Highly porous conductive aerogel acts as a template to load Pt to achieve better performance. Ma et al. [68] used a nitrogen-doped graphene aerogel and its composite with single-walled carbon nanotube (SWCNT) as a Pt-free counter electrode. The aerogels are synthesized from GO-SWCNT colloidal solution using a weak reducing agent glutathione. The dispersion of GO-SWCNT becomes sticky and finally transforms into a hydrogel (Fig. 8.10). After freeze-drying, the solvent is removed to produce graphene/SWCNT aerogel. This composite aerogel is nitrogen-doped with glutathione to improve its electro-catalytic activity. Then, the N-doped aerogels are

Fig. 8.10 Photograph of SWCNTs (**a**) and SWCNTs/GO (**b**) showing dispersion properties in aqueous solution, and of hydrogels (**c**), aerogels (**d**), and CEs (**e**)

ground to a powder and the counter electrodes are fabricated via the spin-coating method [68]. Functionalised graphene sheets alone can function as catalytic counter electrode in DSSCs with a performance similar to Pt [69]. By solution method, oxygen functionalities in graphene sheets can be tuned which in turn results in the increase in material's catalytic activity.

A Nickel oxide (NiO)/graphene composite is utilized as photoelectrode in p-type DSSCs. NiO/rGO composite is prepared and electrochemical impedance spectra demonstrate faster hole transport and less charge recombination in NiO/graphene films compared to plain NiO electrode [70]. When NiO/graphene composite films are used as the photocathode, the injected holes can be transferred more rapidly from NiO nanoparticles into the graphene sheets, which serve as a shuttle to efficiently transport the holes to the external circuit, due to favorable energy level alignment and the strong interaction between NiO particle and graphene. Khurana et al. [71] demonstrate an enhanced photovoltaic performance of DSSC which is fabricated using ZnO–graphene nanocomposite as the counter electrode. Synthesis of ZnO–graphene nanocomposite was performed by adding the graphene oxide solution to other part of zinc nitrate solution with stirring at 100 °C. The solution was made alkaline by addition of sodium hydroxide with continuous stirring. A significant increase in the short circuit current was observed in ZnO–graphene-based DSSC as compared to ZnO nanoparticles based DSSC [71].

Solar cells based on perovskite absorbers are emerging as a promising technology which has the potential of high efficiency and low fabrication cost. Currently two types of architectures are adopted for perovskite solar cells such as mesostructured and planar heterojunction [72–74]. Mesostructured perovskite solar cells consist of a FTO coated glass, a hole blocking layer, mesoporous scaffold, perovskite, hole transporting layer, and the metal contact as given in Fig. 8.11a. Hole blocking layer is a compact layer of TiO_2 of thickness 100 nm. The mesoporous layer of few micron thickness acts as a scaffold for perovskite as well as the electron transport layer. Metal oxides like TiO_2, ZnO, alumina, etc., are used as mesoporous scaffold [72, 75–78]. Longer hole diffusion length than electron diffusion length and charge recombination are the critical factors that limit the electron transport in TiO_2-based electron transporting layer. Figure 8.11b shows the cross-sectional SEM image of solar architecture fabricated with rGP/TiO_2. Due to their excellent conductivity and very high charge extraction rate, graphene derivatives combined with TiO_2 can facilitate electron charge transport. Wang et al. achieved one of the highest efficiencies of 15.6% by utilizing graphene as the charge collection layer in a mesostructured perovskite solar cell [79]. They employed a low-temperature processed nanocomposite of graphene and TiO_2 anatase nanoparticles as the electron collection layer. Han et al. [80] employed solution-processed reduced graphene oxide (rGO)/mesoporous TiO_2 nanocomposite as electron transporting layer in mesostructred perovskite solar cell. rGO/TiO_2 composite shows higher photon conversion efficiency in the visible region. rGO reduces the interfacial resistance and improves charge collection efficiency [80]. Thus graphene composite with metal oxides offer tremendous improvement in the device performance by providing a low energy conduction path for the photogenerated charge carriers.

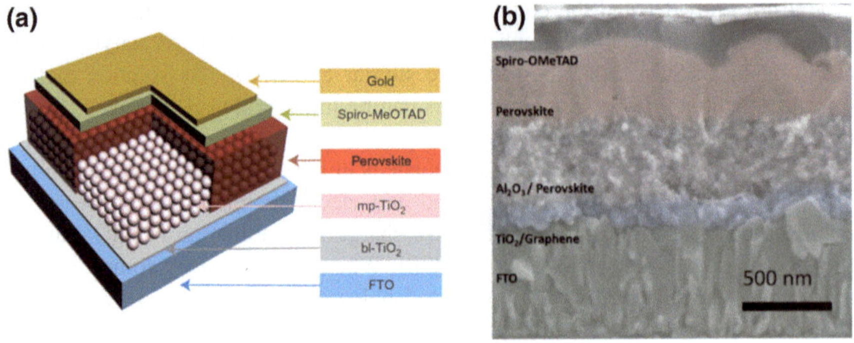

Fig. 8.11 **a** Schematic illustration of perovskite solar cell configuration with dense perovskite covering on the top of mesoporous scaffold. **b** Cross-sectional SEM micrograph of solar cell architecture

8.4 Li-Ion Battery Applications

Lithium-ion batteries are considered to be the most important electrochemical energy storage systems for various applications like mobile communication devices, portable electronic devices, electric/hybrid vehicles, etc., due to their high energy density, high voltage, and light weight [81, 82]. Usually graphite and metal oxides are used as the anode material, which actually determine the battery performance. Even though graphite is a good electrode in Li-ion batteries, the low theoretical capacity limits the use of graphite as electrode material to satisfy the high demand for batteries with higher capacity. The slow Li-ion diffusion, low electron transport, and high charge–discharge rates also limit the uses of Li-ion batteries [83, 84]. A lot of work has focused on the improvement of electrode transport properties of Li-ion batteries by using conducting metal oxides, conductive paints especially TiO_2-based conductive coating and graphene aerogels [85–87].

Donghai Wang et al. synthesized a TiO_2-graphene self-assembled nanocomposite for better lithium-ion insertion. Their studies reveal that it shows double charge capacity compared to that of TiO_2 alone [88]. Bocheng Qiu et al. tried to grow TiO_2 nanocrystals on graphene aerogels through a one-step hydrothermal method. For the successful in situ growth of TiO_2 nanocrystals on graphene aerogels (GAs) surfaces, they have used glucose as the facet-controlling agent and the connection linker [89]. These materials show excellent specific capacity, outstanding lithium storage performance, etc. Studies reveal that there is still lot of scope for the improvement of Li-ion battery performance using sol-gel graphene materials as electrodes.

SnO_2 is important in Li-ion battery research due to its high theoretical reversible Li + storage capacity (782 mAh g^{-1}) compared to that of graphite (372 mAh g^{-1}) [90]. To improve the charge–discharge capacity of the Li-ion battery, Junfei Liang et al. synthesized SnO_2 decorated reduced graphene oxide. They adopted a one-step facile route to synthesize SnO_2-rGO [SG] hybrid. Sn^{4+} salt is taken as the precursor

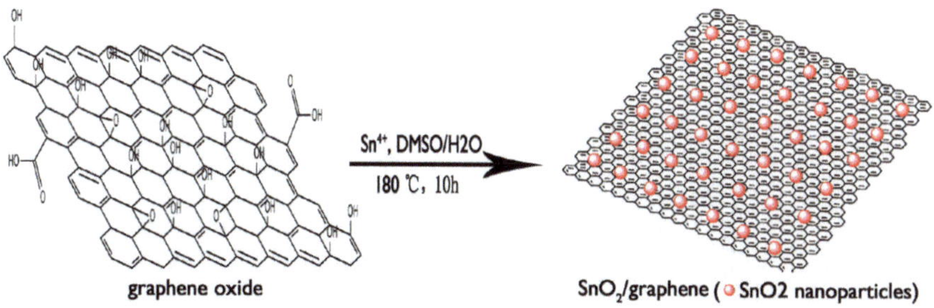

Fig. 8.12 Schematic illustration for the synthesis of SG nanocomposites

for SnO_2 and it is mixed with GO in DMSO/H_2O solvent system. The use of DMSO/H_2O enables the reduction of GO and at the same time it acts as solvent to produce SnO_2 from Sn^{4+} salt. Actually the SnO_2 nanoparticles formed are uniformly decorated on the in situ reduced GO. In the typical procedure, the GO solution is stirred with $SnCl_4{\cdot}5H_2O$ solution, and refluxed at 180 °C for 10 h to thermally reduce GO to rGO and at the same time generation of SnO_2. Figure 8.12 shows the schematic illustration of SnO_2 decorated rGO (SG nanocomposite). The theoretical capacity of SG nanocomposite is 675.4 mAh g^{-1}, which is less than that of SnO_2 (782 mAh g^{-1}) and that of graphene (744 mAh g^{-1}). Even though the theoretical capacity of the nanocomposite is reduced, the charge capacity remains 64.1% of the theoretical value even after 20 cycles [91]. Figure 8.13 shows the charge–discharge CV performance of SG nanocomposite.

SnO_2-graphene aerogels as electrode for Li-ion battery is an important new area. Chen et al. [92] synthesized a 3D graphene aerogels as binder less, freestanding, high-performance electrode material for lithium-ion battery (Fig. 8.14). A porous SnO_2-reduced graphene aerogels with excellent cycling performance, environmental compatibility, and simple manufacturing process are prepared by mixing SnO_2 dispersion and GO dispersion in a mass ratio of 1:1. After 24 h ultrasonication ascorbic acid powder is added into the mixture and stirred for 30 min. Three-dimensional graphene hydrogels are produced after drying the dispersion at 75 °C for 4 h. Then it is treated with deionized water for 24 h, and freeze-dried to get graphene aerogels followed by heating at 550 °C for 30 min to facilitate the reduction of GO. As-prepared graphene aerogels can be used as Li-ion battery electrode materials without any further processing or binder addition [92].

The specific capacitance of the as-prepared SnO_2-graphene aerogel electrode is measured at various charge–discharge cycles at a current density of 100 mA g^{-1}. The first charge–discharge cycle shows a discharging capacity of 1295.2 mAh g^{-1} which is reduced to 866.6 mAh g^{-1} in the second cycle and then it is reduced to 783.5 mAh g^{-1} in the 31st cycle. But it is increased to 867.3 mAh g^{-1} in the 99th cycle with a columbic efficiency higher than 96.5% after the third cycle. Actually, the porous and elastic graphene structure prevented the substantial shrinkage and hence improves the battery performance. Figure 8.15 shows the discharge/charge profile of graphene aerogel.

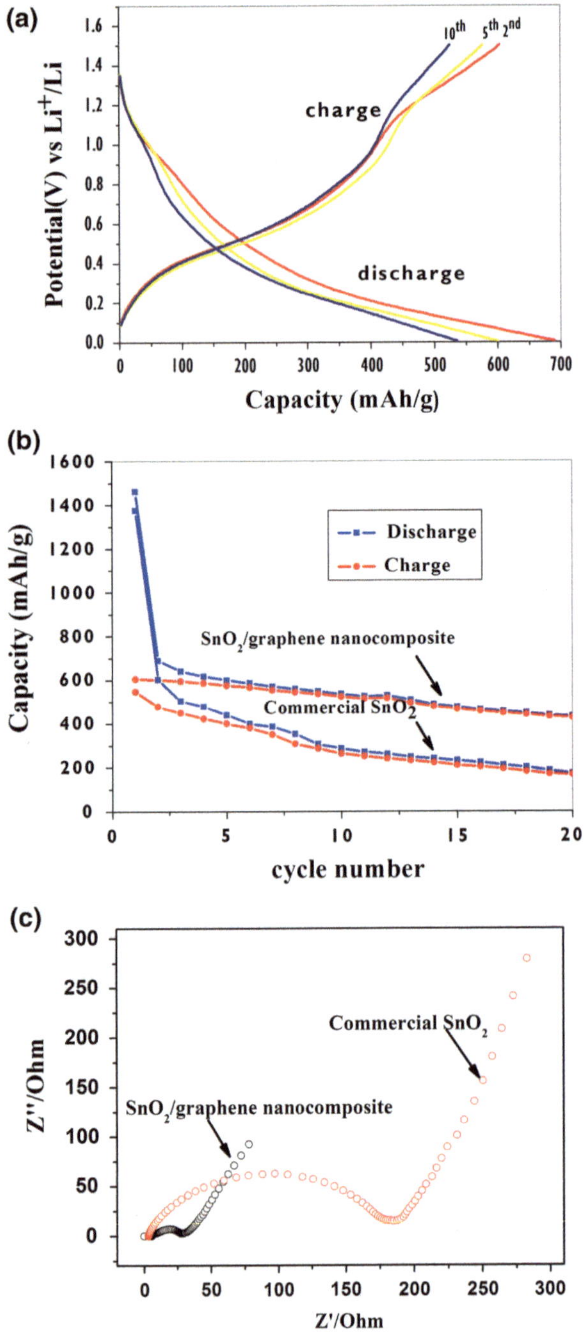

Fig. 8.13 **a** Charge/discharge profile for SG, **b** the cycle performance of SG and commercial SnO_2 nanocrystals, **c** Nyquist plots of SG and commercial SnO_2 nanocrystals obtained by applying a sine wave with amplitude of 5.0 mV over the frequency range from 100 kHz to 0.01 Hz

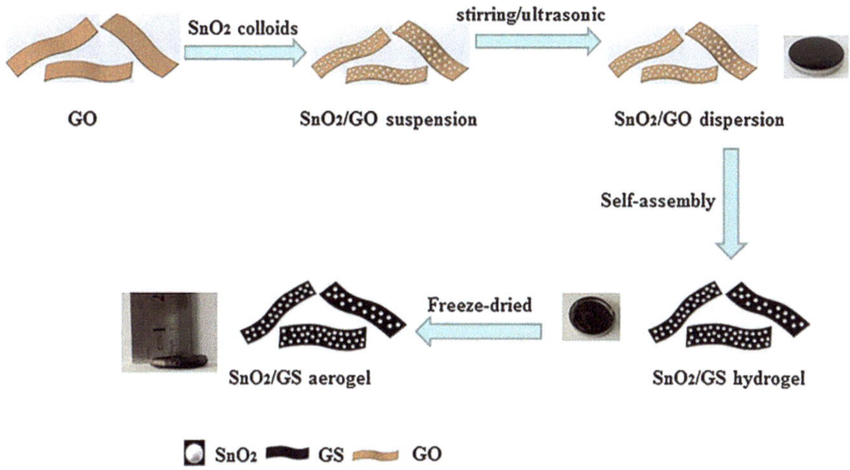

Fig. 8.14 Schematic synthesis procedure of the preparation of the SnO_2-GA and SnO_2-rGA. [Reprinted with permission from Nature publishing group originally published in [92], Copyright © 2016]

Similarly, Gong et al. synthesized 3D SnO_2-graphene aerogels for lithium-ion batteries. A SnO_2 nanoparticle of size 6 nm is encapsulated within graphene aerogel having 3D networks. The composite is prepared by dispersing SnO_2 nanoparticles in GO suspension by ultrasonication followed by heating at 95 °C for 3 h under atmospheric pressure without stirring [93]. Figure 8.16 shows the schematic illustration of synthesis of 3D SnO_2-graphene aerogels. The performance of the battery made with SnO_2-graphene aerogel shows a reversible capacity of 760 mAh g^{-1} even after 50 cycles. It is noteworthy that the structure of the graphene-SnO_2 3D network is preserved even after long cycles.

A 3D porous graphene/Co aerogel is developed [94] for high-performance binder-free lithium-ion battery. The porous graphene/Co aerogel is prepared by the immersion of hydrothermal assembly of GO in cobalt nitrate solution, and followed by freeze-drying and thermal treatment, as shown in Fig. 8.16 [94].

For the evaluation of the electrochemical performance, the Li-ion battery anode is prepared with graphene aerogel, graphene–cobalt aerogel and porous graphene (PG)/Co aerogel. All of the aerogel materials are mixed with PVDF binder for the measurement. The first discharge and charge capacities of the battery measured are 1654.5 and 565.3 mAh g^{-1} for the graphene aerogel, and 2208 and 673.9 mAh g^{-1} for the graphene–cobalt electrode, which is further increased to a discharge capacity of 2455.7 mAh g^{-1} and a reversible charge capacity of 896.9 mAh g^{-1} for porous graphene/cobalt aerogel as shown in Fig. 8.17. The cyclic voltammograms (CV) of the G, G/Co and PG/Co aerogels also follow the same trend.

The studies on the carbon-coated MoO_2–graphene aerogel hybrid as anode material for lithium-ion battery are reported by Zhou et al [95]. The aerogel

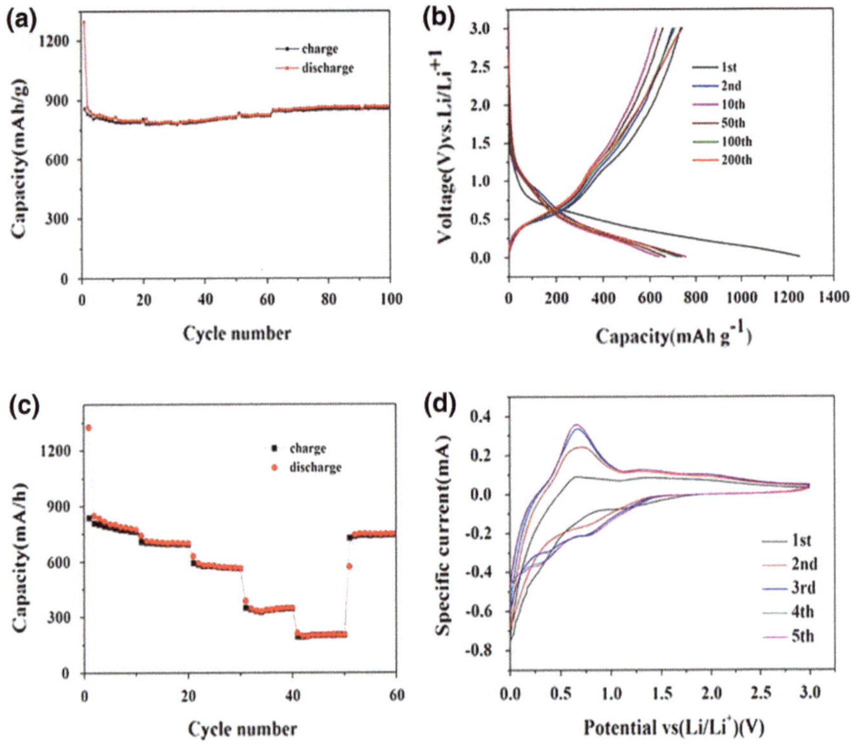

Fig. 8.15 **a** Cycling test for the graphene sheet at the current density of 100 mAh g^{-1}; **b** Discharge/charge profiles (1st, 2nd, 10th, 50th, 100th, 200th cycle) of the as-prepared graphene sheet at the current density of 200 mAh g^{-1} as a LIB electrode. **c** Cycle performance of the graphene sheet as a LIB anode at different current densities. **d** CV curves of the graphene sheet as a LIB electrode. Reprinted with permission from Nature publishing group originally published in [92] Copyrighy © 2016]

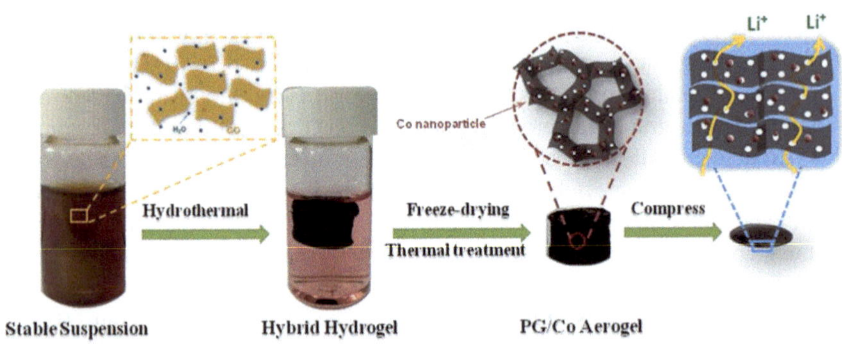

Fig. 8.16 Schematic of the preparation process of the 3D hierarchical porous PG/Co aerogel and the compressed PG/Co electrode

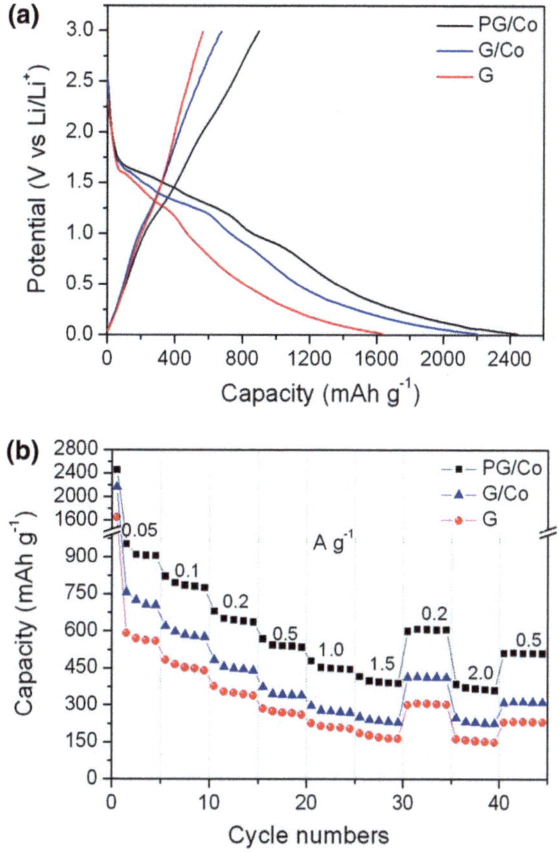

Fig. 8.17 **a** The initial charge/discharge curves of G, G/Co and PG/Co aerogels before compression at a current density of 50 mA g^{-1}. **b** Discharge capacities of the three electrodes between 0.01 and 3.0 V with increasing current densities

electrode is prepared by the ultrasonication of ammonium molybdate tetrahydrate and ascorbic acid was added in GO solution, followed by the addition of glucose. Then the solution is heated at 180 °C for 15 h followed by freeze-drying for 30 h and annealing at 600 °C with a heating rate of 1 °C/min in argon for 5 h. The electrochemical performance of the electrode is excellent and the charge and discharge capacity of the aerogel is 1102.3 and 1515.5 mAh g^{-1} in the first cycle, with a columbic efficiency of 72.7% [95].

8.5 Conclusions

Discovering new ways to produce sustainable energy is a global issue in the current scenario. Many researchers are working on graphene-based materials for developing new energy devices because of its excellent electrical conductivity and high surface area which will favor the development of alternative energy sources. Recently, there is a growing interest in developing graphene aerogels and metaloxide incorporated graphene from sol-gel processes for energy devices such as batteries, supercapacitors, and solar cells. This chapter gives an insight into various graphene-incorporated sol-gel materials for lithium-ion batteries, supercapacitors, and dye sensitized solar cells. It covers the synthesis, important properties, and electrochemical performance of graphene sol-gel materials like graphene aerogels, TiO_2/graphene, ZnO graphene, MnO_2 graphene, SnO_2/graphene, Cobalt oxide/graphene, MoO_2/graphene, etc. All the studies reveal that there is still lot of scope for the improvement of the performance for all these energy devices using sol-gel graphene materials.

References

1. Choi, H.-J., Jung, S.-M., Seo, J.-M., Chang, D.W., Dai, L., Baek, J.-B.: Graphene for energy conversion and storage in fuel cells and supercapacitors. Nano. Energy **1**(4), 534–551 (2012). doi:10.1016/j.nanoen.2012.05.001
2. Morozov, S.V., Novoselov, K.S., Katsnelson, M.I., Schedin, F., Elias, D.C., Jaszczak, J.A., Geim, A.K.: Giant Intrinsic carrier mobilities in graphene and its bilayer. Phys. Rev. Lett. **100**(1), 016602 (2008)
3. Nair, R.R., Blake, P., Grigorenko, A.N., Novoselov, K.S., Booth, T.J., Stauber, T., Peres, N. M.R., Geim, A.K.: Fine structure constant defines visual transparency of graphene. Science **320**(5881), 1308–1309 (2008)
4. Balandin, A.A., Ghosh, S., Bao, W., Calizo, I., Teweldebrhan, D., Miao, F., Lau, C.N.: superior thermal conductivity of single-layer graphene. Nano Lett. **8**(3), 902–907 (2008). doi:10.1021/nl0731872
5. Zhang, Y., Tan, Y.-W., Stormer, H.L., Kim, P.: Experimental observation of the quantum Hall effect and Berry's phase in graphene. Nature **438**(7065), 201–204 (2005). http://www.nature.com/nature/journal/v438/n7065/suppinfo/nature04235_S1.html
6. Pirruccio, G., Martín Moreno, L., Lozano, G., Gómez Rivas, J.: Coherent and broadband enhanced optical absorption in graphene. ACS Nano **7**(6), 4810–4817 (2013). doi:10.1021/nn4012253
7. Castro Neto, A.H., Guinea, F., Peres, N.M.R., Novoselov, K.S., Geim, A.K.: The electronic properties of graphene. Rev. Mod. Phys. **81**(1), 109–162 (2009)
8. Chen, D., Tang, L., Li, J.: Graphene-based materials in electrochemistry. Chem. Soc. Rev. **39**(8), 3157–3180 (2010). doi:10.1039/b923596e
9. Pumera, M.: Graphene-based nanomaterials and their electrochemistry. Chem. Soc. Rev. **39** (11), 4146–4157 (2010). doi:10.1039/c002690p
10. Zhang, L.L., Zhao, X.S.: Carbon-based materials as supercapacitor electrodes. Chem. Soc. Rev. **38**(9), 2520–2531 (2009). doi:10.1039/b813846j
11. Lee, S.K., Kim, H., Shim, B.S.: Graphene: an emerging material for biological tissue engineering. Carbon lett. Korean Carbon Soc. **14**(2), (2013). doi:10.5714/CL.2013.14.2.063 (10.5714/CL.2013.5714.5712.5063)
12. Geim, A.K., Novoselov, K.S.: The rise of graphene. Nat. Mater. **6**(3), 183–191 (2007)
13. Xia, J., Chen, F., Li, J., Tao, N.: Measurement of the quantum capacitance of graphene. Nat. Nano **4**(8), 505–509 (2009). http://www.nature.com/nnano/journal/v4/n8/suppinfo/nnano.2009.177_S1.html
14. Wang, H., Hao, Q., Yang, X., Lu, L., Wang, X.: A nanostructured graphene/polyaniline hybrid material for supercapacitors. Nanoscale **2**(10), 2164–2170 (2010). doi:10.1039/c0nr00224k
15. Yu, G., Hu, L., Vosgueritchian, M., Wang, H., Xie, X., McDonough, J.R., Cui, X., Cui, Y., Bao, Z.: Solution-processed graphene/MnO2 nanostructured textiles for high-performance electrochemical capacitors. Nano Lett. **11**(7), 2905–2911 (2011). doi:10.1021/nl2013828

16. Yan, J., Fan, Z., Wei, T., Qian, W., Zhang, M., Wei, F.: Fast and reversible surface redox reaction of graphene–MnO2 composites as supercapacitor electrodes. Carbon **48**(13), 3825–3833 (2010). doi:10.1016/j.carbon.2010.06.047
17. Wu, Z.-S., Wang, D.-W., Ren, W., Zhao, J., Zhou, G., Li, F., Cheng, H.-M.: Anchoring hydrous RuO2 on graphene sheets for high-performance electrochemical capacitors. Adv. Funct. Mater. **20**(20), 3595–3602 (2010). doi:10.1002/adfm.201001054
18. Liu, C., Yu, Z., Neff, D., Zhamu, A., Jang, B.Z.: Graphene-based supercapacitor with an ultrahigh energy density. Nano Lett. **10**(12), 4863–4868 (2010). doi:10.1021/nl102661q
19. Wang, C.-C., Chen, H.-C., Lu, S.-Y.: Manganese oxide/graphene aerogel composites as an outstanding supercapacitor electrode material. Chem. A Eur. J. **20**(2), 517–523 (2014). doi:10.1002/chem.201303483
20. Chen, S., Zhu, J., Wu, X., Han, Q., Wang, X.: Graphene oxide—MnO_2 nanocomposites for supercapacitors. ACS Nano **4**(5), 2822–2830 (2010)
21. Hu, C.-C., Chen, W.-C., Chang, K.-H.: How to achieve maximum utilization of hydrous ruthenium oxide for supercapacitors. J. Electrochem. Soc. **151**(2), A281–A290 (2004). doi:10.1149/1.1639020
22. Pekala, R.W.: Organic aerogels from the polycondensation of resorcinol with formaldehyde. J. Mater. Sci. **24**(9), 3221–3227 (1989). doi:10.1007/bf01139044
23. Lin, C., Ritter, J.A.: Effect of synthesis pH on the structure of carbon xerogels. Carbon **35**(9), 1271–1278 (1997). doi:10.1016/S0008-6223(97)00069-9
24. Schmidt-Mende, L., MacManus-Driscoll, J.L.: ZnO—nanostructures, defects, and devices. Mater. Today **10**(5), 40–48 (2007). doi:10.1016/S1369-7021(07)70078-0
25. Xu, S., Cheng, C., Guo, W., He, Y., Huang, R., Du, S., Wang, N.: Tuning the optical and electrical properties of hydrothermally grown ZnO nanowires by sealed post annealing treatment. Solid State Commun. **160**, 41–46 (2013). doi:10.1016/j.ssc.2013.02.009
26. Look, D.C., Hemsky, J.W., Sizelove, J.R.: Residual native shallow donor in ZnO. Phys. Rev. Lett. **82**(12), 2552–2555 (1999)
27. Gu, H., Yang, Y., Tian, J., Shi, G.: Photochemical synthesis of noble metal (Ag, Pd, Au, Pt) on graphene/ZnO multihybrid nanoarchitectures as electrocatalysis for H2O2 reduction. ACS Appl. Mater. Interfaces **5**(14), 6762–6768 (2013). doi:10.1021/am401738k
28. Lee, J.M., Pyun, Y.B., Yi, J., Choung, J.W., Park, W.I.: ZnO nanorod—graphene hybrid architectures for multifunctional conductors. J. Phy. Chem. C **113**(44), 19134–19138 (2009). doi:10.1021/jp9078713
29. Nguyen, T.L., Michael, M., Mulvaney, P.: Synthesis of highly crystalline CdSe@ZnO nanocrystals via monolayer-by-monolayer epitaxial shell deposition. Chem. Mater. **26**(14), 4274–4279 (2014). doi:10.1021/cm501858s
30. Son, D.I., Kwon, B.W., Park, D.H., Seo, W.-S., Yi, Y., Angadi, B., Lee, C.L., Choi, W.K.: Emissive ZnO–graphene quantum dots for white-light-emitting diodes. Nat. Nano **7**, 465–471 (2012)
31. Tak, Y., Hong, S.J., Lee, J.S., Yong, K.: Fabrication of ZnO/CdS core/shell nanowire arrays for efficient solar energy conversion. J. Mater. Chem. **19**(33), 5945–5951 (2009). doi:10.1039/b904993b
32. Tian, J., Liu, S., Li, H., Wang, L., Zhang, Y., Luo, Y., Asiri, A.M., Al-Youbi, A.O., Sun, X.: One-step preparation of ZnO nanoparticle-decorated reduced graphene oxide composites and their application to photocurrent generation. RSC Adv. **2**(4), 1318–1321 (2012)
33. Wang, C.-Y., Adhikari, S.: ZnO-CNT composite nanotubes as nanoresonators. Phys. Lett. A **375**(22), 2171–2175 (2011). doi:10.1016/j.physleta.2011.04.031
34. Williams, G., Kamat, P.V.: Graphene—semiconductor nanocomposites: excited-state interactions between ZnO nanoparticles and graphene oxide. Langmuir **25**(24), 13869–13873 (2009). doi:10.1021/la900905h
35. Zhu, Y., Elim, H.I., Foo, Y.L., Yu, T., Liu, Y., Ji, W., Lee, J.Y., Shen, Z., Wee, A.T.S., Thong, J.T.L., Sow, C.H.: Multiwalled carbon nanotubes beaded with ZnO nanoparticles for ultrafast nonlinear optical switching. Adv. Mater. **18**(5), 587–592 (2006). doi:10.1002/adma.200501918

36. Zhang, C., Zhang, J., Su, Y., Xu, M., Yang, Z., Zhang, Y.: ZnO nanowire/reduced graphene oxide nanocomposites for significantly enhanced photocatalytic degradation of rhodamine 6G. Physica E **56**, 251–255 (2014). doi:10.1016/j.physe.2013.09.020
37. Li, B., Cao, H.: ZnO@graphene composite with enhanced performance for the removal of dye from water. J. Mater. Chem. **21**(10), 3346–3349 (2011)
38. Singh, G., Choudhary, A., Haranath, D., Joshi, A.G., Singh, N., Singh, S., Pasricha, R.: ZnO decorated luminescent graphene as a potential gas sensor at room temperature. Carbon **50**(2), 385–394 (2012). doi:10.1016/j.carbon.2011.08.050
39. Wang, J., Gao, Z., Li, Z., Jiang, Z.: Green synthesis of grapheme nanosheets/ZnO composites and electrochemical properties. J. Solid State Chem. **184**(6), 1421–1427 (2011)
40. Chen, S., Zhu, J., Wang, X.: One-step synthesis of graphene—cobalt hydroxide nanocomposites and their electrochemical properties. J. Phys. Chem. C **114**(27), 11829–11834 (2010)
41. Tielrooij, K.J., Song, J.C.W., Jensen, S.A., Centeno, A., Pesquera, A., Zurutuza Elorza, A., Bonn, M., Levitov, L.S., Koppens, F.H.L.: Photoexcitation cascade and multiple hot-carrier generation in graphene. Nat. Phys. **9**(4), 248–252 (2013). doi:http://www.nature.com/nphys/journal/v9/n4/abs/nphys2564.html#supplementary-information
42. Pan, X., Yang, M.-Q., Xu, Y.-J.: Morphology control, defect engineering and photoactivity tuning of ZnO crystals by graphene oxide—a unique 2D macromolecular surfactant. Phys. Chem. Chem. Phys. **16**(12), 5589–5599 (2014). doi:10.1039/c3cp55038a
43. Peining, Z., Nair, A.S., Shengjie, P., Shengyuan, Y., Ramakrishna, S.: Facile fabrication of TiO2–graphene composite with enhanced photovoltaic and photocatalytic properties by electrospinning. ACS Appl. Mater. Interfaces **4**(2), 581–585 (2012). doi:10.1021/am201448p
44. Shao, D., Yu, M., Sun, H., Hu, T., lian, J., Sawyer, S.: High responsivity, fast ultraviolet photodetector fabricated from ZnO nanoparticle-graphene core-shell structures. Nanoscale **5** (9), 3664–3667 (2013). doi:10.1039/c3nr00369h
45. Lightcap, I.V., Kosel, T.H., Kamat, P.V.: Anchoring semiconductor and metal nanoparticles on a two-dimensional catalyst mat. storing and shuttling electrons with reduced graphene oxide. Nano Lett. **10**(2), 577–583 (2010). doi:10.1021/nl9035109
46. Jiang, B., Tian, C., Pan, Q., Jiang, Z., Wang, J.-Q., Yan, W., Fu, H.: Enhanced photocatalytic activity and electron transfer mechanisms of graphene/TiO2 with exposed 001 facets. J. Phy. Chem. C **115**(48), 23718–23725 (2011). doi:10.1021/jp207624x
47. Krishnamurthy, S., Kamat, P.V.: CdSe–graphene oxide light-harvesting assembly: size-dependent electron transfer and light energy conversion aspects. ChemPhysChem **15** (10), 2129–2135 (2014). doi:10.1002/cphc.201301189
48. Chen, L., Zhou, Y., Tu, W., Li, Z., Bao, C., Dai, H., Yu, T., Liu, J., Zou, Z.: Enhanced photovoltaic performance of a dye-sensitized solar cell using graphene-TiO2 photoanode prepared by a novel in situ simultaneous reduction-hydrolysis technique. Nanoscale **5**(8), 3481–3485 (2013). doi:10.1039/c3nr34059g
49. Mehmood, U., Ahmed, S., Hussein, I.A., Harrabi, K.: Improving the efficiency of dye sensitized solar cells by TiO2-graphene nanocomposite photoanode. Photonics Nanostruct. Fundam. Appl. **16**, 34–42 (2015). doi:10.1016/j.photonics.2015.08.003
50. Song, J.-L., Wang, X.: Effect of incorporation of reduced graphene oxide on ZnO-based dye-sensitized solar cells. Physica E **81**, 14–18 (2016). doi:10.1016/j.physe.2016.02.005
51. Cheng, P., Yang, Z., Wang, H., Cheng, W., Chen, M., Shangguan, W., Ding, G.: TiO2–graphene nanocomposites for photocatalytic hydrogen production from splitting water. Int. J. Hydrogen Energy **37**(3), 2224–2230 (2012). doi:10.1016/j.ijhydene.2011.11.004
52. Zhou, K., Zhu, Y., Yang, X., Jiang, X., Li, C.: Preparation of graphene-TiO2 composites with enhanced photocatalytic activity. New J. Chem. **35**(2), 353–359 (2011). doi:10.1039/c0nj00623h
53. Zhan, Z., Zheng, L., Pan, Y., Sun, G., Li, L.: Self-powered, visible-light photodetector based on thermally reduced graphene oxide-ZnO (rGO-ZnO) hybrid nanostructure. J. Mater. Chem. **22**(6), 2589–2595 (2012)

54. Manga, K.K., Wang, S., Jaiswal, M., Bao, Q., Loh, K.P.: High-gain graphene-titanium oxide photoconductor made from inkjet printable ionic solution. Adv. Mater. **22**(46), 5265–5270 (2010). doi:10.1002/adma.201002939
55. Azarang, M., Shuhaimi, A., Yousefi, R., Jahromi, S.P.: One-pot sol-gel synthesis of reduced graphene oxide uniformly decorated zinc oxide nanoparticles in starch environment for highly efficient photodegradation of Methylene Blue. RSC Adv. **5**(28), 21888–21896 (2015). doi:10.1039/c4ra16767h
56. Kusumawati, Y., Martoprawiro, M.A., Pauporté, T.: Effects of Graphene in Graphene/TiO2 Composite Films Applied to Solar Cell Photoelectrode. J. Phys. Chem. C **118**(19), 9974–9981 (2014). doi:10.1021/jp502385p
57. Sun, S., Gao, L., Liu, Y.: Enhanced dye-sensitized solar cell using graphene-TiO2 photoanode prepared by heterogeneous coagulation. Appl. Phys. Lett. **96**(8), 083113 (2010). doi:10.1063/1.3318466
58. Yang, N., Zhai, J., Wang, D., Chen, Y., Jiang, L.: Two-dimensional graphene bridges enhanced photoinduced charge transport in dye-sensitized solar cells. ACS Nano **4**(2), 887–894 (2010). doi:10.1021/nn901660v
59. Lim, S.P., Pandikumar, A., Huang, N.M., Lim, H.N.: Reduced graphene oxide–titania nanocomposite-modified photoanode for efficient dye-sensitized solar cells. Int. J. Energy Res. **39**(6), 812–824 (2015). doi:10.1002/er.3307
60. Kim, S.-B., Park, J.-Y., Kim, C.-S., Okuyama, K., Lee, S.-E., Jang, H.-D., Kim, T.-O.: Effects of graphene in dye-sensitized solar cells based on nitrogen-doped TiO2 composite. J. Phys. Chem. C **119**(29), 16552–16559 (2015). doi:10.1021/acs.jpcc.5b02309
61. Cheng, G., Akhtar, M.S., Yang, O.B., Stadler, F.J.: Novel preparation of anatase TiO2@reduced graphene oxide hybrids for high-performance dye-sensitized solar cells. ACS Appl. Mater. Interfaces **5**(14), 6635–6642 (2013). doi:10.1021/am4013374
62. Fan, J., Liu, S., Yu, J.: Enhanced photovoltaic performance of dye-sensitized solar cells based on TiO2 nanosheets/graphene composite films. J. Mater. Chem. **22**(33), 17027–17036 (2012). doi:10.1039/c2jm33104g
63. Zhu, P., Nair, A.S., Shengjie, P., Shengyuan, Y., Ramakrishna, S.: Facile fabrication of TiO2–graphene composite with enhanced photovoltaic and photocatalytic properties by electrospinning. ACS Appl. Mater. Interfaces **4**(2), 581–585 (2012). doi:10.1021/am201448p
64. Anish Madhavan, A., Kalluri, S., Chacko, D.K., Arun, T.A., Nagarajan, S., Subramanian, K. R.V., Sreekumaran Nair, A., Nair, S.V., Balakrishnan, A.: Electrical and optical properties of electrospun TiO2-graphene composite nanofibers and its application as DSSC photo-anodes. RSC Adv. **2**(33), 13032–13037 (2012). doi:10.1039/c2ra22091a
65. Anta, J.A., Guillén, E., Tena-Zaera, R.: ZnO-based dye-sensitized solar cells. J. Phys. Chem. C **116**(21), 11413–11425 (2012). doi:10.1021/jp3010025
66. Wang, H., Hu, Y.H.: Graphene as a counter electrode material for dye-sensitized solar cells. Energy Environ. Sci. **5**(8), 8182–8188 (2012). doi:10.1039/c2ee21905k
67. Cheng, W.-Y., Wang, C.-C., Lu, S.-Y.: Graphene aerogels as a highly efficient counter electrode material for dye-sensitized solar cells. Carbon **54**, 291–299 (2013). doi:10.1016/j.carbon.2012.11.041
68. Ma, J., Li, C., Yu, F., Chen, J.: 3 D Single-walled carbon nanotube/graphene aerogels as Pt-free transparent counter electrodes for high efficiency dye-sensitized solar cells. ChemSusChem **7**(12), 3304–3311 (2014). doi:10.1002/cssc.201403062
69. Roy-Mayhew, J.D., Bozym, D.J., Punckt, C., Aksay, I.A.: Functionalized graphene as a catalytic counter electrode in dye-sensitized solar cells. ACS Nano **4**(10), 6203–6211 (2010). doi:10.1021/nn1016428
70. Yang, H., Guai, G.H., Guo, C., Song, Q., Jiang, S.P., Wang, Y., Zhang, W., Li, C.M.: NiO/graphene composite for enhanced charge separation and collection in p-type dye sensitized solar cell. J. Phys. Chem. C **115**(24), 12209–12215 (2011). doi:10.1021/jp201178a

71. Khurana, G., Sahoo, S., Barik, S.K., Katiyar, R.S.: Improved photovoltaic performance of dye sensitized solar cell using ZnO–graphene nano-composites. J. Alloys Compd. **578**, 257–260 (2013). doi:10.1016/j.jallcom.2013.05.080
72. Listorti, A., Juarez-Perez, E.J., Frontera, C., Roiati, V., Garcia-Andrade, L., Colella, S., Rizzo, A., Ortiz, P., Mora-Sero, I.: Effect of mesostructured layer upon crystalline properties and device performance on perovskite solar cells. J. Phys. Chem. Lett. **6**(9), 1628–1637 (2015). doi:10.1021/acs.jpclett.5b00483
73. Zhang, L.Q., Zhang, X.W., Yin, Z.G., Jiang, Q., Liu, X., Meng, J.H., Zhao, Y.J., Wang, H. L.: Highly efficient and stable planar heterojunction perovskite solar cells via a low temperature solution process. J. Mater. Chem. A **3**(23), 12133–12138 (2015). doi:10.1039/c5ta01898f
74. Zhao, Y., Nardes, A.M., Zhu, K.: Solid-state mesostructured perovskite CH3NH3PbI3 solar cells: charge transport, recombination, and diffusion length. J. Phys. Chem. Lett. **5**(3), 490–494 (2014). doi:10.1021/jz500003v
75. Ball, J.M., Lee, M.M., Hey, A., Snaith, H.J.: Low-temperature processed meso-superstructured to thin-film perovskite solar cells. Energy Environ. Sci. **6**(6), 1739–1743 (2013). doi:10.1039/c3ee40810h
76. Lee, M.M., Teuscher, J., Miyasaka, T., Murakami, T.N., Snaith, H.J.: Efficient hybrid solar cells based on meso-superstructured organometal halide perovskites. Science **338**(6107), 643–647 (2012)
77. Matas Adams, A., Marin-Beloqui, J.M., Stoica, G., Palomares, E.: The influence of the mesoporous TiO2 scaffold on the performance of methyl ammonium lead iodide (MAPI) perovskite solar cells: charge injection, charge recombination and solar cell efficiency relationship. J. Mater. Chem. A **3**(44), 22154–22161 (2015). doi:10.1039/c5ta06041a
78. Son, D.-Y., Im, J.-H., Kim, H.-S., Park, N.-G.: 11% Efficient perovskite solar cell based on ZnO nanorods: an effective charge collection system. J. Phys. Chem. C **118**(30), 16567–16573 (2014). doi:10.1021/jp412407j
79. Wang, J.T.-W., Ball, J.M., Barea, E.M., Abate, A., Alexander-Webber, J.A., Huang, J., Saliba, M., Mora-Sero, I., Bisquert, J., Snaith, H.J., Nicholas, R.J.: Low-temperature processed electron collection layers of graphene/TiO2 nanocomposites in thin film perovskite solar cells. Nano. Lett. **14**(2), 724–730 (2014b). doi:10.1021/nl403997a
80. Han, G.S., Song, Y.H., Jin, Y.U., Lee, J.-W., Park, N.-G., Kang, B.K., Lee, J.-K., Cho, I.S., Yoon, D.H., Jung, H.S.: reduced graphene oxide/mesoporous TiO2 nanocomposite based perovskite solar cells. ACS Appl.Mater. Interfaces **7**(42), 23521–23526 (2015). doi:10.1021/acsami.5b06171
81. Kang, K.S., Meng, Y.S., Breger, J., Grey, C.P., Ceder, G.: Electrodes with high power and high capacity for rechargeable lithium batteries. Science **311**, 977–980 (2006)
82. Chan, C.K., Peng, H.L., Liu, G., Mcllwrath, K., Zhang, X.F., Huggins, R.A., Nat, C.Y.: High-performance lithium battery anodes using silicon nanowires. Nanotechnology **3**, 31–35 (2008)
83. Maier, J.: Nanoionics: ion transport and electrochemical storage in confined systems. Nat. Mater. **4**, 805–815 (2005)
84. Tarascon, J.M., Armand, M.: Issues and challenges facing rechargeable lithium batteries. Nature, **414,** 359–367 (2001)
85. Zhou, Y.K., Cao, L., Zhang, F.B., He, B.L., Li, H.L.: Lithium insertion into TiO2 nanotube prepared by the hydrothermal process. J. Electrochem. Soc. **150**,1246–1249 (2003)
86. Baudrin, E., Cassaignon, S., Koesch, M., Jolivet, J.P., Dupont, L., Tarascon, J.M.: Structural evolution during the reaction of Li with nano-sized rutile type TiO2 at room temperature. Electrochem. Commun. **9**, 337–342 (2007)
87. Hu, Y.S., Kienle, L., Guo, Y.G., Maier, J.: High lithium electroactivity of nanometer-sized rutile TiO2. Adv. Mater. **18**, 1421–1426 (2006)
88. Wang, D., Choi, D., Li, J., Yang, Z., Z., Nie, Kou, R., Hu, D., Wang, C., Saraf, L.V., Zhang, J., Aksay, I.A., Liu, J.: Self-assembled TiO2–graphene hybrid nanostructures for enhanced Li-Ion insertion. ACS Nano **3**(4), 907–914 (2009)

89. Qiu, B., Xing, M., Zhang, J.: Mesoporous TiO2 nanocrystals grown in situ on graphene aerogels for high photocatalysis and lithium-ion batteries. J. Am. Chem. Soc. **136**, 5852–5855 (2014)
90. Idota, Y., Kubota, T., Matsufuji, A., Maekawa, Y., Miyasaka, T.: A high-capacity lithium-ion-storage material. Science **276**, 1395–1397 (1997)
91. Liang, J., Wei, W., Zhong, D., Yang, Q., Li, L., Guo, L.: One-step in situ synthesis of sno2/graphene nanocomposites and its application as an anode material for li-ion batteries. ACS Appl. Mater. Interfaces **4**(1), 454–459 (2012). doi:10.1021/am201541s
92. Chen, Z., Li, H., Tian, R., Duan, H., Guo, Y., Chen, Y., Zhou, J., Zhang, C., Dugnani, R., Liu, H.: Three dimensional Graphene aerogels as binder-less, freestanding, elastic and high-performance electrodes for lithium-ion batteries. Sci. Rep. **6**, 7365 (2016). doi:10.1038/srep27365. http://www.nature.com/articles/srep27365#supplementary-information
93. Gong, C., Zhang, Y., Yao, M., Wei, Y., Li, Q., Liu, B., Liu, R., Yao, Z., Cui, T., Zou, B., Liu, B.: Green synthesis of 3D SnO2/graphene aerogels and their application in lithium-ion batteries. RSC Adv. **5**(50), 39746–39751 (2015). doi:10.1039/c5ra05711f
94. Cao, H., Zhou, X., Deng, W., Liu, Z.: A compressible and hierarchical porous graphene/Co composite aerogel for lithium-ion batteries with high gravimetric/volumetric capacity. J. Mater. Chem. A **4**(16), 6021–6028 (2016). doi:10.1039/c6ta00064a
95. Zhou, Y., Liu, Q., Liu, D., Xie, H., Wu, G., Huang, W., Tian, Y., He, Q., Khalil, A., Haleem, Y.A., Xiang, T., Chu, W., Zou, C., Song, L.: Carbon-coated MoO2 dispersed in three-dimensional graphene aerogel for lithium-ion battery. Electrochimica Acta **174**, 8–14 (2015). doi:10.1016/j.electacta.2015.05.153
96. Yoo, J.J., Balakrishnan, K., Huang, J., Meunier, V., Sumpter, B.G., Srivastava, A., Conway, M., Mohana Reddy, A.L., Yu, J., Vajtai, R., Ajayan, P.M.: Ultrathin Planar Graphene Supercapacitors. Nano. Lett. **11**(4), 1423–1427 (2011). doi:10.1021/nl200225j
97. Du, Q., Zheng, M., Zhang, L., Wang, Y., Chen, J., Xue, L., Dai, W., Ji, G., Cao, J.: Preparation of functionalized graphene sheets by a low-temperature thermal exfoliation approach and their electrochemical supercapacitive behaviors. Electrochimica Acta **55**(12), 3897–3903 (2010). doi:10.1016/j.electacta.2010.01.089
98. Jeong, H.M., Lee, J.W., Shin, W.H., Choi, Y.J., Shin, H.J., Kang, J.K., Choi, J.W.: Nitrogen-doped graphene for high-performance ultracapacitors and the importance of nitrogen-doped sites at basal planes. Nano. Lett. **11**(6), 2472–2477 (2011). doi:10.1021/nl2009058
99. Fan, Z., Yan, J., Zhi, L., Zhang, Q., Wei, T., Feng, J., Zhang, M., Qian, W., Wei, F.: A three-dimensional carbon nanotube/graphene sandwich and its application as electrode in supercapacitors. Adv. Mater. **22**(33), 3723–3728 (2010). doi:10.1002/adma.201001029
100. Guo, C.X., Li, C.M.: A self-assembled hierarchical nanostructure comprising carbon spheres and graphene nanosheets for enhanced supercapacitor performance. Energy Environ Sci **4** (11), 4504–4507 (2011). doi:10.1039/c1ee01676h
101. Alvi, F., Ram, M.K., Basnayaka, P.A., Stefanakos, E., Goswami, Y., Kumar, A.: Graphene–polyethylenedioxythiophene conducting polymer nanocomposite based supercapacitor. Electrochimica Acta **56**(25), 9406–9412 (2011). doi:10.1016/j.electacta.2011.08.024
102. Xia, X., Tu, J., Mai, Y., Chen, R., Wang, X., Gu, C., Zhao, X.: Graphene sheet/porous NiO hybrid film for supercapacitor applications. Chem. Eur. J. **17**(39), 10898–10905 (2011) doi:10.1002/chem.201100727
103. Qu, Q., Yang, S., Feng, X.: 2D sandwich-like sheets of iron oxide grown on graphene as high energy anode material for supercapacitors. Adv. Mater. **23**(46), 5574–5580. (2011) doi:10.1002/adma.201103042
104. Yu, G., Hu, L., Liu, N., Wang, H., Vosgueritchian, M., Yang, Y., Cui, Y., Bao, Z.: Enhancing the supercapacitor performance of graphene/MnO2 nanostructured electrodes by conductive wrapping. Nano Lett. **11**, 4438–4442 (2011)

Chapter 9
Sol-Gel Synthesis of Titanium Dioxide

Sanjay Gopal Ullattil and Pradeepan Periyat

9.1 Sol and Gel

A sol is a stable suspension of colloidal solid particles in a liquid phase [1]. These solid particles are denser than the surrounding liquid and small enough for the forces of dispersion to be greater than that of gravity. Sols are classified as lyophobic and lyophilic. When the solvent particle interaction is relatively weak, it is called a lyophobic sol and if this interaction is relatively strong, a lyophilic sol. Sols become stable when the forces between two particles are repulsive and the contributing forces from the surface will prevent the agglomeration and hence coagulation. Once the charge on the particle is removed by another additive, the system collapses (flocculation) leading to gel formation.

A gel is a porous three-dimensional inter-connected solid network and is transparent due to smaller sized particles [2]. In other words gel is a two-component system of semi-solid nature in liquid. Gelation is a process in which a sol or a solution transforms into a gel and the continuity of solid structure gives the gel elasticity. These solid particles present in the gel can be amorphous, crystalline solids or macromolecules. When the solid network is made up of colloidal sol particles, the gel is known to be colloidal and when the solid network is made up of sub-colloidal chemical units, then gel is polymeric. The main difference between the colloidal and polymeric gels is, in the former, sol–gel transition which is caused by a physicochemical effect and in the latter, by chemical bonding in contrast to a chemical reaction such as polycondensation.

S.G. Ullattil
Department of Chemistry, University of Calicut, Thenhipalam 673635, Kerala, India

P. Periyat (✉)
Department of Chemistry, Central University of Kerala, Kasaragod 671314, Kerala, India
e-mail: pperiyat@cukerala.ac.in

S.C. Pillai and S. Hehir (eds.), *Sol-Gel Materials for Energy, Environment and Electronic Applications*, Advances in Sol-Gel Derived Materials and Technologies,
DOI 10.1007/978-3-319-50144-4_9

Sols are thermodynamically unstable, due to their large surface area to volume ratio. However, they can be stabilized by introducing an energy barrier on the solid particles presented in the sol. This stabilization process can be done either by electrostatic stabilization or steric stabilization [3].

9.1.1 Electrostatic Stabilization of the Sol

The net force between the particles in a sol is assumed to be the sum of attractive Van der Waals forces and electrostatic repulsion created by charges adsorbed on the particles [4]. Van der Waals interaction arises between two microscopic particles placed in liquid medium. When solid particles dispersed in a liquid medium, which also contains an electrolyte, some specific ions often preferentially adsorbed on the surface of the particles. A surface charge is developed in solution adjacent to the charged surface. The overlap of these diffused layers surrounding two sol particles produce repulsion. The sol will be stable if repulsive forces due to the surface charges exceed the van der Waals force of attraction. The repulsive barrier depends upon the two types of ions that make up the double layer: charge determining ions that control the charge on the surface of the particles and counter ions that are present in the solution in the vicinity of the particle. The counter ions act as screen charges of the potential determining ions and the potential difference between the two layers is called 'zeta potential'. The stability of a colloid is also related to the magnitude of zeta potential. The potential at which value of zeta potential is zero is called isoelectric point and the pH at which the particle is neutrally charged is called point of zero charge (PZC) [5].

9.1.2 Steric Stabilization of the Sol

Steric stabilization can be achieved by adsorbing sterically crowded organic molecules on the surface of the colloidal particles. This adsorbed organic layer constitutes a steric barrier, which discourages the close approach of the two particles in terms of both enthalpy and entropy. To provide an effective barrier, the adsorbed layer must meet the following requirements [6]:

1. Surface of the particle should be completely covered.
2. The organic molecules should be selected such that the energy barrier created is greater than or large enough to overcome the attractive van der Waals forces.

The electrostatic stabilization can be combined with steric stabilization, thus causing stabilization by electrostatic effects.

9.2 Sol–Gel Process

Sol–gel process is a versatile solution method for making ceramic glass materials [7]. In general, the sol–gel process involves the transition of a system from a liquid 'sol' which is mostly in colloidal form converted into solid 'gel' phase. Using this process, it is possible to fabricate a wide variety of ceramic or glass materials in the form of ultra-fine or spherical-shaped powders, thin film coating, ceramic fibres, microporous inorganic membranes, monolithic ceramics and glasses or extremely porous aerogel materials [8].

The starting materials used in the preparation of the 'sol' are usually metal organic compounds such as metal alkoxide/inorganic metal salts. In a series of hydrolysis and polymerization, reactions form a colloidal suspension or a sol. Further processing of the sol enables one to make ceramic materials in different forms. When the sol is cast into a mould, a wet 'gel' will form. With further drying and heat treatment, the gel is converted into dense ceramic materials. If the liquid in a wet gel is removed under a supercritical condition, a highly porous and extremely low-density material called aerogel is obtained [9]. As the viscosity of a sol is adjusted into a proper viscosity range, ceramic fibres can be drawn from the sol [10]. Ultra-fine and uniform ceramic powders are formed by precipitation [11], spray hydrolysis [12] or emulsion techniques [13]. The details of the sol–gel process are shown in Fig. 9.1.

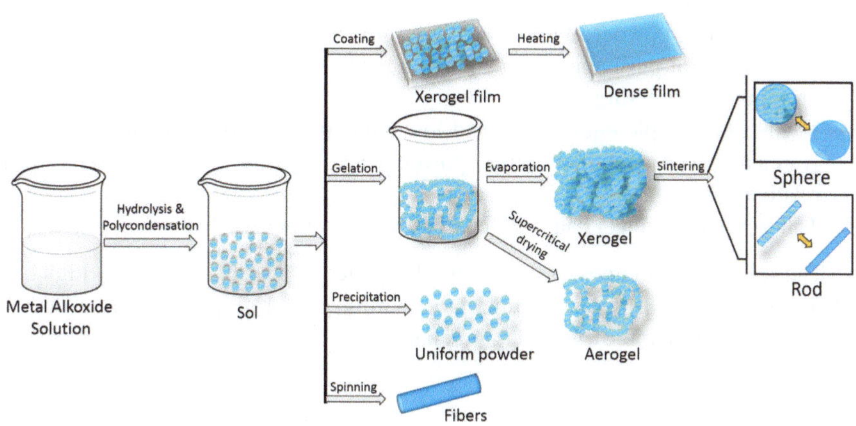

Fig. 9.1 Schematic representation of sol–gel processing

9.2.1 Various Steps in Sol–Gel Processing [14]

The steps involved in sol–gel processing are as follows:

1. The preparation of homogenous solution of liquid precursor in organic solvent miscible with water/other common reagents.
2. Conversion of homogeneous solution into a sol, by treatment with a suitable reagent. Here molecular-level manipulations are possible like adjustment of the hydrolysis conditions to obtain precisely controlled particle sizes, adding dopants to bring about special properties in the desired product, etc.
3. Conversion of the sol into gel by self-polymerization and condensation method. This sol–gel conversion can lead to a numerous number of applications in the industrial/catalysis area.
4. Shaping the gel into desired shape such as thin films, fibres, spheres, etc.
5. Sintering the gel into ceramic materials with desired shape [15].

9.2.2 Advantages of Sol–Gel Processing

The followings are the advantages of sol–gel process [2, 4]:

1. The increased chemical homogeneity in multi-component systems as well as in doped systems.
2. Production of powder materials having high surface area.
3. High chemical purity can be maintained for the products due to the absence of grinding and pressing steps.
4. Needs only simple equipments and low temperature of preparation.
5. A range of product like fibres, powders and coating can be prepared with relative ease.
6. Tailor-made products can be prepared by controlling the features.

The sol–gel method possess some disadvantages also such as relatively high cost of precursors, long processing time, large shrinkage during processing and possibility of the formation of hard agglomerates.

9.3 Titanium Dioxide (TiO_2)

TiO_2 is well recognized as a valuable material with applications in paints [16], as filler in paper [17], textiles [18], photocatalysis [19], etc. TiO_2 exists in three crystal modifications, known as anatase (tetragonal, $a = b = 3.782$ Å, $c = 9.502$ Å), rutile (tetragonal, $a = b = 4.854$ Å, $c = 2.953$ Å) and brookite (rhombohedral, $a = 5.436$ Å, $b = 9.166$ Å, $c = 5.135$ Å) [20]. In all the three TiO_2 structures, the

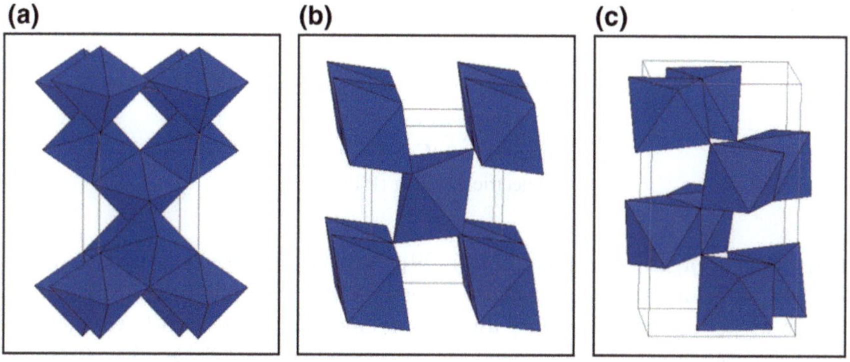

Fig. 9.2 Structure of **a** anatase, **b** rutile and **c** brookite [23]

stacking of the octahedral results in threefold coordinated oxygen atoms [21]. Thermodynamically, the rutile structure is the stable one [22]. In all the three crystalline forms each of Ti^{4+} ions is surrounded by an irregular octahedron of oxide ions, but the number of edges shared by the octahedron increases from, two in rutile to three in brookite to four in anatase. The structures of anatase, rutile and brookite are shown in Fig. 9.2.

Both in rutile and anatase, the six oxide ions that surround the titanium ions can be grouped into two [24]. The two oxygen atoms are farthest from titanium ions and the other four oxide ions are relatively closer to titanium ion. In rutile these distances are 2.01 and 1.92 Å, respectively, and in anatase they are 1.95 and 1.91 Å, respectively. Anatase and rutile structures are the important forms of TiO_2 reported in the literature, and Table 9.1 shows the structural properties of anatase and rutile.

9.3.1 *TiO_2 as Photocatalyst*

TiO_2 is a semiconductor, with band gap lying in the UV range and transparent in the visible region. This makes it active for photocatalytic reactions like degradation of organic compounds [26]. In 1972, Fujishima and Honda discovered the photocatalytic splitting of water electrodes [27]. This event marked the beginning of a new area in heterogeneous photocatalysis. They achieved UV light-induced water cleavage using a TiO_2 photo anode in combination with a platinum counter electrode soaked in an electrode aqueous solution.

TiO_2 has a band gap of 3.2 eV and this band gap can only utilize light with wavelength below 400 nm, and thus only UV light is suitable since it has wavelength range of 200–400 nm [28]. When penetrable photon energy is applied, the electrons in the TiO_2 valence band (VB) will be excited causing the electron jump to the conduction band (CB) resulting in a positive hole in the valence band. The radical electron has a very high oxidizing energy and it is able to breakdown

Table 9.1 Properties of TiO_2 [25]

Crystal form	Anatase	Rutile
Density (g/cm^3)	3.87	4.24
Hardness (Mohr's scale)	5–6	6–7
Crystal structure	Tetragonal	Tetragonal
Specific heat (Cal °C^{-1} g^{-1})	0.177	0.17
Dielectric constant (RT)	48	114

inorganic and organic chemical bonds [29]. Among the three polymorphic phases (anatase, rutile and brookite), anatase is found to be better photocatalyst because it has the higher adsorption affinity towards organic compounds and lower electron–hole recombination rate compared to other polymorphs of TiO_2 [30]. CdS with smaller band gap energy can able to absorb the visible light instead of UV to carry out the photocatalytic reaction [31]. However, CdS is not used for environmental cleanup due to its high toxicity. TiO_2 is stable, eco-friendly and easy to synthesize. Hence, it is widely used for environmental cleanup, waste water treatment to remove toxic chemical dyes [32].

When a photon with an energy hv exceeds the energy of the band gap of TiO_2 (>3.2 eV), an electron is promoted from VB to CB leaving behind a hole (h^+) [33]. A portion of this photo excited electron–hole pair diffuses to the surface of the catalytic particle and takes part in the chemical reaction with the adsorbed donor (D) or acceptor (A) molecules [34]. The holes can oxidize donor molecules, whereas the conduction band electrons can reduce appropriate electron acceptor molecules [35]:

$$D + h^+ \rightarrow D^+.$$
$$A + e^- \rightarrow A^-.$$

A characteristic feature of TiO_2 semiconducting metal oxides is the strong oxidation power of their holes. They can react in a one-electron oxidation step with water to produce the highly reactive hydroxyl radical (OH). Both the holes and hydroxyl radicals are very powerful oxidants, which can be used to oxidize most organic contaminants:

$$H_2O + h^+ \rightarrow OH + H^+$$

In general, air oxygen acts as electron acceptor by forming the superoxide ion O^{2-}:

$$O^{2+} + e^- \rightarrow O^{2-}.$$

Superoxide ions are also highly reactive particles, which are able to oxidize organic materials [36].

9.4 Sol–Gel Synthesis of TiO_2

Depending on the titanium metal precursor type, two types of sol–gel process can be identified to produce TiO_2 semiconductors: (i) an alcohol-based process where the starting precursor is metal alkoxide and (ii) an aqueous-based process where the starting precursor is inorganic metal salt.

9.4.1 Alcohol-Based Process

The important metal alkoxide precursors of TiO_2 are $Ti(OC_2H_5)$, $Ti(OC_3H_7)_4$ and $Ti(OC_4H_9)_4$. In these alkoxides, there exists a metal–oxygen bond and due to large difference in electro-negativity between Ti and O, the bond becomes highly polar and extremely reactive. Hence by addition of water, simultaneous hydrolysis and condensation reactions proceed ultimately leading to the formation of a gel [37].

9.4.2 Aqueous-Based Process

The important precursors for the aqueous-based process are $TiOSO_4$ [38] and $TiCl_4$ [39]. Aqueous-based sol–gel preparation can also be done using a hydrolysis and condensation method. The slow controlled nucleation and growth of crystals at an elevated temperature form the particles that can also be used [40]. Steps involved in the aqueous-based sol–gel process are precipitation and peptisation. Precipitation involves the hydrolysis of the inorganic metal salt, which hydrolyses rapidly to get a gelatinous precipitate by the addition of base, followed by washing to remove excess electrolyte. Peptisation is generally the process of re-dispersing a colloid which will prevent coagulation [37]. Peptisation actually refers to a direct disintegration of substance into particles of colloid size by an added agent known as peptising agent [41]. It is very difficult to peptise a colloid that has been fallen into a primary potential minimum.

9.5 High-Temperature Stability of TiO_2

Anatase and rutile are two important polymorphs of TiO_2 [42–44]. The room-temperature phase is anatase and high-temperature phase is rutile. Anatase transforms irreversibly and exothermically into rutile in the range of 400–1200 °C [45, 46] depending on parameters such as method of preparation, grain size, morphology, degree of agglomeration, nature of impurities and reaction atmosphere.

The anatase to rutile transformation involves an overall contraction of oxygen and a movement of ions so that a co-operative rearrangement of Ti^{4+} and O^{2-} occurs. The transformation implies that two of six Ti–O bonds of anatase structure break to form a rutile structure. The removal of oxygen ions, which generates lattice vacancies, accelerates the transformation and inhibits the formation of interstitial titanium. As a representative example the decomposition of $TiOSO_4$ to anatase and its transformation to rutile is represented in Scheme 9.1 [21].

The rate of transformation of pure TiO_2 in the anatase modification was found to be slow below 885 K and extremely rapid above 1000 K. The transition is a nucleation growth process and follows first-order kinetics, with activation energy of 90 kCal/mol [47–52]. The impurities (additive or dopants) that have most pronounced inhibiting action are chloride, sulphate and fluoride ions, whereas those that accelerate the transformation include Sb, Fe, V and Zn [53–56]. Ions having lower oxidation states (Li^+, Na^+, K^+, Mg^{2+}, Ca^{2+}, Sr^{2+}, Ba^{2+}, Al^{3+}, Y^{3+}, La^{3+}, Er^{3+}, Ti^{4+}, Co^{2+}, Ni^{2+}, Cu^{2+} and Zn^{2+}) and the same having higher oxidation states such as W^{6+} and Mo^{6+} reduce vacancy concentration and will retard the reaction [57, 58]. The effect of reaction atmosphere such as hydrogen, static air, oxygen, argon, nitrogen and chlorine affect the phase transformation to different extents. Lida and Ozaki[58] as well as Shannon [59] found that transformation rate of anatase to rutile in a hydrogen atmosphere is greater than in air (Fig. 9.3).

The enhancement or inhibiting effect of impurities (additive or dopants) on the anatase to rutile transformation also depends on their ability to enter into the TiO_2 lattice thereby creating oxygen vacancies or interstitial Ti^{3+} ions. The impurities that can increase the concentration of oxygen vacancies would also tend to promote the transformation, while some impurities will retard the transformation by increasing the lattice defect concentration of Ti interstitials in TiO_2 [60].

Oxides of Al, La, Yb, Cu and Ce mixed with anatase TiO_2 increase the photocatalytic efficiency but decrease the anatase phase stability. Both alumina and silica as single additives exert an inhibitory effect on the transformation of anatase to rutile [61]. For example, Suyama and Kata reported the inhibiting effect on the anatase to rutile transformation by silica additive and found to be that the transition temperature was improved to 980–1030 °C, depending on the amount of silica added and homogeneity of silica in TiO_2 [62]. The reduction in anatase/anatase contact points by adjacent silica particles resulted in the suppression of the nucleation of rutile particles. For the up-to-date detailed report on the high-temperature stability of anatase TiO_2, interested readers are referred to a more specific review [63].

Another important parameter which controls the phase transition temperature is the ionic radius of impurity cations (dopants) added [43]. The impurity (dopant) has no effect on the amorphous gel to anatase transformation temperature, but influences the anatase to rutile transformation. When the dopant ion size is less than that of titanium, anatase phase will be stabilized to higher temperature and if dopant size is higher anatase phase will be destabilized [64]. Depending on the ionic radius of impurity (dopant) compared with radius of titanium, the dopant can be introduced either substitutionally or interstitially.

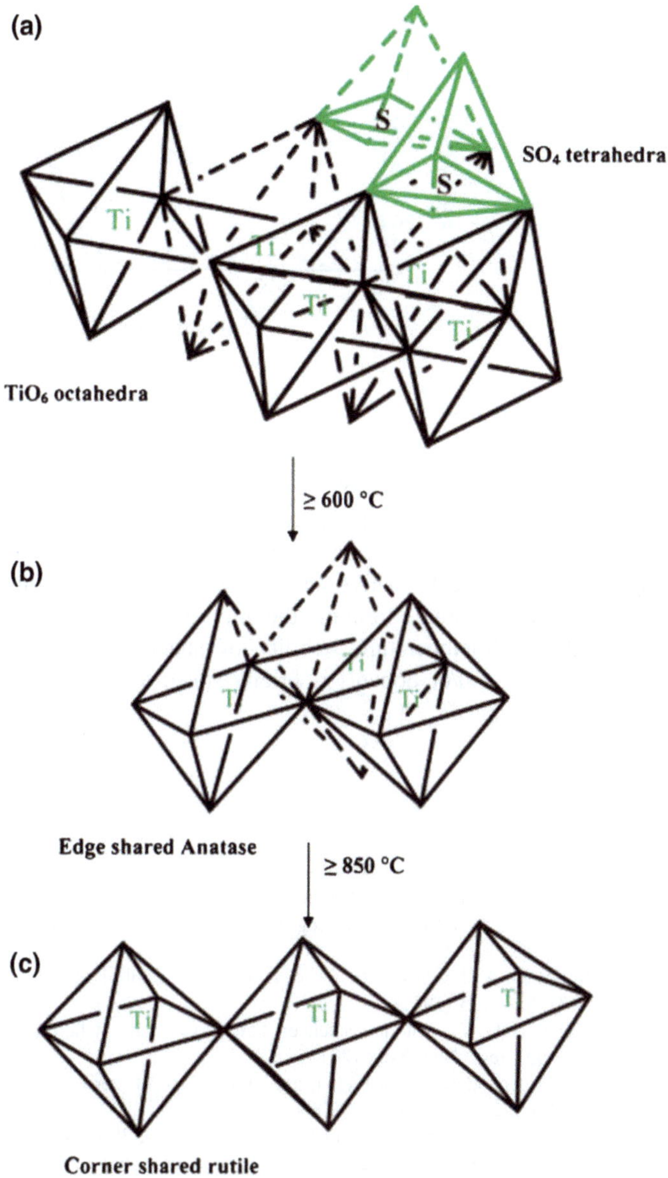

Scheme 9.1 Schematic representation of decomposition of $TiOSO_4$ to anatase and its transformation to rutile (**a** $TiOSO_4$; **b** anatase TiO_2; **c** rutile TiO_2) [21]. Reprinted with permission from Ref. [21]. Copyright 2008, American Chemical Society

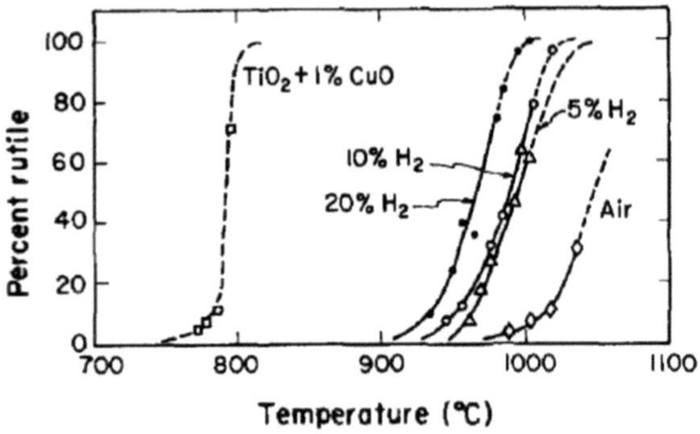

Fig. 9.3 Percent transformation to rutile versus temperature [58]. Reprinted with permission from Ref. [58]. Copyright 1999, Wiley-VCH

9.6 Conclusion

In this chapter, the basic principles of sol–gel processing and their advantages/disadvantages are explicitly stated, concentrating on the synthesis of nanocrystalline TiO_2. From, three different crystal forms of TiO_2, anatase, rutile and brookite, more photocatalytically active anatase phase is usually converted into rutile phase at low temperature, limiting the high-temperature applications of anatase. Hence, the high-temperature stability of anatase TiO_2 using various impurities like addictive and/or dopants is briefly stated and their photocatalytic efficiencies are also explained.

References

1. Roy, R.: Ceramics by the solution-sol-gel route. Science **238**(4834), 1664–1669 (1987)
2. Pierre, A.C.: Introduction to Sol-Gel Processing, vol. 1. Springer Science & Business Media (2013)
3. Napper, D.H., Netschey, A.: Studies of the steric stabilization of colloidal particles. J. Colloid Interface Sci. **37**(3), 528–535 (1971)
4. Brinker, C.J., Scherer, G.W.: Sol-Gel Science: The Physics and Chemistry of Sol-Gel Processing. Academic Press, New York (2013)
5. Kosmulski, M.: PH-dependent surface charging and points of zero charge II. Update. J. Colloid Interface Sci. **275**(1), 214–224 (2004)
6. Napper, D.H.: Steric stabilization. J. Colloid Interface Sci. **58**(2), 390–407 (1977)
7. Xu, Q., Anderson, M.A.: Sol–gel route to synthesis of microporous ceramic membranes: preparation and characterization of microporous TiO_2 and ZrO_2 xerogels. J. Am. Ceram. Soc. **77**(7), 1939–1945 (1994)
8. Feinle, A., Elsaesser, M.S., Hüsing, N.: Sol–gel synthesis of monolithic materials with hierarchical porosity. Chem. Soc. Rev. (2016)
9. Hong, I.: VOCs degradation performance of TiO_2 aerogel photocatalyst prepared in SCF drying. J. Ind. Eng. Chem. **12**(6), 918–925 (2006)
10. Zhang, Y., Ding, Y., Gao, J., Yang, J.: Mullite fibres prepared by sol–gel method using polyvinyl butyral. J. Eur. Ceram. Soc. **29**(6), 1101–1107 (2009)

11. Chen, J.F., Wang, Y.H., Guo, F., Wang, X.M., Zheng, C.: Synthesis of nanoparticles with novel technology: high-gravity reactive precipitation. Ind. Eng. Chem. Res. **39**(4), 948–954 (2000)
12. Zhou, M., Yu, J., Liu, S., Zhai, P., Huang, B.: Spray-hydrolytic synthesis of highly photoactive mesoporous anatase nanospheres for the photocatalytic degradation of toluene in air. Appl. Catal. B **89**(1), 160–166 (2009)
13. Imhof, A., Pine, D.J.: Uniform macroporous ceramics and plastics by emulsion templating. Adv. Mater. **10**(9), 697–700 (1998)
14. Woodhead, J.L., Segal, D.L.: Sol-gel processing (1984)
15. Kakihana, M.: Invited review "sol-gel" preparation of high temperature superconducting oxides. J. Sol-Gel Sci. Technol. **6**(1), 7–55 (1996)
16. Ceresa, E.M., Burlamacchi, L., Visca, M.: An ESR study on the photoreactivity of TiO_2 pigments. J. Mater. Sci. **18**(1), 289–294 (1983)
17. Park, J.K., Kim, J.K., Kim, H.K.: TiO_2-SiO_2 composite filler for thin paper. J. Mater. Process. Technol. **186**(1), 367–369 (2007)
18. Montazer, M., Seifollahzadeh, S.: Enhanced self-cleaning, antibacterial and UV protection properties of nano TiO_2 treated textile through enzymatic pretreatment. Photochem. Photobiol. **87**(4), 877–883 (2011)
19. Ullattil, S.G., Periyat, P.: A 'one pot' gel combustion strategy towards Ti^{3+} self-doped 'black' anatase TiO_{2-x} solar photocatalyst. J. Mater. Chem. A **4**(16), 5854–5858 (2016)
20. Etacheri, V., Di Valentin, C., Schneider, J., Bahnemann, D., Pillai, S.C.: Visible-light activation of TiO_2 photocatalysts: advances in theory and experiments. J. Photochem. Photobiol., C **25**, 1–29 (2015)
21. Periyat, P., Pillai, S.C., McCormack, D.E., Colreavy, J., Hinder, S.J.: Improved high-temperature stability and sun-light-driven photocatalytic activity of sulfur-doped anatase TiO_2. J. Phys. Chem. C **112**(20), 7644–7652 (2008)
22. Carp, O., Huisman, C.L., Reller, A.: Photoinduced reactivity of titanium dioxide. Prog. Solid State Chem. **32**(1), 33–177 (2004)
23. http://ruby.colorado.edu/~smyth/min/tio2.html(2016). Accessed on 10 June 2016
24. Linsebigler, A.L., Lu, G., Yates Jr., J.T.: Photocatalysis on TiO_2 surfaces: principles, mechanisms, and selected results. Chem. Rev. **95**(3), 735–758 (1995)
25. Patton, T.C.: Pigment hand book, vol. 1. In: Properties and Economics. Wiley, Inc., New York (1973)
26. Tang, W.Z., Zhang, Z., An, H., Quintana, M.O., Torres, D.F.: TiO2/UV photodegradation of azo dyes in aqueous solutions. Environ. Technol. **18**(1), 1–12 (1997)
27. Fujishima, A., Honda, K.: Photolysis-decomposition of water at the surface of an irradiated semiconductor. Nature **238**(5385), 37–38 (1972)
28. Hattori, A., Tada, H.: High photocatalytic activity of F-doped TiO_2 film on glass. J. Sol-Gel Sci. Technol. **22**(1–2), 47–52 (2001)
29. Schneider, J., Matsuoka, M., Takeuchi, M., Zhang, J., Horiuchi, Y., Anpo, M., Bahnemann, D.W.: Understanding TiO_2 photocatalysis: mechanisms and materials. Chem. Rev. **114**(19), 9919–9986 (2014)
30. Luttrell, T., Halpegamage, S., Tao, J., Kramer, A., Sutter, E., Batzill, M.: Why is anatase a better photocatalyst than rutile?—Model studies on epitaxial TiO_2 films. Sci. Rep. **4** (2014)
31. Smith, A.M., Nie, S.: Semiconductor nanocrystals: structure, properties, and band gap engineering. Acc. Chem. Res. **43**(2), 190–200 (2009)
32. Pelaez, M., Nolan, N.T. Pillai, S.C., Seery, M.K., Falaras, P., Kontos, A.G., Dunlop, P.S., Hamilton, J.W., Byrne, J.A., O'shea, K., Entezari, M.H.: A review on the visible light active titanium dioxide photocatalysts for environmental applications. Appl. Catal. B: Environ. **125**, 331–349 (2012)
33. Kumar, S.G., Devi, L.G.: Review on modified TiO_2 photocatalysis under UV/visible light: selected results and related mechanisms on interfacial charge carrier transfer dynamics. J. Phys. Chem. A **115**(46), 13211–13241 (2011)

34. Fujishima, A., Rao, T.N., Tryk, D.A.: Titanium dioxide photocatalysis. J. Photochem. Photobiol., C **1**(1), 1–21 (2000)
35. Hoffmann, M.R., Martin, S.T., Choi, W., Bahnemann, D.W.: Environmental applications of semiconductor photocatalysis. Chem. Rev. **95**(1), 69–96 (1995)
36. Ullattil, S.G., Periyat, P.: Green microwave switching from oxygen rich yellow anatase to oxygen vacancy rich black anatase TiO_2 solar photocatalyst using Mn (II) as 'anatase phase purifier'. Nanoscale **7**(45), 19184–19192 (2015)
37. Schubert, U.: Chemical modification of titanium alkoxides for sol–gel processing. J. Mater. Chem. **15**(35–36), 3701–3715 (2005)
38. Periyat, P., Baiju, K.V., Mukundan, P., Pillai, P.K., Warrier, K.G.K.: Aqueous colloidal sol–gel route to synthesize nanosized ceria-doped titania having high surface area and increased anatase phase stability. J. Sol-Gel Sci. Technol. **43**(3), 299–304 (2007)
39. Baiju, K.V., Periyat, P., Wunderlich, W., Pillai, P.K., Mukundan, P., Warrier, K.G.K.: Enhanced photoactivity of neodymium doped mesoporous titania synthesized through aqueous sol–gel method. J. Sol-Gel Sci. Technol. **43**(3), 283–290 (2007)
40. Periyat, P., Saeed, P.A., Ullattil, S.G.: Anatase titania nanorods by pseudo-inorganic templating. Mater. Sci. Semicond. Process. **31**, 658–665 (2015)
41. Varma, H.K., Mani, T.V., Damodaran, A.D., Warrier, K.G., Balachandran, U.: Characteristics of alumina powders prepared by spray-drying of boehmite sol. J. Am. Ceram. Soc. **77**(6), 1597–1600 (1994)
42. Kim, D.W., Kim, T.G., Hong, K.S.: Origin of a shrinkage anomaly in anatase. J. Am. Ceram. Soc. **81**(6), 1692–1694 (1998)
43. Penn, R.L., Banfield, J.F.: Oriented attachment and growth, twinning, polytypism, and formation of metastable phases: Insights from nanocrystalline TiO_2. Am. Mineral. **83**(9–10), 1077–1082 (1998)
44. Clark, R.J.H.: The Chemistry of Titanium and Vanadium. Elsevier, Amsterdam (1968)
45. Shannon, R.D., Pask, J.A.: Kinetics of the anatase-rutile transformation. J. Am. Ceram. Soc. **48**(8), 391–398 (1965)
46. Banfield, J.F., Bischoff, B.L., Anderson, M.A.: TiO_2 accessory minerals: coarsening, and transformation kinetics in pure and doped synthetic nanocrystalline materials. Chem. Geol. **110**(1), 211–231 (1993)
47. Schubert, U., Pierre, A.C.: Introduction to sol-gel processing. Angew. Chem. Int. Ed. **37**(23), 3324–3325 (1998)
48. Hishita, S., Mutoh, I., Koumoto, K., Yanagida, H.: Inhibition mechanism of the anatase-rutile phase transformation by rare earth oxides. Ceram. Int. **9**(2), 61–67 (1983)
49. Gribb, A.A., Banfield, J.F.: Particle size effects on transformation kinetics and phase stability in nanocrystalline TiO_2. Am. Mineral. **82**(7–8), 717–728 (1997)
50. Kumar, K.N.P.: Growth of rutile crystallites during the initial stage of anatase-to-rutile transformation in pure titania and in titania-alumina nanocomposites. Scr. Metall. Mater. **32** (6), 873–877 (1995)
51. Spurr, R.A., Myers, H.: Quantitative analysis of anatase-rutile mixtures with an X-ray diffractometer. Anal. Chem. **29**(5), 760–762 (1957)
52. Penn, R.L., Banfield, J.F.: Formation of rutile nuclei at anatase 112 twin interfaces and the phase transformation mechanism in nanocrystalline titania. Am. Mineral. **84**(5–6), 871–876 (1999)
53. Eppler, R.A..: Effect of antimony oxide on the anatase-rutile transformation in titanium dioxide. J. Am. Ceram. Soc. **70**(4) (1987)
54. Iida, Y., Ozaki, S.: Grain growth and phase transformation of titanium oxide during calcination. J. Am. Ceram. Soc. **44**(3), 120–127 (1961)
55. Zhang, Y.H., Reller, A.: Phase transformation and grain growth of doped nanosized titania. Mater. Sci. Eng., C **19**(1), 323–326 (2002)
56. MacKenzie, K.J.D.: Calcination of titania: IV, effect of additives on the anatase rutile transformation. Trans. J. Br. Ceram. Soc. **74**(2), 29–34 (1975)

57. Vargas, S., Arroyo, R., Haro, E., Rodriguez, R.: Effects of cationic dopants on the phase transition temperature of titania prepared by the sol-gel method. J. Mater. Res. **14**(10), 3932–3937 (1999)
58. Iida, Y., Ozaki, S.: Grain growth and phase transformation of titanium oxide during calcination. J. Am. Ceram. Soc. **44**(3), 120–127 (1961)
59. Shannon, R.D.: Phase transformation studies in TIO_2 supporting different defect mechanisms in vacuum-reduced and hydrogen-reduced rutile. J. Appl. Phys. **35**(11), 3414–3416 (1964)
60. Yang, J., Ferreira, J.M.F.: Inhibitory effect of the Al_2O_3–SiO_2 mixed additives on the anatase–rutile phase transformation. Mater. Lett. **36**(5), 320–324 (1998)
61. Periyat, P., Baiju, K.V., Mukundan, P., Pillai, P.K., Warrier, K.G.K.: High temperature stable mesoporous anatase TiO_2 photocatalyst achieved by silica addition. Appl. Catal. A **349**(1), 13–19 (2008)
62. Suyama, Y., Kato, A.: The inhibitory effect of SiO_2 on the anatase-rutile transition of TiO_2. J. Ceram. Soc. Jpn. **86**(3), 119–125 (1978)
63. Periyat, P., Naufal, B., Ullattil, S.G.: A review on high temperature stable anatase TiO_2 photocatalysts. Mater. Sci. Forum **855**, 78–93 (2016)
64. Rodriguez-Talavera, R., Vargas, S., Arroyo-Murillo, R., Montiel-Campos, R., Haro-Poniatowski, E.: Modification of the phase transition temperatures in titania doped with various cations. J. Mater. Res. **12**(02), 439–443 (1997)

Chapter 10
Sol-Gel Lanthanum Phosphate: A Versatile Ceramic Material for Diverse Functional Applications

Sankar Sasidharan, Rajesh Komban, Shijina Nambiar, Balagopal N. Nair, M. Padmanabhan, Krishna G. Warrier and U.S. Hareesh

10.1 Synthesis of Lanthanum Phosphate ($LaPO_4$)—Literature Review

Lanthanum phosphate synthesis gathered attention after Hikichi et al. [1] established the phase stability in 1987. Considerable number of reports are available on the synthesis of lanthanum phosphate employing techniques like solid-state synthesis, precipitation, hydrothermal, solid–liquid reaction, colloidal, microwave assisted, ultrasonication and sol-gel [1–25]. The starting compounds in nearly all the above techniques are rare earth nitrates, carbonates or acetates and the phosphates obtained are in submicron or nanometre range. Ruigang et al [2] synthesized lanthanum phosphate powders by direct solid–liquid reaction of lanthanum oxide and phosphoric acid at the La: P ratio of 1:1. Phase pure and fine grain-sized $LaPO_4$ powders could be obtained by calcining the as-received powders at temperature higher than 600 °C. Lanthanum oxide dissolved in dilute phosphoric acid in the La to P atomic ratio 1:1

S. Sasidharan · S. Nambiar · K.G. Warrier (✉) · U.S. Hareesh (✉)
Materials Science and Technology Division, CSIR—National Institute for Interdisciplinary Science and Technology, Thiruvananthapuram 695019, Kerala, India
e-mail: warrierkgk@yahoo.com

U.S. Hareesh
e-mail: hareesh@niist.res.in

R. Komban
Centrum for Angewandte Nanotechnologie (CAN) GmbH, Hamburg, Germany

B.N. Nair
R&D Centre, Noritake Co. Ltd, Miyoshi-Aichi, Japan

M. Padmanabhan
Department of Chemistry, Amrita Vishwa Vidyapeetham University, Kollam, India

S.C. Pillai and S. Hehir (eds.), *Sol-Gel Materials for Energy, Environment and Electronic Applications*, Advances in Sol-Gel Derived Materials and Technologies, DOI 10.1007/978-3-319-50144-4_10

provided a direct reaction between lanthanum oxide and phosphoric acid to yield no by-products other than water:

$$\mathbf{La_2O_3 + 2H_3PO_4 \rightarrow 2LaPO_4 + 3H_2O.}$$

Crystallization from boiling phosphoric acid solution was used by Kijowska [3] to synthesize Ln phosphates (Ln = La, Ce, Pr, Nd, Sm, Eu, Gd, Tb, Dy, Ho, Er, Tm, Yb, Lu and Y). Formation of hexagonal $LnPO_4.H_2O$ (La, Ce, Pr, Nd, Sm, Eu, Gd, Tb), tetragonal $LnPO_4.H_2O$ (Ho, Er, Tm, Yb, Lu), isostructural with YPO_4 and orthorhombic $DyPO_4.H_2O$ was observed.

A facile solution–precipitation process for the preparation of complex $LaPO_4$ nanostructures via oriented attachment of small nanoparticles was adopted by Chai et al. [4]. They could obtain nanostructures with different shapes, including spherical, flake-like, star-shaped and short nano-rods when the molar ratio of La^{3+}: H_3PO_4 was adjusted or when additional ionic surfactants were used (Fig. 10.1).

Rare earth phosphate powders (Re = La, Ce or Y) were synthesized from an aqueous precipitation method by Lucas et al. [5], by adding rare earth chloride solution to either phosphoric acid or diammonium hydrogen phosphate. The pH was maintained at a constant value by the addition of ammonium hydroxide solution. After the complete addition of the rare earth solution, the suspension was aged for various times and the resulting precipitate was washed with distilled water, and dried at 80 °C. The needle-like precipitate crystallized in the hexagonal structure of rhabdophane $RePO_4$ $0.5H_2O$ with a specific surface area of 80 m^2/g (Fig. 10.2).

Lanthanide orthophosphate crystals ($LnPO_4$) crystallized in hexagonal or tetragonal structure depending on the ionic radius of the lanthanide ions. $LnPO_4$ (Ln = La → Dy) nano-wires/nano-rods crystallized in the hexagonal form, while (Ho → Lu, Y) $LnPO_4$ materials were of hexagonal structure when synthesized using low-temperature hydrothermal conditions. The X-ray analysis indicated that the intensity of the (200) peak was much stronger than those of the other peaks, and was different from bulk hexagonal $LnPO_4$ (Ln = La → Dy). This indicated that the obtained $LnPO_4$ (Ln = La → Dy) nano-wires/nano-rods grew preferentially along the [001] direction (the c axis). Based on the crystal structure of $LaPO_4$, Fang et al.

Fig. 10.1 SEM images and TEM insets of $LaPO_4$ nanostructures obtained in a typical synthesis with different volumes of H_3PO_4–EG (New J. Chem. **2009,** 33, 1657)

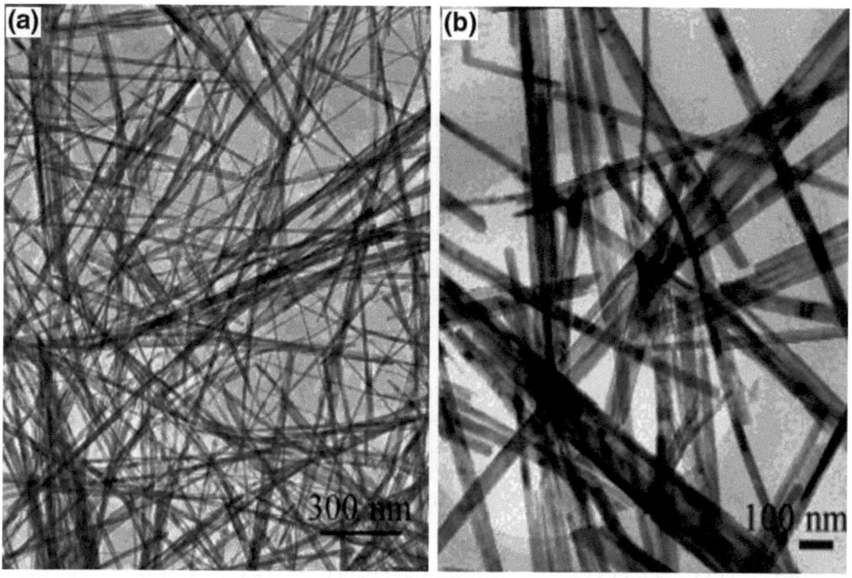

Fig. 10.2 TEM images of the obtained hexagonal $LaPO_4$ (J. Am. Chem. Soc. **2003**, 125, 16025)

[6] suggested that the growth of $LaPO_4$ nano-wires were highly anisotropic along the c axis.

Buhler [7] used microwave irradiation for synthesizing highly luminescent $LaPO_4$: Ce, Tb nanocrystals based on the use of ionic liquids (ILs) as reaction media (Fig. 10.3). During the synthesis, a solution of rare earth chlorides in ionic liquid of tributyl methyl ammonium triflylimide (IL) and a co-solvent (for example, ethanol, dimethyl formamide (DMF), dimethyl sulfoxide (DMSO) or pyridine) was added to the phosphate precursor (phosphoric acid), dissolved in the IL and a co-solvent, at 70 °C. The ionic liquids were selected from a variety of candidates for their extra-ordinary thermal and chemical stability. Precipitation of $LaPO_4$: Ce, Tb particles occurs immediately, as indicated by the appearance of a slight opalescence. The crystallinity of the preformed nanoparticles was enhanced by heating the dispersion to 300 °C within 10 s in a microwave oven. Transmission electron microscopy (TEM) images revealed spherical to slightly ellipsoidal particles with sizes of 9–12 nm and corresponding electron diffraction pattern and the lattice fringes observed in high-resolution images indicated that the particles are crystalline.

Fang et al. [15] developed a room-temperature synthetic protocol to obtain high yields of crystalline lanthanide phosphate nano-wires using a rotating tube microfluidic platform. They made use of the high surface energy single-walled carbon nanotubes (SWCNTs) as nucleation centres to grow, stabilize and isolate well-defined nanocrystals monomers before the fusion process into 1D nano-wires (Fig. 10.4) it was demonstrated that quantum yields close to the value of the

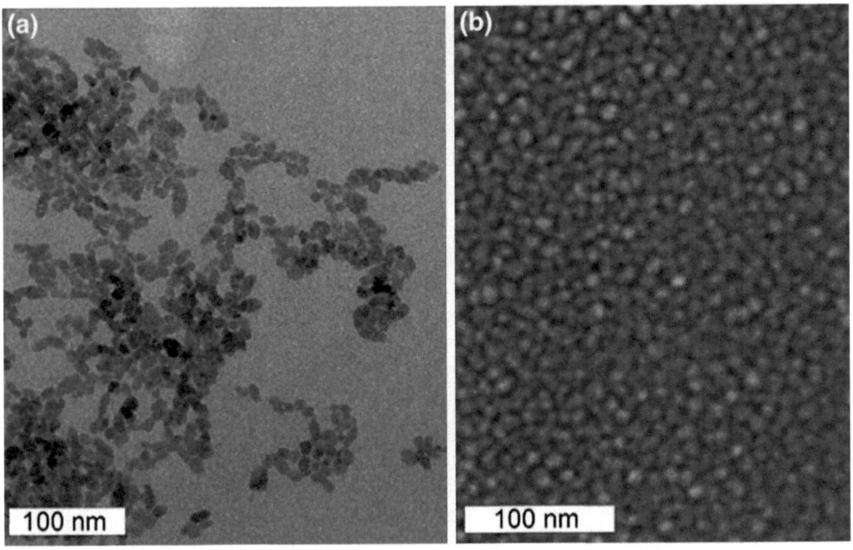

Fig. 10.3 **a** TEM and **b** SEM images of as-prepared $LaPO_4$: Ce,Tb nanocrystals (Angew. Chem. Int. Ed. **2006**, 45, 4864)

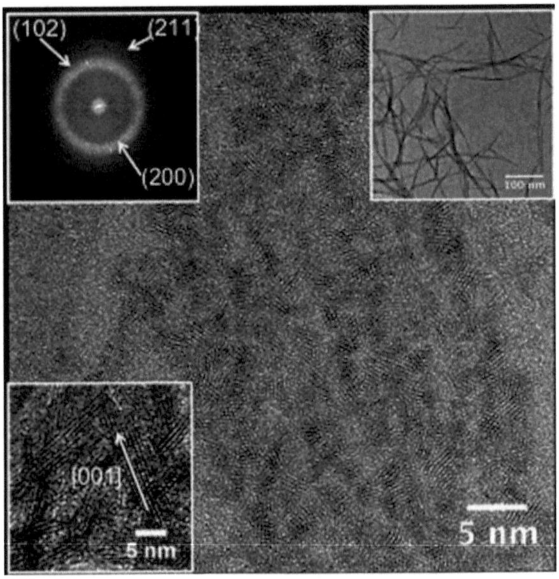

Fig. 10.4 TEM micrographs of 1D nano-wires of $LaPO_4$: Ce: Tb (Dalton Trans., **2011**, 40, 3122)

nano-crystalline phosphor material can be achieved without the use of a passivating shell in the nanoscale.

Matraszek et al. [16], in a typical *Pechini* process synthesised lanthanum phosphate by mixing citric acid and ethylene glycol with solutions of lanthanide

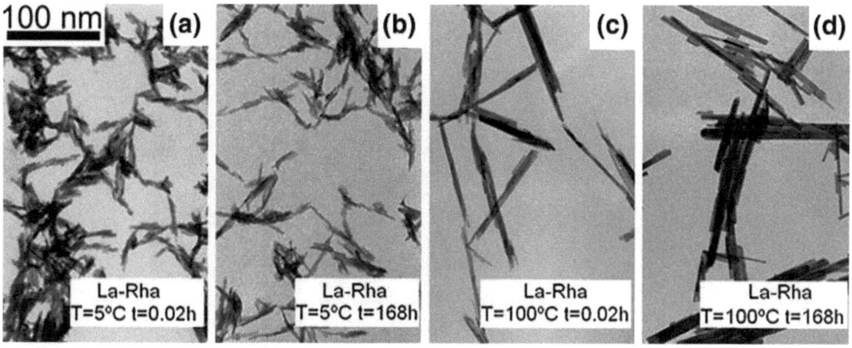

Fig. 10.5 TEM micrographs of La—rhabdophane synthesized at different temperatures (J. Nanopart. Res. (**2011**) 13, 4049)

nitrates in varying molar ratios; citric acid: La^{3+}—2:1, citric acid: ethylene glycol—1:1. The obtained solutions after concentration to a viscous liquid were mixed with $NH_4H_2PO_4$ in appropriate molar ratio under stirring. The dried gels were calcined in a furnace preheated at 500 °C. The lanthanum, cerium and praseodymium orthophosphates prepared by the above method were characterized by much smaller average crystallite size (about 16 nm) in comparison with that of the orthophosphates produced through precipitation from a phosphoric acid solution at a temperature of 80 °C. High specific surface area and increased mesopores were obtained. A direct precipitation technique used by T.R. Herrero resulted in rhabdophane $LaPO_4$ was synthesized by mixing an aqueous solution of either 0.1 M La $(NO_3)_3.9H_2O$ or (Nd $(NO_3)_3.9H_2O$ for $NdPO_4$) with an aqueous solution containing 0.1 M KH_2PO_4 and 0.09 M KOH (Fig. 10.5) [17].

More recently Yan li et al. [24] successfully synthesized luminescent hydrogel $LaPO_4$: Eu^{3+} using in situ polymerization. The $LaPO_4$ hydrogel showed very high emission intensities than those obtained from conventional methods. Xiao [25] synthesized hydrated and dehydrated $Eu3^+$/$Tb3^+$-doped $REPO_4$ (RE5Y, La and Gd) nanophosphors via a solvo-thermal technology wherein they altered the temperature and solvents. X-ray powder diffraction and scanning electron microscopy revealed varying structures and morphologies (Fig. 10.6), compared to the conventionally derived ones.

10.2 Sol-Gel Approach for Synthesis of $LaPO_4$ and Composites

Fisher et al. [26] successfully employed the Sol-Gel template-assisted synthesis method to produce $LaPO_4$: Eu nanotubes with sizes ranging from 20 to 200 nm (Fig. 10.7). Powder X-ray diffraction data indicated that the untemplated dried bulk powder was a mix of hexagonal and monazite phases, whereas the dried and

Fig. 10.6 Scanning electron microscope image of hydrothermal synthesis (J. Am. Ceram. Soc. (**2010**) 93, 2195)

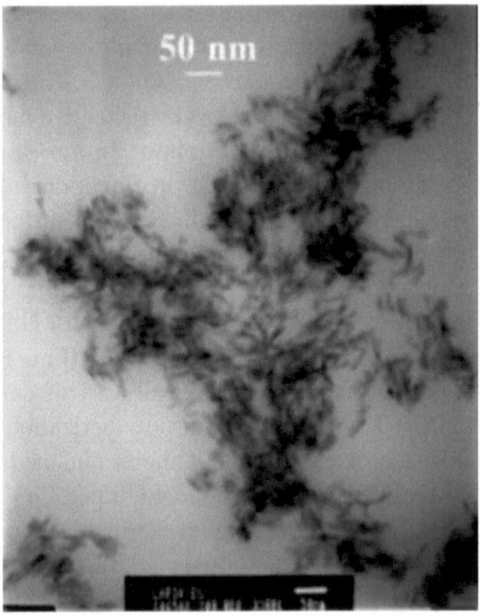

Fig. 10.7 TEM image of $LaPO_4$: Eu bulk powder dried at 70 °C, showing rod-like morphology (J. Phys. Chem. C **2008**, 112, 1901)

calcined bulk powder was of pure monazite phase. Guo et al. [27] prepared lanthanum phosphate using a sol-gel process in both ethanolic and aqueous media. The $LaPO_4$ gels prepared using this technique gave macro-crystalline $LaPO_4$.

Rajesh et al. [18] successfully employed room-temperature aqueous Sol-Gel process for synthesis of rod-shaped nano-crystalline lanthanum phosphate. Transmission electron microscopy revealed that the sol and gel particles have rod-shaped morphology and comparable particle sizes (Fig. 10.8). Precursor gel calcined at 400 °C possessed high specific surface area of 100 m^2/g, when measured using Brunauer–Emmett–Teller (BET) nitrogen adsorption analysis.

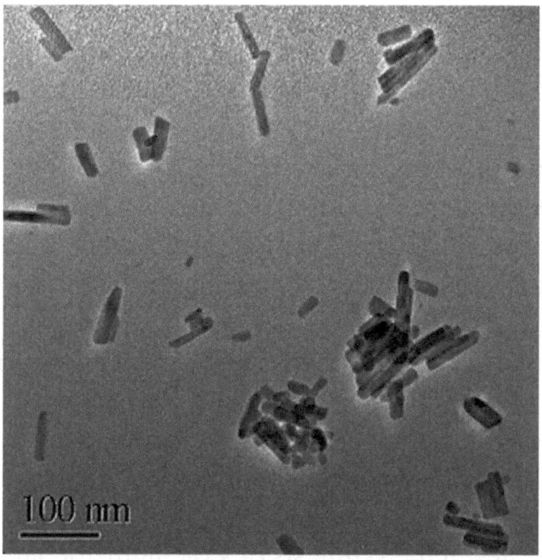

Fig. 10.8 TEM of the lanthanum phosphate gel calcined at 70 °C (Adv. Func. Mater. **2007**, 17, 1682)

Analysis of the reported literature available for the synthesis of lanthanum phosphate revealed that most of the methods involve multiple steps, high synthesis temperature/pressure and the products formed were in the size ranges of submicron to microns. Sol-Gel approach carried out in aqueous medium involving simple precipitation–peptization–gelation process was not widely reported and therefore was attempted for the synthesis of stoichiometric LaP- and LaP-based composites.

10.3 Lanthanum Phosphate Nanocomposites and Coatings — a Case Study

The efficient sol-gel synthesis of $LaPO_4$, and the development of nanocomposites and nano-coatings are reported in this study. The stability, hydrophobicity, non-reactivity with molten metals, densification behaviour and transparency are reported.

10.3.1 Sol-Gel Preparation of $LaPO_4$

The synthesis of stoichiometric $LaPO_4$, and composites is immensely important, as the non-stoichiometry will certainly have a deleterious effect on its high-temperature properties. Further, stable particulate sols derived through hydrolysis–peptization route, through strict control of surface charge, are required for functional transparent coatings. However, the testified synthesis techniques, apart from the sol-gel technique for lanthanum phosphate, resulted in submicron to micron sized particles. The Sol-Gel processing thus evolves as a low-temperature wet chemical method for the synthesis of stoichiometric nano-lanthanum phosphate. Synthesis of lanthanum phosphate and its composites using the room-temperature Sol-Gel process could be the method for control of stoichiometry and purity that is essential for developing materials for high-temperature as well as functional applications. Here a modified Sol-Gel processing in which precipitation–peptization method is followed by electrostatic stabilization is adopted for obtaining the lanthanum phosphate particles [18, 28].

Lanthanum phosphate ($LaPO_4$) sol which is obtained at a pH of 1.8–1.9 (Sankar et al. [28]) obtained using sol-gel processing was allowed to self-assemble in ammonia atmosphere. The $LaPO_4$ gel obtained can be dried and heated to various temperatures, while the sol obtained at earlier stages could be used for coating purposes.

One of the main issues in the synthesis of stoichiometric lanthanum phosphate is the removal of anions from the precipitation medium. Any excess phosphate remaining in the precipitate may seriously affect the temperature stability of the $LaPO_4$. Addition of ammonia solution (NH_4OH) facilitates the formation of ammonium salts with anions which can be easily removed through filtration/centrifugation/washing as well as through easy decomposition during thermal treatment.

The preparation of $LaPO_4$-based nanocomposites was carried out using aqueous Sol-Gel route where intimate mixing of the constituents is guaranteed by regulating the zeta potential (surface charge) in the precursor. The homogenous mixing of the two phases was made sure by dispersing the precursor salts solution of the second-phase additives in the sol of lanthanum phosphate prepared. This case study involves the use of zirconia and yttria as the additives, since most of the properties of these are identical with that of $LaPO_4$ and $LaPO_4$ has emerged as an excellent interphase as well as reinforcement material for ceramic composites over the years [29–34].

For the synthesis of $LaPO_4$-**x**wt% nano-composite, corresponding nitrate (Y and Zr) was initially dissolved in minimum amount of distilled water and then added to 0.05 M lanthanum phosphate sol prepared as mentioned above under constant stirring condition. After stirring for homogeneous mixing of lanthanum phosphate sol and the nitrate solution, the precipitate and $LaPO_4$ were flocculated slowly by adjusting the pH to ~ 8 using ammonia solutions. The precipitate obtained was later dried and calcined to obtain the composite. The composite formation was confirmed by carrying out thermal decomposition studies by heating to a temperature of 1000 °C using a TG–DTA analyzer.

10.3.1.1 Preparation of $LaPO_4$ Nano-coatings

The coatings were obtained starting from stock sols of lanthanum phosphate. The sol thus obtained was coated over surfaces and annealed to different temperatures

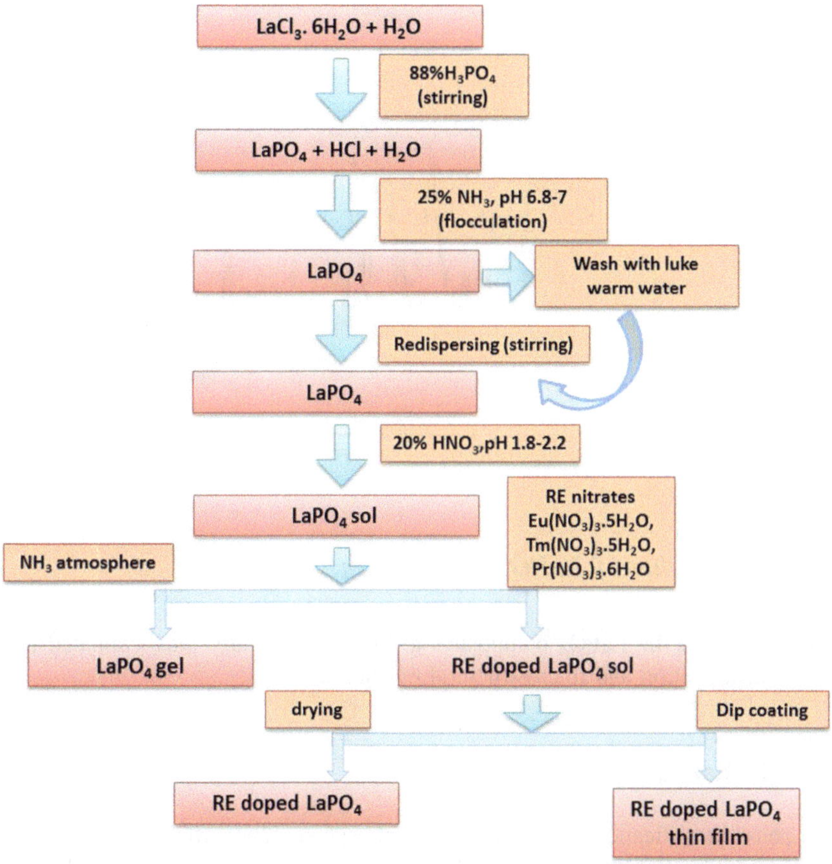

Fig. 10.9 Flow chart of preparation of $LaPO_4$ gel and $LaPO_4$ sol-based coatings

(100–400 °C) to prepare the coatings. The $LaPO_4$ nano-coatings were made on glass substrate using a dip coater with different advancing and receding rates and soaking times. Parameters were optimized to obtain a good adherent coating.

The coated glass slides were dried at 50 °C for 5 h and then annealed at a temperature range of 100–400 °C. The morphology of the samples was observed under a scanning electron microscope. Coatings based on $LaPO_4$ as well as Eu-, Tm- and Pr-doped $LaPO_4$ were prepared using dip coating technique. For the doped sol synthesis, rare earth nitrates such as Eu $(NO_3)_3.5H_2O$, Tm $(NO_3)_3.5H_2O$, Pr $(NO_3)_3.6H_2O$ were used. This precursor sol was used for the coating of substrates (Fig. 10.9).

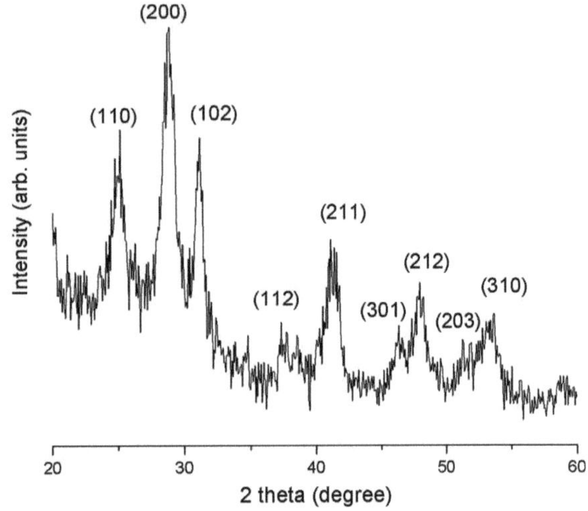

Fig. 10.10 Powder X-ray diffraction pattern of 70 °C dried lanthanum phosphate J. Sol-Gel Sci. Technol (**2011**) 58,195)

10.3.2 Stability of $LaPO_4$ Composites

The stability of the $LaPO_4$ composites at high temperatures was evaluated using phase identification and microstructural evaluation.

10.3.2.1 $LaPO_4$ Phase Transformation

The phase identification of the lanthanum phosphate nanoparticles was done from the X-ray diffraction pattern obtained for the gel powder dried at 70 °C. The peaks are well indexed and found matching with hexagonal rhabdophane phase pattern with JCPDS file no: 46-1439. The (200) peak intensity is much higher than the (102) peak intensity and is in accordance with the growth patterns reported for unidirectional growth of lanthanum phosphate nano-rods differentiating from bulk hexagonal lanthanum phosphate.

The phase patterns obtained for $LaPO_4$ particles heat treated at various temperatures are shown in Fig. 10.11. The $LaPO_4$ changes from hydrated form (Fig. 10.10) to hexagonal phase (A) on heating at 400 °C and the XRD pattern obtained matches with the JCPDS file no: 04-0635. The particles appeared to possess good crystalline nature. The further transformation from hexagonal phase to monoclinic phase (B) could be seen around 800 °C (JCPDS file no.35-0731). From the pattern obtained for sample calcined at 1200 °C, the complete transformation to the stable monoclinic phase for the lanthanum phosphate was observed.

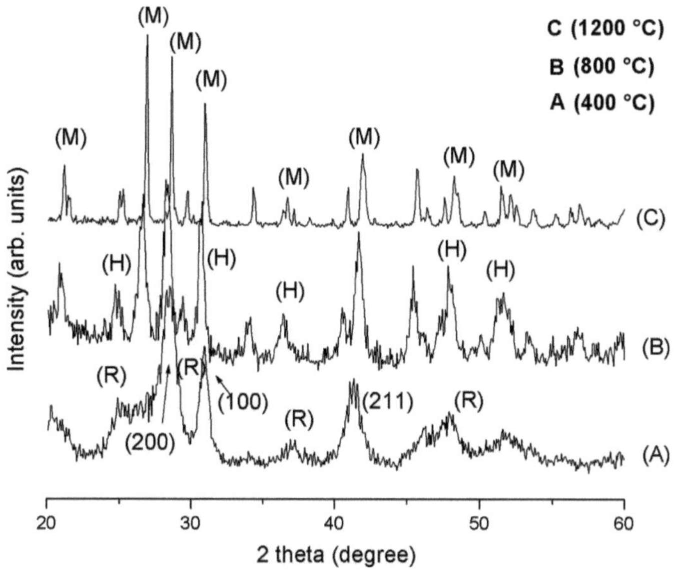

Fig. 10.11 PXRD patterns of lanthanum phosphate gel powder calcined at 400, 800 and 1200 °C (J. Sol-Gel Sci. Technol (**2011**) 58,195)

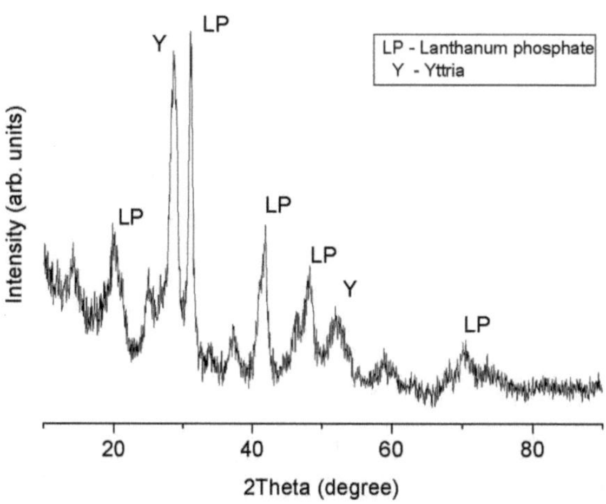

Fig. 10.12 X-ray diffraction pattern for $LaPO_4$–Y_2O_3 composite precursor heated at 600 °C (Mater. Res. Bull. (**2012**) 47,1835)

10.3.2.2 Lanthanum Phosphate Composite Phase Transformation

X-ray diffraction pattern obtained for $LaPO_4$–Y_2O_3 composite gel precursor heated at 600 °C is shown in Fig. 10.12 and it could be seen that the peaks are well indexed with that of monoclinic $LaPO_4$, JCPDS file no.46-1326 and that of cubic

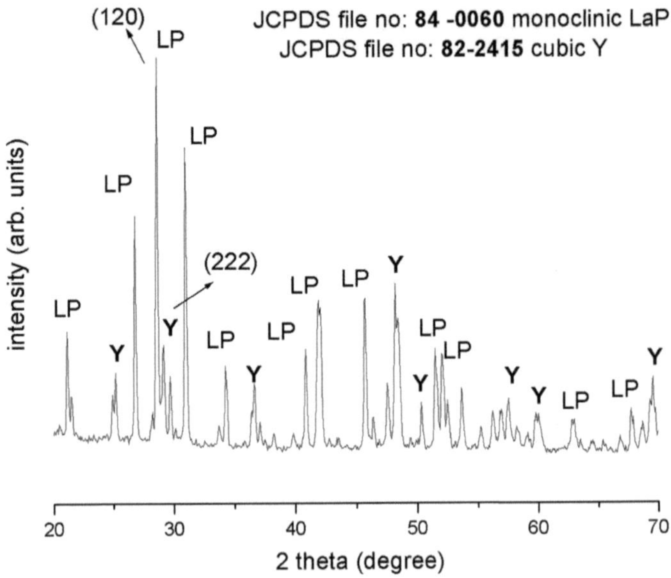

Fig. 10.13 X-ray diffraction pattern for $LaPO_4$–Y_2O_3 nanocomposite calcined at 1300 °C (Mater. Res. Bull. (**2012**) 47, 1835)

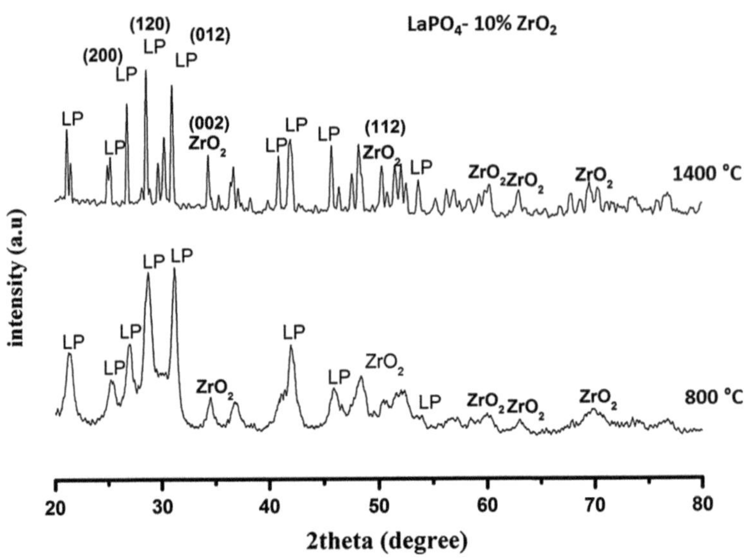

Fig. 10.14 X-ray diffraction pattern for $LaPO_4$–ZrO_2 composite precursor heated at 800 °C and further calcined at 1400 °C

Fig. 10.15 TEM images of as-precipitated $LaPO_4$ rods (J. Sol-Gel Sci. Technol (**2011**) 58,195)

Y_2O_3, JCPDS file no.44-0399. The phases are really distinct and there seems to be no reaction between the constituents. The results are in accordance with previous reports indicating that the formation of cubic yttrium oxide formation occurs when the hydroxide is heated at 500 °C. The composite when heated to higher temperatures of 1300 °C, corresponding X-ray diffraction pattern obtained (Fig. 10.13), shows that there is no reaction between the $LaPO_4$ and Y_2O_3 and the composite is phase pure with the diffraction peaks well indexed with monoclinic $LaPO_4$ (JCPDS file no.84-0600) and cubic Y_2O_3 (JCPDS file no.82-2415).

In the case of the $LaPO_4$–ZrO_2 composite, the precursor was heated to a temperature of 800 °C and the corresponding X-ray diffraction patterns were recorded as shown in Fig. 10.14. The peaks obtained are well indexed with monoclinic $LaPO_4$ (JCPDS file no. 35-0731) and tetragonal ZrO_2 (JCPDS file no. 81-1544) for the sample heated at 800 °C. The diffraction peaks are indexed to (200), (120) and (012) crystal planes of monoclinic $LaPO_4$ appearing at 2θ values of 26°, 28° and 30°. The low-temperature formation of the tetragonal phase of the zirconia particles could be attributed to the lower particle size obtained by the Sol-Gel method adopted. The composite was further heated to a temperature of 1400 °C, and the results showed no reaction between the constituents as well as absence of any new phases (Fig. 10.14). This shows the greater possibility of using the composite prepared at high-temperature applications.

10.3.3 Morphological Evolution and Features

10.3.3.1 Lanthanum Phosphate

Initially, the morphology of the as-precipitated $LaPO_4$ from lanthanum salt using phosphoric acid was examined using TEM (Fig. 10.15). The micrographs show an elongated and very long rod-like morphology of the particles during this

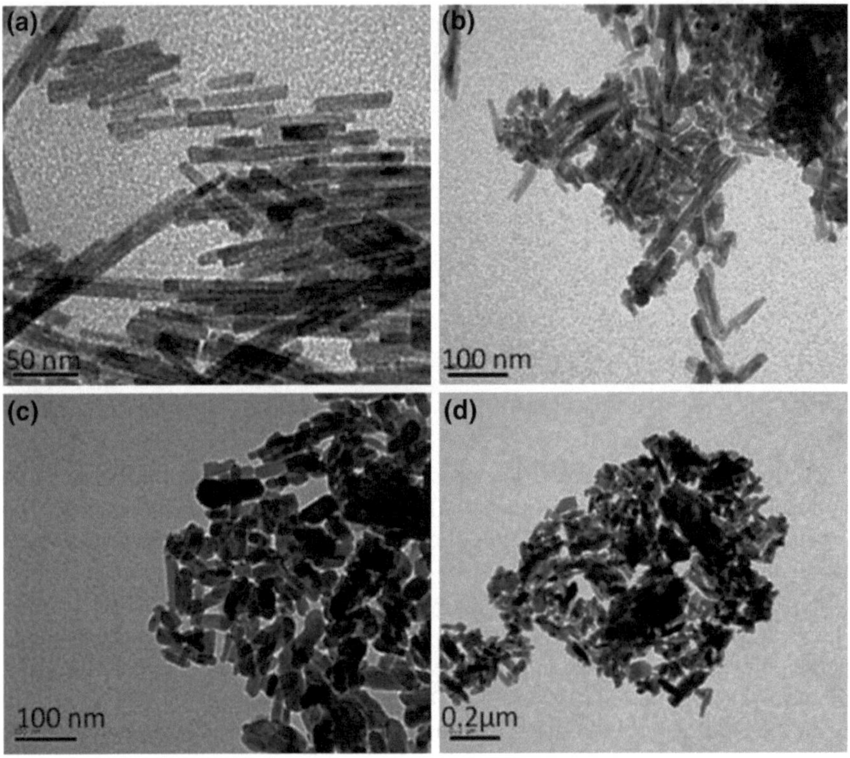

Fig. 10.16 TEM micrographs of $LaPO_4$ dried at (**a**) 70 °C, (**b**) 100 °C, (**c**) and (**d**) 800 °C. Use new pictures obtained from recent studies (J. Sol-Gel Sci. Technol (**2011**) 58,195)

precipitation process. A few of the rods obtained are seem to be agglomerated. TEM micrographs show the unidirectional growth of lanthanum phosphate particles in the precipitated state itself. This unidirectional precipitation of $LaPO_4$ could be due to the spherical diffusion model growth as explained by Fang et al. [19] and the structure can also be envisaged as infinite linear chains extending along the c axis and such a structure resulting in growth along one direction.

TEM micrographs of the dried and calcined samples (Fig. 10.16) reveal the rod-like morphology obtained for the phosphate particles. The samples were dried at 70 and 100 °C and later calcined at 800 °C for the morphological analysis. We could see an average length of ~50 nm and diameter ~15 nm after heat treatment of particles at 800 °C. This shows that the morphology is more or less retained even at higher temperatures (Fig. 10.16).

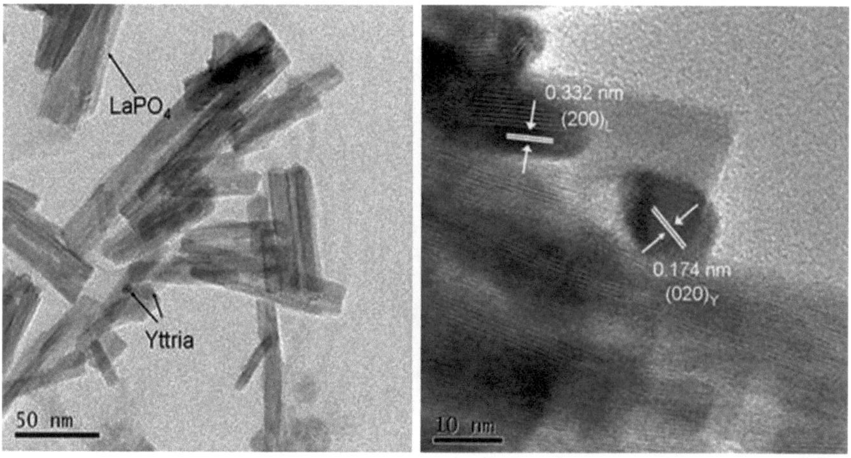

Fig. 10.17 TEM micrographs of lanthanum phosphate–yttria composite calcined at 600 °C

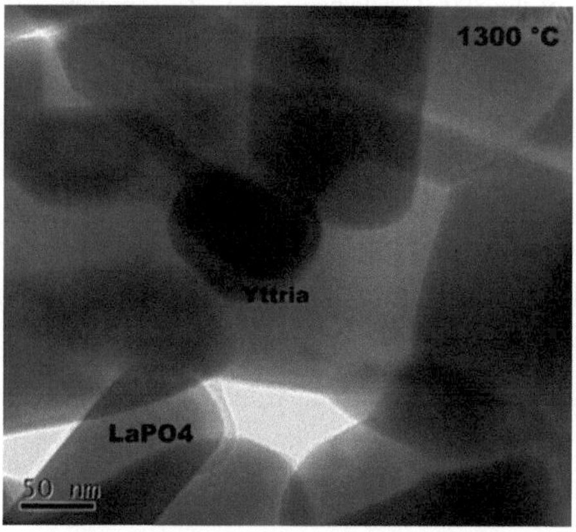

Fig. 10.18 HRTEM micrograph of $LaPO_4$-20 wt%Y_2O_3 composite calcined at 1300 °C

10.3.3.2 Lanthanum Phosphate Composites

The $LaPO_4$-20 wt%Y_2O_3 composite precursor powder heated to 600 °C was observed under transmission electron microscope (Fig. 10.17). The corresponding micrographs show that $LaPO_4$ possesses rod-shape morphology and yttria was found to be spherical in shape. $LaPO_4$ nano-rods having ~70–80 nm length and of ~5 nm width were found along with yttria particles having average diameter of ~10 nm. The EDAX spectrum (Fig. 10.19) shows the phases of $LaPO_4$ and ZrO_2 during the formation of $LaPO_4$-ZrO_2 composite. The rod morphology of $LaPO_4$ and spherical nature of ZrO_2 nanoparticles is evident from the corresponding TEM micrographs.

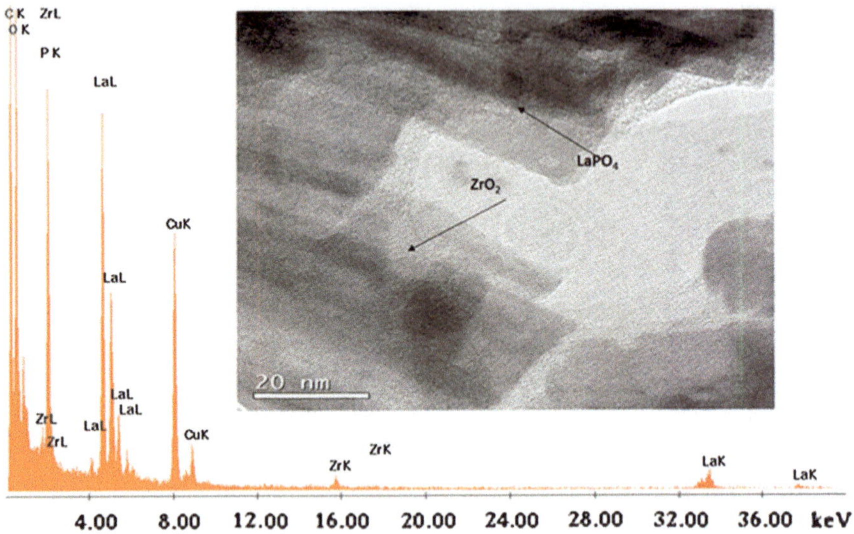

Fig. 10.19 TEM image of the 800 °C calcined $LaPO_4$–ZrO_2 precursor powder with corresponding EDAX spectrum

TEM micrograph of lanthanum phosphate–yttria nanocomposite calcined at 1300 °C is presented in Fig. 10.18. The micrograph clearly indicates the presence of two phases with its corresponding planes. As evident from phase identification studies which is supported by TEM micrographs, the composites synthesized are phase pure at higher temperatures and could be used for high-temperature applications (Fig. 10.19).

10.3.4 Densification Behaviour

In order to analyse the densification behaviour of $LaPO_4$ and the composites, the powders were uniaxially pressed to obtain pellets samples which were then subjected to TMA (dilatometric) measurement at a heating rate of 10 °C/min to analyse the sintering behaviour as well as evaluate the coefficient of thermal expansion. Figure 10.20 shows the sintering behaviour of $LaPO_4$. It could be seen that the densification occurs in three steps with a shrinkage onset at 800 °C. The densification shrinkage is completed at 1300 °C (Fig. 10.20b). Density as high as >99% was obtained on sintering at 1300 °C. This low-temperature sinterability of the sol–gel-derived $LaPO_4$ compared to conventional methods is certainly due to fine particle size and good surface area, which enhances the powder diffusion characteristics. The coefficient of thermal expansion (Fig. 10.20a) measured for $LaPO_4$ was found to be 8.6×10^{-6}/°C in the temperature range of 100—1300 °C which is

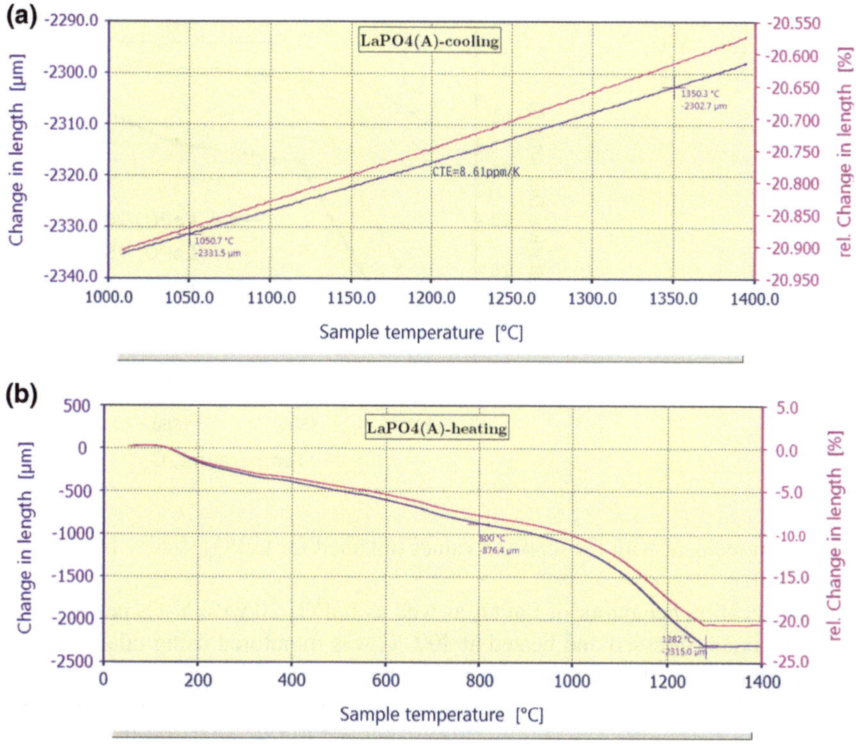

Fig. 10.20 Thermo-mechanical analysis of $LaPO_4$ **a** coefficient of thermal expansion and **b** sintering behaviour of $LaPO_4$

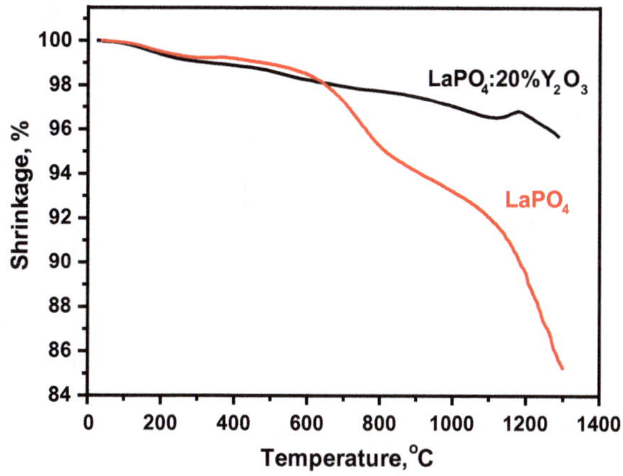

Fig. 10.21 Shrinkage schedule obtained for $LaPO_4$-20% Y_2O_3 composite and $LaPO_4$

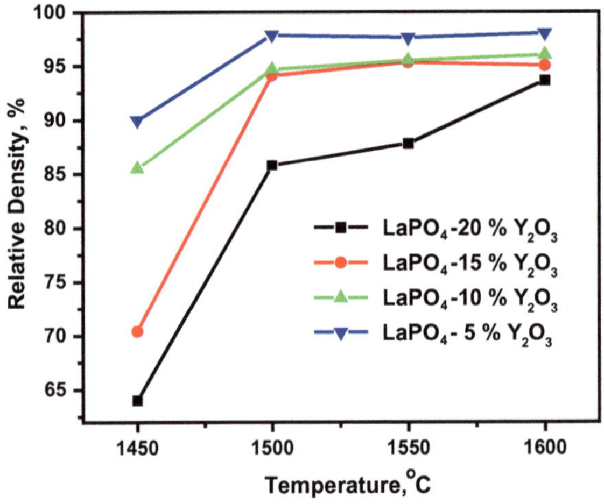

Fig. 10.22 Plot of relative density of $LaPO_4$–Y_2O_3 composite versus temperature

in good agreement with the reported values obtained for $LaPO_4$ (9.6×10^{-6}/°C, Du et al. [45]).

Densification behaviour of $LaPO_4$ as well as $LaPO_4$-20 wt% Y_2O_3 pellets which were uniaxially pressed and heated at 400 °C was monitored using dilatometry up to a temperature of 1400 °C and the results are shown in Fig. 10.21. The shrinkage profiles indicated that the rate of shrinkage in composite is less compared to pure $LaPO_4$. For pure $LaPO_4$, sintering is completed at 1300 °C while for the composite ($LaPO_4$-20%Y_2O_3) higher temperatures are essential to get acceptable levels of densification due to the presence of second-phase dispersions.

The relative density of sintered samples was calculated by rule of mixtures. The theoretical density of $LaPO_4$ is 5.06 g/cc and that of yttrium oxide is 5.03 g/cc. The relative densities of different compositions at varying sintering temperatures are represented in Fig. 10.22. It could be observed that the sintered density increased with increasing temperature for all compositions. The relative density of $LaPO_4$-5 wt% Y_2O_3 was found to be 98.6%, whereas $LaPO_4$-20 wt% Y_2O_3 has 93% density at a sintering temperature of 1600 °C. Pure $LaPO_4$ attained sintered density of 99.8 wt% at 1300 °C. The decrease in density of $LaPO_4$ on the addition of Y_2O_3 is due to distribution of fine yttria particles on the grain boundaries of $LaPO_4$ (Fig. 10.23).

The microstructural investigations on sintered pellets at different temperatures were carried out on polished and thermally (50 °C less than the sintering temperature) etched specimens. Figure 10.23 shows the SEM of $LaPO_4$–Y_2O_3 composite sample sintered at 1600 °C with varying concentration of yttria content. Average grain size is found to be ~2 μm for $LaPO_4$. The EDAX spectrum obtained confirmed the presence of yttria along the grain boundaries.

Fig. 10.23 SEM images of $LaPO_4$-Y_2O_3 composites sintered at 1600 °C, **a** 5, **b** 10, **c** 15 and **d** 20 wt% of yttria

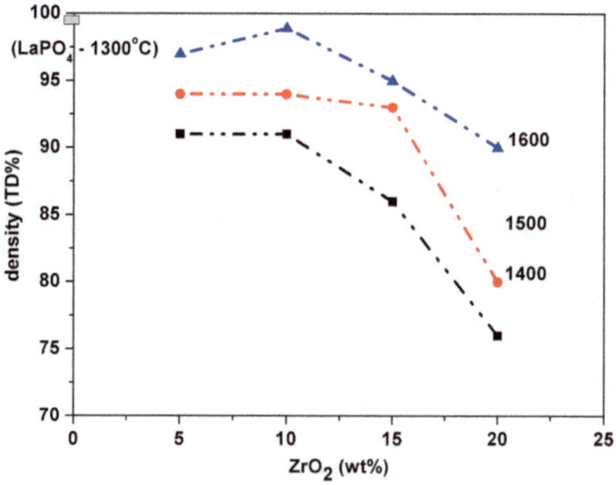

Fig. 10.24 Plot showing densities acquired for the composite samples with respect to sintering temperature and ZrO_2 addition

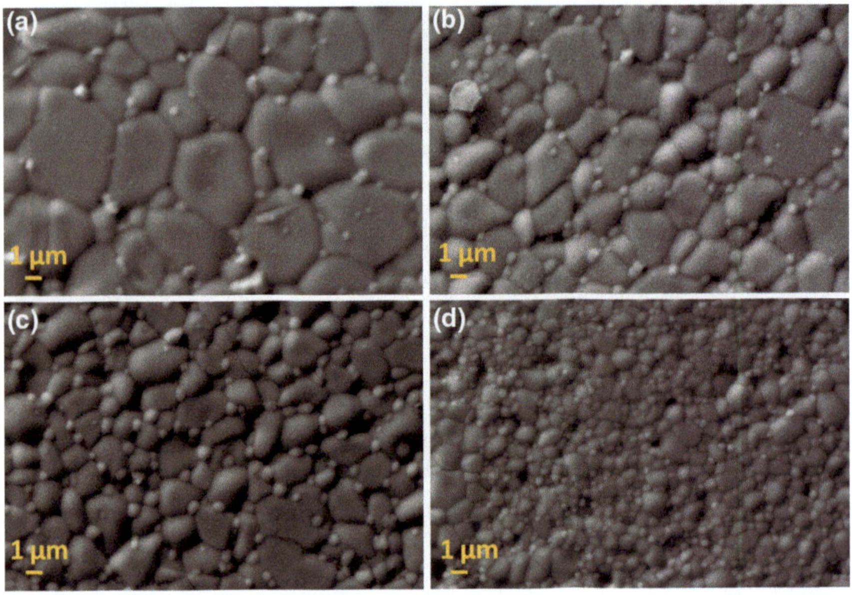

Fig. 10.25 SEM micrographs of $LaPO_4$–ZrO_2 composites sintered at 1600 °C **a** $LaPO_{4-5}$% ZrO_2 **b** $LaPO_4$-10% ZrO_2 **c** $LaPO_4$-15% ZrO_2 **d** $LaPO_4$-20% ZrO_2

10.3.4.1 $LaPO_4$–Zirconia Composite System

Figure 10.24 revealed that on sintering at 1400 °C $LaPO_4$-ZrO_2 composite at samples with 5 and 10 wt% ZrO_2 yielded >90% TD density while $LaPO_4$—20% ZrO_2 had poor sinterability. Sintering at 1600 °C, all the four compositions yielded greater than 90% of the theoretical density. $LaPO_4$-10% ZrO_2 composite yielded density of greater than 98% with 5 and 15% compositions having > 95% TD. Although lanthanum phosphate gets sintered at ~1300 °C, the decrease in density on the addition of ZrO_2 could be attributed to the distribution of fine ZrO_2 particles on the grain boundaries of $LaPO_4$ as evident from the SEM observation (Fig. 10.25).

The pellets were mirror polished, thermally etched and were observed under scanning electron microscope for the microstructural evaluation. The SEM micrographs obtained for composite samples, sintered at 1600 °C, with addition of 5–10% ZrO_2 presented in Fig. 10.25 reveal that the size of $LaPO_4$ grains are smaller compared to that of pure $LaPO_4$ (>4 µm). With the increase in content of ZrO_2 to 20%, the grain growth of $LaPO_4$ has been inhibited to nearly 50% to average grain size of ~1–2 µm as shown in Fig. 10.25d ($LaPO_4$-20% ZrO_2). The distribution of ZrO_2 is evident from the micrographs obtained.

Although there are reports on ZrO_2 acting as grain growth inhibitor, it is hypothesized in this case that ZrO2 will restrict the growth of phosphate grains at their grain boundaries due to the higher thermal expansion coefficient of ZrO_2 in comparison with $LaPO_4$. The reduction in sintered grain size will definitely improve the $LaPO_4$ stability to higher temperatures. Further (as Lin et al. [35] suggests), the presence of La^{3+} induces oxygen vacancies and strain in zirconia which stabilizes tetragonal phase and decreases the concentration of monoclinic phase with poor mechanical properties.

10.3.5 Hydrophobicity of $LaPO_4$ Nano-coatings

Lanthanum phosphate is investigated for its non-wetting characteristics with respect to water in the form of coatings on glass surfaces and as sintered solids, separately. The contact angle provides the inverse measure of wettability of a surface with water. The contact angle (θ), which is the angle at which the liquid–vapour interface meets the solid–liquid interface, is determined by the resultant between adhesive and cohesive forces. Hydrophobic property of a material is of extreme prominence in inducing self-cleaning property to substrates and surfaces and is of particular importance in wettability control. Hydrophobic coatings are expected to aid various industrial usages including anti-wetting, anti-ice, anti-rusting as well as reducing friction resistance due to the very low solid–liquid interaction.

Common ceramic oxides like alumina as well as zirconia are tend to be hydrophilic in nature as hydrogen bonds with interfacial water molecules established in order to fulfil the octet configuration resulting in clathrate or hydrophilic hydration structure (Argyris et al. [36]) leading to the wetting behaviour. An important property that administrates a material's affinity towards water molecules is the surface polarity. The increase in polarity generally leads to increased hydrophilicity. Azimi et al. [37] suggested that rare earth oxides (REOs) which are expected to be polar and hydrophilic have an intrinsic hydrophobic behaviour, primarily due to the electronic structure of the lanthanide series, where unfilled 4f orbitals of the rare earth atoms are shielded from interactions with the surrounding environment by the full octet of electrons in the $5s^2p6$ outer shell. Consequently, such atoms would have a lower tendency to exchange electrons and form hydrogen bond with interfacial water molecules. Considering characteristic electronic structure of rare earth atoms, lanthanum phosphate is expected to be hydrophobic in nature and initial result (Table 10.1) validates this argument.

Water Contact angle data measured for the $LaPO_4$ powder samples heated for 3 h duration at different temperatures ranging from 70 to 800 °C are given in Table 10.1. The contact angle values for $LaPO_4$ particles dried at 70 °C and calcined at 400 °C are nearly identical while there is a marginal decrease in the value for particles heated at 800 °C. The inherent hydrophobic character of $LaPO_4$ is the

Table 10.1 Contact angle values for $LaPO_4$ dried at different temperatures

$LaPO_4$ samples	Temperature (°C)	Contact angle (°)
Sample I	70	89
Sample II	400	87
Sample III	800	84

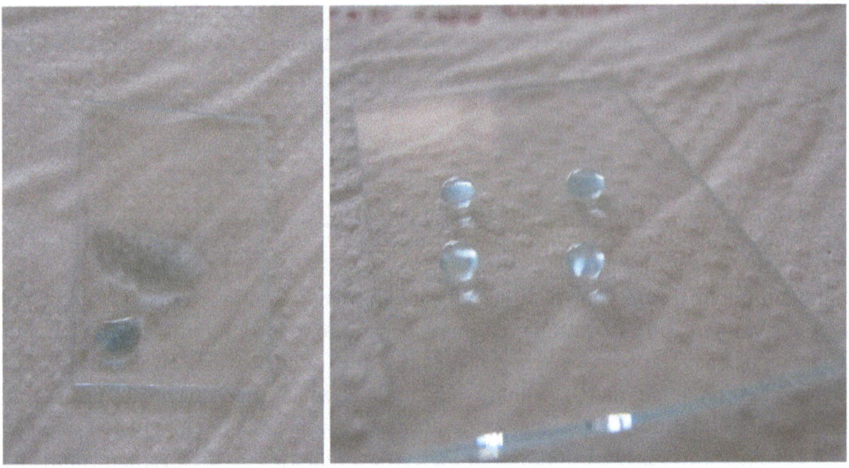

Fig. 10.26 Water droplets on coated and uncoated glass plate

Fig. 10.27 Water droplet on $LaPO_4$: Pr coated glass surface

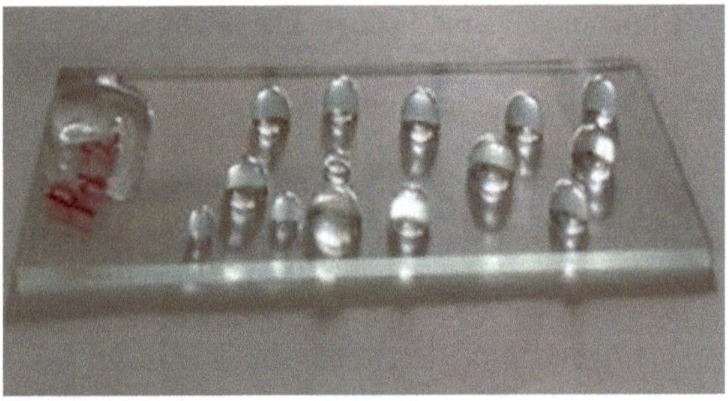

reason or the minimal difference in values at different temperatures. The hydrophobic behaviour possibly arises due to the electronic structure of the $LaPO_4$ where unfilled 4f orbitals of the rare earth atoms are shielded from interactions with the surrounding environment by the full octet of electrons in the $5s^2p^6$ outer shell and also by shielding effects of the bulky PO_4 groups. The values obtained from repeated measurements indicate the possibility of using $LaPO_4$ as a non-wetting inorganic material.

The coatings developed over glass surface are found to be hydrophobic in nature with water droplets assuming spherical shape when compared to the wetting of water drops in uncoated samples (Fig. 10.26). These films were checked for the hydrophobic behaviour using sessile drop method. The thin films showed water contact angle value as high as 120°, which was significantly higher than the water contact angle values of ~20° measured for the uncoated glass substrate. This hydrophobic nature of $LaPO_4$ coatings is explained on the basis of the inherent hydrophobicity arising from the electronic structure as well as the surface roughness

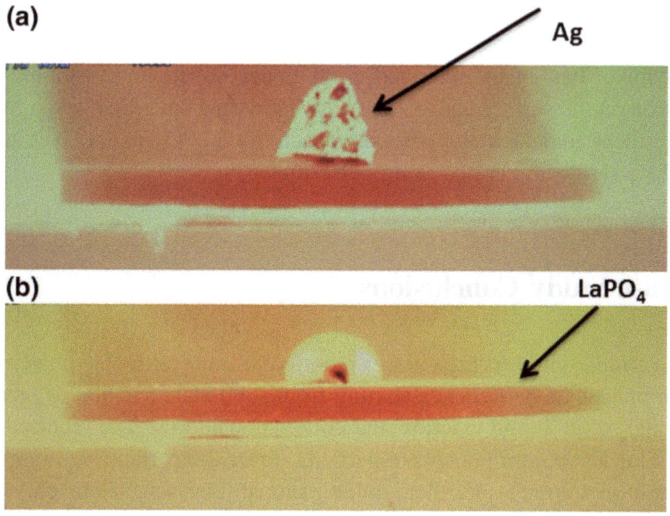

Fig. 10.28 **a** Silver starting to melt (966 °C) **b** Silver melts over the $LaPO_4$ surface (1050 °C) and forms a globule

induced by the nano-rods over the surface of the glass surface. The Re(Rare Earth)-doped $LaPO_4$ sol-coated glass substrates also show hydrophobic behaviour (Fig. 10.27).

10.3.6 Non-reactivity of $LaPO_4$ Nano-coatings

Crucible materials for melting metals should not contaminate the melt. During metal casting operations the moulds/crucibles are usually coated with layers of release agents comprising mainly of inorganic materials like alumina, zirconia, hexagonal BN and graphite. The layers acting as an interfacial coating prevents chemical interaction with mould material and casting metal. Non-reactivity of lanthanum phosphate with the molten metals facilitates the use of lanthanum phosphate-based coatings and monoliths as non-reactive surfaces for molten metals for high-temperature applications such as mould release coatings or crucibles for metal casting. Sintered lanthanum phosphate discs were kept in furnace with metallic silver placed over the disc. The furnace was heated above the melting point of Ag (1050 °C). The IR imaging revealed the formation of a globule of Ag (Fig. 10.28) without spreading indicates the non-wettability of lanthanum phosphate surface to molten metal. Further the contact angle was measured to be >125° showing the non-reactivity of the material under study.

Further, the reactivity with zinc metal was done. Lanthanum phosphate nano-powder synthesized was uniaxially pressed to form pellets which was covered with Zn metal flakes (Fig. 30a) and was heated to the melting point of the metal with a holding time of approximately 20 mts. When cooled, it was found that the $LaPO_4$ is totally non-reactive with the Zn metal flake which was visibly observed as the metal flake was easily removed from pellet on cooling with only a small impregnation on the surface caused by the pellet during softening of the metal flake. This experiment indicated that the non-reactivity $LaPO_4$ pellet over the melted metal surface shows the anti-wetting behaviour.

10.4 Case Study Conclusions

There are various synthesis procedures reported for rare earth phosphate, especially lanthanum phosphate particulates starting from solid-state reaction to precipitation, hydrothermal, solid–liquid reaction and colloidal synthesis. The reported synthesis techniques for lanthanum phosphate particles cover the particle size in micrometre or submicron metre range. Further, stable nanoparticle suspensions and lanthanum phosphate having excellent stoichiometry are not covered. Particulate colloidal suspensions derived through hydrolysis–peptization route stable for long time through strict control of surface charge are required for functional transparent coatings and no detailed report on such synthesis is available. Although high-temperature phase stability of lanthanum phosphate is studied, synthesis and properties of lanthanum phosphate composites are reported seldom. Only a few reports are available on the reactivity of $LaPO_4$ with molten metals. The hydrophobic behaviour of lanthanum phosphate is not all reported.

Synthesis of stoichiometric lanthanum phosphate nanoparticles from lanthanum chloride through an environment friendly aqueous sol-gel method was done and the resultant compositions were characterized. The nano-rod morphology was found to happen in the initial step of the current Sol-Gel process and this morphology is retained even at higher temperatures. The average length of the $LaPO_4$ nanorods after precipitation was ~ 110 nm and which was further reduced to ~ 80 nm. The lanthanum phosphate was compacted and sintered at temperature as low as 1300 °C due to the fine size and was characterized using TGA, FTIR, XRD, high-resolution transmission electron microscopy, scanning electron microscopy and contact angle measurements. The contact angle values obtained in the range of 80–90° for the $LaPO_4$ particles indicate the potential for further applications in hydrophobic non-wetting coatings and composites. Lanthanum phosphate-based yttria ($LaPO_4$–Y_2O_3) and zirconia composites ($LaPO_4$–ZrO_2) are successfully prepared extending the precipitation–peptization mechanism. Both yttria and zirconia contents of 5–20 wt% were used to synthesize the composites and with $LaPO_4$-10% ZrO_2 and $LaPO_4$–5% Y_2O_3 helped in achieving >98% TD obtained on sintering at 1600 °C. The stability of the composites at higher temperature was confirmed using phase identification as well as microstructural evaluation which showed no reaction between the constituent phases. Microstructural evaluation confirmed the grain growth inhibition of lanthanum phosphate by the addition of ZrO_2. [38] The excellent insulation property obtained for the composite could prove to be advantageous in developing $LaPO_4$–ZrO_2-based composites as thermally insulating material and for high-temperature application including thermal barrier coatings and prove advantageous over the existing oxide-based TBC materials. Lanthanum phosphate is thus identified as an all inorganic and versatile material for developing hydrophobic and non-reactive surfaces [39]. Coatings and surfaces with multi-functional properties ranging from transparency to hydrophobicity could be achieved using $LaPO_4$. The reason for the enhanced hydrophobicity is related to structure of rare earths in general where the interaction with OH sites is prevented due to the shielding of 'f' shell by full octet filled outer shell, unlike in many oxides like aluminium oxide and zirconium oxide. Non-reactive crucibles for metal melting applications and coatings that are anti-wetting to molten metals can be developed for foundry and other metallurgical applications using $LaPO_4$. $LaPO_4$ coatings had rough surface due to elongated particle morphology and hence hydrophobicity due to the surface texture could also be made available in addition to the intrinsic hydrophobicity.

10.5 $LaPO_4$-Future Perspectives

Rare earth phosphate, lanthanum phosphate in particular, is characterized by attractive properties like low thermal conductivity (1.30 W/mK at 1000 °C), high-temperature stability (>1900 °C), non-reactivity with many of the high-temperature oxides and specifically non-reactivity and non-wettability with molten metals. Suitably modified and doped rare earth phosphates are known to possess interesting optical properties also. Structurally these phosphates identify with the natural monazites opening up possibilities of containing other metal ions within the structure. The synthesis methods are well reported. Most of the investigations have been concentrated on the major constituent of the rare earths family, lanthanum. Therefore, the family of rare earth phosphates opens up a large spectrum of research and application opportunities.

The low thermal conductivity leads to further investigations on use in thermal barrier coatings, high-temperature fibre-insulating materials and mesoporous lanthanum phosphate. Since lanthanum phosphate is a 'soft ceramic' with low hardness (5 GPa) compared to many high-temperature oxides (Alumina (20 GPa) and Zirconia (10 GPa)), effective use in the form of composites will be the best approach, including that of machinable ceramics, which is much less investigated at present.

The hydrophobic properties of lanthanide phosphates provide us opportunity to investigate applications for high-temperature coatings for even corrosive environments. Hydrophobic properties of lanthanide phosphates coupled with optical properties of metal oxides (titanium oxide, zinc oxide) could lead to composites to self-cleaning surfaces and with specific colour and multifunctionality. Hydrophobic composite coatings combined with alumina or zirconia could be useful for creating meso- and nano-porous layers in ceramic membrane substrates for improved water flux. Preliminary observations made on the structural features of lanthanum phosphate indicate promising possibility for use in immobilizing radioactive wastes imitating natural containment as in the case of monazite mineral. Detailed investigation on the metallophobic properties of lanthanum phosphate may lead to much wider perspectives in the field of metallurgy, foundry, materials processing and surface engineering.

Prospective applications of $LaPO_4$ can be summarized as in the following table:

Composition	Property	Application
Lanthanum phosphate and other mixed rare earth phosphates as coatings and monoliths	Non-wettability with molten metals	Mould release coatings Monoliths for metallurgical applications (For e.g. spray atomization nozzles)
Thin films of lanthanum phosphate	Hydrophobicity	Hydrothermally stable coatings (For e.g. in boilers)
Doped lanthanum phosphate	Luminescence	Biomarkers
Lanthanum phosphate	Low thermal conductivity	Thermal insulation coatings
Lanthanum phosphate	Hydrophobicity	As inorganic fillers to improve hydrophobicity

References

1. Hikichi, Y., Nomura, T.: J. Am. Ceram. Soc. **70**(10), 252 (1987)
2. Ruigang, W., Wei, P., Jian, C., Minghao, F., Zhenzhu, C., Yongming, L.: Mater. Chem. Phy. **79**, 30 (2003)
3. Kijowska, R.: J. Mater. Sci. **38**, 229 (2003)
4. Chai, Z., Gao, L., Wang, C., Zhang, H., Zheng, R., Webley, P.A., Wang, H.: New J. Chem. **33**, 1657 (2009)
5. Lucas, S., Champion, E., Bregiroux, D., Bernache-Assollant, D., Audubert, F.: J. Solid State Chem. **4–5**, 1302 (2004)
6. Fang, Y.P., Xu, A.W., Song, R.Q., Zhang, H.X., You, L.P., Yu, J., Liu, H.Q.: J Am Chem Soc **125**(51), 16025–16034 (2003)
7. Buhler, G., Feldmann, C.: Angew. Chem. Int. Ed. **45**, 4864 (2006)
8. Riwotzki, R., Meysammy, H., Schnablegger, H., Kornowski, A., Hasse, M.: Angew. Chem. Int. Ed. **40**, 573 (2001)
9. Yu, M., Lin, J., Fu, J., Han, Y.C.: Chem. Phys. Lett. **371**(1), 178–183 (2003)
10. Schuetz, P., Caruso, F.: Chem. Mater. **14**, 4509 (2002)
11. Yang, M., You, H., Liu, K., Zheng, Y., Guo, N., Zhang, H.: Inorg. Chem. **49**(11), 4996 (2010)
12. Buissette, V., Moreau, M., Gacoin, T., Boilot, J.P.: Adv. Func. Mater. **16**(3), 351–355 (2006)
13. Shen, J., Sun, L.D., Yan, C.H.: Dalton Trans. **42**, 5687–5697 (2008)
14. Meiser, F., Cortez, C., Caruso, F.: Angew. Chem. **116**, 6080 (2004)

15. Fang, J., Guo, Y., Lu, G., Raston, C.L., Iyer, K.S.: Dalton Trans. **40**, 3122 (2011)
16. Matraszek, A., Radominska, E., Szczygiel, I.: J. Therm. Anal. Calorim. **103**, 813 (2011)
17. Roncal-Herrero, T., Rodrıguez-Blanco, J.D., Oelkers, E.H., Benning, L.G.: J. Nanopart. Res. **13**, 4049 (2011)
18. Rajesh, K., Shajesh, P., Seidel, O., Mukundan, P., Warrier, K.G.K.: Adv. Func Mater. **17**, 1682 (2007)
19. Fang, Y.P., Xu, A.W., Song, R.Q., Zhang, H.X., You, L.P., Yu, J., Liu, H.Q.: J. Am. Chem. Soc. **125**(51), 16025–16034 (2003)
20. Murphy, K.E., Altman, M.B., Wunderlich, B.: J. Appl. Phy. **48**, 4122 (1977)
21. Min, W., Miyahara, D., Yokoi, K., Yamaguchi, T., Daimon, K., Hikichi, Y., Matsubura, T., Ota, T.: Mater. Res. Bull. **36**, 939 (2001)
22. Winter, M.R., Clarke, D.R.: J. Am. Ceram. Soc. **90**(2), 533 (2007)
23. Du, A., Wan, C., Qu, Z., Pan, W.: J. Am. Ceram. Soc. **92**(11), 2687 (2009)
24. Zheng, Y., Wang, Q., Cai, W., Yu, Y.: Op. Mater. **34**(7), 1019–1022 (2012)
25. Xiao, X., Yan, B.: J. Am. Ceram. Soc. **93**(8), 2195 (2010)
26. Fisher, M.J., Wang, W., Dorhout, P.K., Fisher, E.R.: J. Phys. Chem. C **112**, 2008 (1901)
27. Guo, Y., Wozniacki, P., Barkatt, A.: J. Mater. Res. **11**(3), 639 (1996)
28. Sankar, S., Warrier, K.G.: J. Sol-Gel. Sci. Technol. **58**, 195 (2011)
29. Marshall, D.B., Morgan, P.E.D., Housley, R.M., Cheung, J.T.: J. Am. Ceram. Soc. **81**, 951 (1998)
30. Morgan, P.E.D., Marshall, D.B.: a-structural materials properties microstructure and processing. Mater. Sci. Eng. **162**, 15 (1993)
31. Mawdsley, J.R., Kovar, D., Halloran, J.W.: J. Am. Ceram. Soc. **83**, 802 (2000)
32. Davis, J.B., Marshall, D.B., Morgan, P.E.D.: J. Eur. Ceram. Soc. **20**, 583 (2000)
33. a)Chawlaa, K.K., Liu, H., Janczak, R.J., Sambasivan, S.: J. Eur. Ceram. Soc. **20**, 551 (2000). b) Majeed, A., Vijayaraghavan, L., Malhotra, S.K., Krishnamurthy R.: Int. J. Mach. Tools Manuf.**48**, 40 (2008)
34. Ergun, C.: J. Mater. Process. Technol. **199**, 178 (2008)
35. J-Tsiar, L., H-Yang, Lu.: Ceram. Int. **14**, 251 (1988)
36. Argyris, D., Ashby, P.D., Striolo, A.: ACS Nano **5**, 2215 (2011)
37. Azimi, G., Dhiman, R., Kwon, H.M., Paxson, A.T., Varanasi, K.K.: Nat. Mater. **12**(4), 315–320 (2013)
38. Shijina, K., Sankar, S., Midhun, M., Firozkhan, M., Nair, B.N., Warrier, K.G., Hareesh, U.N. S.: New J. Chem. **40**, 5333 (2016)
39. Sankar, S., Nair, B.N., Suzuki, T., Anilkumar, G.M., Padmanabhan, M., Hareesh, U.N.S., Warrier, K.G.: Sci. Rep. **6**, 22732 (2016)
40. Lucas, S., Champion, E., Assllant, D.B., Leroy, G.: J. Solid State Chem. **1774**, 1312 (2004)
41. Kumar, S.R., Pillai, S.C., Hareesh, U.S., Mukundan, P., Warrier, K.G.K.: Mater. Letters **43**, 286 (2000)
42. Prasad, K., Pinjari, D.V., Pandit, A.B., Mhaske, S.T.: Ultrason. Sonochem. **18**, 1128 (2011)
43. Ren, X., Guo, S., Zhao, M., Pan, W.: J. Mater. Sci. **49**, 2243 (2014)
44. Cao, X.Q., Vassen, R., Stoever, D.: J. Eur. Ceram. Soc. **24**, 1 (2004)
45. Du, A., Wan, C., Qu, Z., Pan, W.: J. Am. Ceram. Soc. **92**, 2687 (2009)
46. Heimann, R. B. Plasma-spray coating: principles and applications. VCH VerlagsgesellschaftmbH (1996)
47. Pragatheeswaran, A., Ananthapadmanabhan, P.V., Chakravarthy, Y., Bhandari, S., Thiyagarajan, T.K., Tiwari, N., Saha, T.K., Ramachandran, K.: Ceram. Int. **40**(7), 10441–10446 (2014)
48. Schindler, D.W.: Science **4139**, 897 (1974)
49. Wang, R., Pan, W., Chen, J., Jiang, M., Luo, Y., Fang, M.: Ceram. Inter. **29**, 19 (2003)
50. Zheng, Q., Wang, X., Liu, X., Yin, Y.: Adv. Mater. Res. **79**, 2227 (2009)
51. Ergun, C., Liu, H., Webster, T.J.: Biomed. Mater. Res. Part A. **89**, 727 (2009)
52. Zhou, Z.J., Yang, Z.F., Yuan, Q.M., Li, X.H.: J. Rare Earths **20**(3), 197–203 (2002)
53. Min, W., Daimon, K., Matsubara, T., Hikichi, Y.: Mater. Res. Bull. **37**, 1107 (2002)

54. Rajesh, K., Baiju, K.V., Jayasankar, M., Warrier, K.G.: J. Am. Ceram. Soc. **91**, 2415 (2008)
55. Hay, R.S., Boakye, E., Petry, M.D.: J. Eur. Ceram. Soc. **20**, 589 (2000)
56. Boakye, E.E., Mogilevsky, P., Parthasarathy, T.A., Hay, R.S., Welter, J., Kerans, R.J.: J. Am. Ceram. Soc. **89**(11), 3475–3480 (2006)
57. Zhu, X., Lu, Z., Wei, B., Huang, X., Zhang, Y., Su, W.: J. Power Sources **196**, 729 (2011)
58. Yihua, H., Dongliang, J., Jingxian, Z., Qingling, L.: J. Am. Ceram. Soc. **92**, 2883 (2009)
59. Lloyd, I.K.: Ceram. Trans. **221**, 115 (2010)
60. Wu, J., Xie, G., Lin, J., Lan, Z., Huang, M., Huang, Y.: J. Power Sources **195**, 6937 (2010)
61. Curtis, C.E.: J. Am. Ceram. Soc. **40**, 274 (1957)
62. Erdei, S., Roy, R., Harshe, G., Juwhar, H., Agrawal, I.D., Ainger, F.W., White, W.B.: Mater. Res. Bull. **30**, 745 (1995)
63. Alangi, N., Mukherjee, J., Anupama, P., Verma, M.K., Chakravarthy, Y., Padmanabhan, P.V. A., Das, A.K., Gantayet, L.M.: J. Nucl. Mater. **410**, 39 (2011)
64. Fang, Y.P., Xu, A.W., Song, R.Q., Zhang, H.X., You, L.P., Yu, J., Liu, H.Q.: J. Am. Chem. Soc. **125**(51), 16025–16034 (2003)
65. Du, A., Wan, C., Qu, Z., Pan, W.: J. Am. Ceram. Soc. **92**(11), 2687 (2009)
66. J-Tsiar, L., H-Yang L.: Ceram Int. **14**, 251 (1988)
67. Mullica, D.F., Milligan, W.O., Grossie, David A., Beall, G.W., Boatner, L.A.: Inorg. Chim. Acta **95**, 231 (1984)
68. Ananthapadmanabhan, P.V., Sreekumar, K.P., Thiyagarajan, T.K., Satpute, R.U., Krishnan, K., Kulkarni, N.K., Kutty, T.R.G.: Mater. Chem. Phy. **113**(1), 417 (2009)

Index

S.C. Pillai and S. Hehir (eds.), *Sol-Gel Materials for Energy, Environment and Electronic Applications*, Advances in Sol-Gel Derived Materials and Technologies,
DOI 10.1007/978-3-319-50144-4

MIX
Papier aus verantwortungsvollen Quellen
Paper from responsible sources
FSC® C105338

FSC
www.fsc.org

If you have any concerns about our products,
you can contact us on
ProductSafety@springernature.com

In case Publisher is established outside the EU,
the EU authorized representative is:
Springer Nature Customer Service Center GmbH
Europaplatz 3, 69115 Heidelberg, Germany

Printed by Libri Plureos GmbH
in Hamburg, Germany